Springer Remote Sensing/Photogrammetry

More information about this series at http://www.springer.com/series/10182

Faisal Hossain
Editor

Earth Science Satellite Applications

Current and Future Prospects

 Springer

Editor
Faisal Hossain
University of Washington
Seattle, WA
USA

ISSN 2198-0721   ISSN 2198-073X (electronic)
Springer Remote Sensing/Photogrammetry
ISBN 978-3-319-33436-3   ISBN 978-3-319-33438-7 (eBook)
DOI 10.1007/978-3-319-33438-7

Library of Congress Control Number: 2016936681

© Springer International Publishing Switzerland 2016
This work is subject to copyright. All rights are reserved by the Publisher, whether the whole or part of the material is concerned, specifically the rights of translation, reprinting, reuse of illustrations, recitation, broadcasting, reproduction on microfilms or in any other physical way, and transmission or information storage and retrieval, electronic adaptation, computer software, or by similar or dissimilar methodology now known or hereafter developed.
The use of general descriptive names, registered names, trademarks, service marks, etc. in this publication does not imply, even in the absence of a specific statement, that such names are exempt from the relevant protective laws and regulations and therefore free for general use.
The publisher, the authors and the editors are safe to assume that the advice and information in this book are believed to be true and accurate at the date of publication. Neither the publisher nor the authors or the editors give a warranty, express or implied, with respect to the material contained herein or for any errors or omissions that may have been made.

Printed on acid-free paper

This Springer imprint is published by Springer Nature
The registered company is Springer International Publishing AG Switzerland

Preface

We present here the book titled '*Earth Science Satellite Applications: Current and Future Prospects.*' When talking about prospects of satellite applications, it is important to talk about capacity, or capacity building for handling satellite (or Earth Observation) systems. Capacity building using Earth Observing (EO) systems and data (i.e., from orbital and non-orbital platforms) to enable societal applications may be defined as the network of soft and hard components comprising human, non-human, technical, non-technical, hardware, and software dimensions necessary to successfully cross the gap between science and research and societal application taken from "Hossain et al. (2016)".

In today's world, it has become quite clear that the capacity building community of scientists and stakeholders need to be better prepared to take advantage of the rapidly emerging, abundant scientific output and remote sensing data from satellite missions by converting them into decision-making products for end users.

An organization is said to have resilient capacity when it can retain and continue to build capacity in the face of unexpected shocks or stresses. Shocks may also include extreme events such as disasters and losing key staff with technical and institutional knowledge. *So how do we change this course and take full advantage of satellite Earth observational capability towards a more sustainable, safer future in the coming decades?*

To address this key question and strengthen the global societal applications and capacity building community's voice, a 3-day workshop was held in Tacoma (Washington) during June 23–25, 2015, in anticipation of the 2017–2027 NRC Decadal Survey. This edited book is a result of this workshop. The workshop was sponsored by the NASA Applied Sciences Program as an E2 Topical Workshop, Symposium, and Conference (TWSC) event. It brought together experts from the applied sciences community already engaged in capacity building across various themes for the stakeholder community; NASA Applied Sciences and Capacity Building programmatic personnel; and several international stakeholder agencies with a history of using and a need for EO systems and data. The workshop aimed to debate issues to formulate a vision and a path forward for the NASA Applied Sciences and Capacity Building community.

First, we provide in this book international perspectives from around the world on the value of EO/satellite systems for societal systems and the hurdles to achieving full potential of earth observations. Such perspectives have been gathered from various corners of the world such as Asia, Africa, and Americas. The book also provides a thematic breakdown of societal applications such as disasters, water resources, health, and ecosystems. A program to engage the application community early in the process for making planned satellite missions for societally relevant is also discussed. Finally, the book provides a real-world and optimistic scenario of how one stakeholder agency was able to benefit from the engagement with scientific community and take advantage of satellite system for flood management.

This book would not have been possible without the active support of Dr. Nancy Searby, Program Manager of NASA Capacity Building Program, Lawrence Friedl, Dan Irwin (SERVIR), Ashutosh Limaye (SERVIR), International Center of Integrated Mountain Development (ICIMOD), Asian Disaster Preparedness Center (ADPC), Regional Center for Mapping of Resources (RCMRD), and Greg Miller (University of Washington). In addition, we also acknowledge the role of numerous participants of the workshop (many of them now authors of various chapters) who provided valuable input to the discussion.

We hope that readers benefit from this book and ponder with scientists and stakeholders alike how best to move forward with building durable capacity of satellite EO systems in the future.

Faisal Hossain
On behalf of the Capacity Building and Satellite Application Community

Reference

Hossain, F., Serrat-Capdevila, A., Granger, S., Thomas, A., Saah, D., Ganz, D., et al. (2016, In press). A global capacity building vision for societal applications of Earth observing systems and data: Key questions and recommendations. *Bulletin of the American Meteorological Society*. doi:10.1175/BAMS-D-15-00198.1, http://journals.ametsoc.org/doi/pdf/10.1175/BAMS-D-15-00198.1.

Contents

Part I International Perspectives of Satellite Earth Observation Data for Societal Benefit

The Role of Earth Observation for Managing Biodiversity and Disasters in Mesoamerica: Past, Present, and Future 3
Victor H. Ramos and Africa I. Flores

Reform Earth Observation Science and Applications to Transform Hindu Kush Himalayan Livelihoods—Services-Based Vision 2030 27
M.S.R. Murthy, Deo Raj Gurung, Faisal Mueen Qamer,
Sagar Bajracharya, Hammad Gilani, Kabir Uddin, Mir Matin,
Birendra Bajracharya, Eric Anderson and Ashutosh Limaye

Integrating Earth Observation Systems and Data into Disaster Preparedness in the Lower Mekong: Experiences from the Asian Disaster Preparedness Center 63
David J. Ganz, Peeranan Towashiraporn, N.M.S.I. Arambepola,
Anisur Rahman, Aslam Perwaiz, Senaka Basnayake, Rishiraj Dutta
and Anggraini Dewi

Land Cover Mapping for Green House Gas Inventories in Eastern and Southern Africa Using Landsat and High Resolution Imagery: Approach and Lessons Learnt 85
Phoebe Oduor, Jaffer Ababu, Robinson Mugo, Hussein Farah,
Africa Flores, Ashutosh Limaye, Dan Irwin and Gwen Artis

Part II Thematic Perspectives of Satellite Earth Observation Data for Societal Benefit

Role of Earth Observation Data in Disaster Response and Recovery: From Science to Capacity Building 119
Guy Schumann, Dalia Kirschbaum, Eric Anderson and Kashif Rashid

Applying Earth Observations to Water Resources Challenges 147
Christine M. Lee, Aleix Serrat-Capdevila, Naveed Iqbal,
Muhammad Ashraf, Benjamin Zaitchik, John Bolten, Forrest Melton
and Bradley Doorn

**Use of Remotely Sensed Climate and Environmental Information
for Air Quality and Public Health Applications** 173
William Crosson, Ali Akanda, Pietro Ceccato, Sue M. Estes,
John A. Haynes, David Saah, Thomas Buchholz, Yu-Shuo Chang,
Stephen Connor, Tufa Dinku, Travis Freed, John Gunn,
Andrew Kruczkiewicz, Jerrod Lessel, Jason Moghaddas, Tadashi Moody,
Gary Roller, David Schmidt, Bruce Springsteen, Alexandra Sweeney
and Madeleine C. Thomson

**Satellite Remote Sensing in Support of Fisheries Management
in Global Oceans** 207
Dovi Kacev and Rebecca L. Lewison

**Improving NASA's Earth Observation Systems and Data Programs
Through the Engagement of Mission Early Adopters** 223
Vanessa M. Escobar, Margaret Srinivasan and Sabrina Delgado Arias

**Application of Satellite Radar Altimeter Data in Operational
Flood Forecasting of Bangladesh** 269
Amirul Hossain and Md. Arifuzzaman Bhuiyan

Part I
International Perspectives of Satellite Earth Observation Data for Societal Benefit

The Role of Earth Observation for Managing Biodiversity and Disasters in Mesoamerica: Past, Present, and Future

Victor H. Ramos and Africa I. Flores

Abstract Mesoamerica is a term used sometimes in cultural context, but in this article we are using it to name the land bridge between North and South America, made up of the Southern Mexico States (Chiapas, Quintana Roo, Yucatán, Campeche y Tabasco), Guatemala, Belize, El Salvador, Honduras, Nicaragua, Costa Rica, and Panama. With an area of approximately 755,000 square kilometers, it is one of the most heterogeneous regions of the world in terms of elevation, land forms, climate, natural ecosystems and human populations. In the general context given, the potential of Earth observation (EO) to assist the management of natural resources, biodiversity and disasters in the region is clear. In this chapter, we discuss the current state of EO applications and future perspectives related to land-use change, ecosystem dynamics and biodiversity and solid-earth hazards. We hope that this contribution can identify current and future challenges related to obtaining the biggest societal benefit of EO and suggest actions to take advantage of anticipated innovations and data availability.

1 Introduction

Mesoamerica is a term used sometimes in cultural context, but in this article we are using it to name the land bridge between North and South America, made up of the Southern Mexico States (Chiapas, Quintana Roo, Yucatán, Campeche y Tabasco),

---

V.H. Ramos (✉)
National Council of Protected Areas, Guatemala-Monitoring and Evaluation Center (CONAP-CEMEC), Antiguo Hospital de San Benito, Petén, Guatemala, USA
e-mail: vhramos@wcs.org

V.H. Ramos
Wildlife Conservation Society (WCS), Avenida 15 de Marzo, Calle Fraternidad, Flores, Petén, Guatemala, USA

A.I. Flores (✉)
University of Alabama in Huntsville, Huntsville, USA
e-mail: Africa.flores@nasa.gov

A.I. Flores
NASA-SERVIR Science Coordination Office, Huntsville, AL, USA

© Springer International Publishing Switzerland 2016
F. Hossain (ed.), *Earth Science Satellite Applications*,
Springer Remote Sensing/Photogrammetry, DOI 10.1007/978-3-319-33438-7_1

Guatemala, Belize, El Salvador, Honduras, Nicaragua, Costa Rica, and Panama. With an area of approximately 755,000 km^2, it is one of the most heterogeneous regions of the world in terms of elevation, land forms, climate, natural ecosystems, and human populations.

With an estimated population of 58 million inhabitants by 2015, of which close to one third are under 15 years of age, the population is expected to grow to 69 million inhabitants by 2030 (Celade 2004, 2014; CONAPO 2014). Population growth dynamics vary from country to country with extremes in Guatemala, which is projected to have between 2015 and 2030 a 1.87 %/year increase in population and El Salvador that is projected to have 0.39 %/year increase for the same period. Ethnic diversity is another characteristic of human populations in Mesoamerica, with over 60 different ethnic/linguistic groups (Center for Support of Native Lands, National Geographic 2002). Around one third of the population lives with less than US$4/day and therefore they are considered poor, although the trend appears to be the reduction of this number (World Bank 2013).

The region has also been the scene of violent conflicts, including civil wars in Guatemala, El Salvador and Nicaragua that concluded by the end of the 1990s decade. More recently, and in connection with common crime, gangs, and organized crime, four of the eight countries in Mesoamerica (Honduras, El Salvador, Belize, and Guatemala) have been included in the top ten ranking of murder rates per 100,000/inhabitants (UNODC 2013). The aggregate effects of poverty and violence are the believed cause of massive migrations, highlighted recently by the crises of unaccompanied children migrating from Central America and Mexico into the U.S. (Pierce 2015).

Mesoamerica is one of the original proposed global biodiversity hotspots and contains 24,000 species of plants, 2859 species of vertebrates and is one of the leading regions of the world regarding the presence of endemic species (Myers et al. 2000). The northern part of the region is one of the eight centers of origin of domesticated plants, for world crops like maize, beans, peppers, tomato, cotton, and cacao (Vavilov and Freier 1951). Around 28 % of all the terrestrial area is under some kind of formal protection (UNEP-WCMC 2014) but there is concern regarding the limitations of the effectiveness of protected areas in conserving biodiversity, preventing deforestation and degradation(Blackman et al. 2015), which is often interpreted as the results of economical and social issues and the general scarcity of resources to manage and administer protected areas.

Deforestation is one of the biggest threats for biodiversity and protected areas in Mesoamerica. An estimated 5 million hectares of forest have been lost between 2000 and 2013 in the eight countries of the region (Hansen et al. 2013). Deforestation has been historically linked to the promotion of colonization by governments, spontaneous migration and more recently to the increasing importance of agro-industrial crops like oil palm (Rival and Levang 2015). Fire in the wild-lands is closely related to agricultural expansion because of the almost universal use of fire as a clearing tool and is the other big threat to biodiversity. Severe fire events occurred in 1998, 2003, and 2005 when hundreds of thousands of hectares of forests and savannas burned for several months and smoke of those burnings reached the U.S. (Wang et al. 2006).

Mesoamerica is also a region vulnerable to disasters that cause periodically important human and material losses. Statistics between 1990 and 2014 indicate that there have been 36 earthquakes, 104 floods, 17 landslides, 82 storms, and 21 volcanic events. These disasters have caused more than 28 thousand deaths, left homeless close to one million people and caused US$50 billion in damages (Guha-Sapir et al. 2009).

In the general context given, the potential of Earth Observation (EO) to assist the management of natural resources, biodiversity, and disasters in the region is clear. In this chapter, we discuss the current state of EO applications and future perspectives related to Land-use Change, Ecosystem Dynamics and Biodiversity, and Solid-Earth Hazards. We hope that this contribution can identify current and future challenges related to obtaining the biggest societal benefit of EO and suggest actions to take advantage of anticipated innovations and data availability.

2 Current State

2.1 Land-Use Change, Ecosystem Dynamics, and Biodiversity

Wide availability and free access of remote sensing data along with advances in communications technology have made possible, relatively recently, an exponential growth of the use of EO data. The turning point in this growth of applications was probably the decision to make freely available all Landsat data in the U.S. Geological Survey (USGS) archive by 2008, when distribution went from a few thousand scenes by year to millions by year (Turner et al. 2015). Mesoamerica is no exception to that trend and increasingly more uses and applications are present in government, private, and academic contexts.

To characterize the recent progress and current state in the use of EO data in applications related to land use change, ecosystems, and biodiversity carried on in Mesoamerica we selected examples at national and regional scales. Information on 19 cases has been collected, 11 of them focused at the national level, and 8 are regional or multicountry level. A summary of these cases is presented in Table 1.

The earliest analysis included in Table 1 (Vreugdenhil et al. 2002) was the Map of Ecosystems of Central America and had the primary objective to describe the ecosystems to establish a baseline of biodiversity in the region and serve as a basis for biological monitoring programs. The map was also intended to be used as a reference in the design of the Mesoamerican Biological Corridor. The map was derived using visual interpretation of Landsat scenes with local experts in each country, helping to delineate ecosystems with a strong emphasis on field verification and ecosystem data collection. The map was perhaps the first time EO data was used at the regional scale, and still today is used even if the acquisition date of the Landsat scenes ranged between 1997 and 1999.

As expected, the most often used EO data comes from the Landsat Project (63 %), followed by the Moderate Resolution Imaging Spectroradiometer (MODIS)

Table 1 Selected EO applications in Mesoamerica on land-use change, ecosystem dynamics, and biodiversity

Description	Country coverage	Sensors	Acquisition date	Method	Institution	References
Map of the ecosystems of Central America	GT, BZ, HN, SV, NI, CR, PA	Landsat	1997–1999	Visual interpretation of remote sensing data, field verification	World Bank and the Comisión Centroamericana de Ambiente y Desarrollo (CCAD)	Vreugdenhil et al. (2002)
Changes in land cover and deforestation in Central America 1990–2008	GT, BZ, HN, SV, NI, CR, PA	AVHRR, Landsat, MODIS, Spot Vegetation, MERIS, Alos Palsar	1981–2010	Data compilation, digital classification of remote sensing data, field verification	Centro del Agua del Trópico Húmedo para América Latina y el Caribe (CATHALAC)	Cherrington et al. (2011)
Ocean Algal bloom monitoring for Mesoamerica	MX, GT, BZ, HN, SV, NI, CR, PA	MODIS	2005–present	SeaDAS (SeaWiFS Data Analysis System) image processing	Mesoamerican Regional Visualization and Monitoring System (SERVIR)	Graves et al. (2007)
Potential Impacts of climate change on Biodiversity in Central America, Mexico, Dominican Republic	MX, GT, BZ, HN, SV, NI, CR, PA, DO	Inputs related to landuse derived from Landsat	1997–2005	Data integration, climate change anomalies derivation	Centro del Agua del Trópico Húmedo para América Latina y el Caribe (CATHALAC)	Anderson et al. (2008)
Using EO data to improve REDD+ policy	MX, GT, BZ, HN, SV, NI, CR, PA, DO	Landsat, MODIS, SRTM, TRMM	1995–present	Data integration, decision tools, digital classification of remote sensing data	Resources for the Future, Inc and Mesoamerican Regional Visualization and Monitoring System (SERVIR)	Blackman (2013)

(continued)

Table 1 (continued)

Description	Country coverage	Sensors	Acquisition date	Method	Institution	References
Operational program for the detection of hot-spots using remote sensing techniques	MX, GT, BZ, HN, SV, NI, CR, PA	AVHRR, MODIS	1998–present	AVHRR and MODIS fire detection algorithms adapted to the region	Comisión Nacional para el Conocimiento y Uso de la Biodiversidad (CONABIO)	Flasse and Ceccato (1996) Giglio et al. (2003) Ressl et al. (2009)
MAD-MEX: Automatic Wall-to-Wall Land Cover Monitoring for the Mexican REDD-MRV Program Using All Landsat Data	MX	Landsat	1993–2008	Digital classification of remote sensing data	Comisión Nacional para el Conocimiento y Uso de la Biodiversidad (CONABIO), Comisión Nacional Forestal (CONAFOR), Instituto Nacional de Estadística y Geografía (INEGI)	Gebhardt et al. (2014)
A National, Detailed Map of Forest Aboveground Carbon Stocks in Mexico	MX	Landsat, Alos Palsar, SRTM	2000–2008	Modeling of the relationship between fused optical and SAR data and field above ground biomass measurements	Comisión Nacional Forestal (CONAFOR), Woods Hole Research Center (WHRC)	Cartus et al. (2014)
Mangroves of Mexico: extent, distribution, and monitoring	MX	Aerial photos, SPOT, Landsat	1970–2010	Visual interpretation of aerial photos, digital classification of remote sensing data	Comisión Nacional para el Conocimiento y Uso de la Biodiversidad (CONABIO)	Rodríguez-Zúñiga et al. (2013)
A novel Satellite-based Ocean Monitoring System for Mexico	MX, GT, BZ, HN, SV, NI, CR, PA	MODIS	2002–present	Algorithms to obtain surface temperature, and ocean color products, chlorophyll-a concentration, total suspended matter concentration, chlorophyll fluorescence and other variables	Comisión Nacional para el Conocimiento y Uso de la Biodiversidad (CONABIO)	Cerdeira López (2011)

(continued)

Table 1 (continued)

Description	Country coverage	Sensors	Acquisition date	Method	Institution	References
Dynamics of forest cover in Guatemala	GT	Landsat, Aster	2000–2010	Digital classification of remote sensing data	Instituto Nacional de Bosques (INAB), Consejo Nacional de Áreas Protegidas (CONAP), Universidad del Valle de Guatemala (UVG), Ministerio de Ambiente y Recursos Naturales (MARN), Universidad Rafael Landivar	INAB, CONAP, UVG, URL (2012)
Geospatial Information System for Fire Management (SIGMA-I) in Guatemala	GT	MODIS, Landsat, TRMM	1998–2009	Data compilation, digital classification of remote sensing data	Consejo Nacional de Áreas Protegidas (CONAP), Instituto Nacional de Bosques (INAB), Coordinadora Nacional para la Reducción de Desastres (CONRED), Ministerio de Ambiente y Recursos Naturales (MARN)	CONAP, INAB, CONRED, MARN (2010)
Forest cover and deforestation in Belize: 1980–2010	BZ	Landsat	1979–2010	Digital classification of remote sensing data	Centro del Agua del Trópico Húmedo para América Latina y el Caribe (CATHALAC), Land Information Centre, Forest Department	Cherrington et al. (2010)
A MODIS generated land cover mapping of Honduras: a base-line layout to create a national monitoring center	HN	MODIS	2009	Digital classification of remote sensing data	University of Utah, Escuela Nacional de Ciencias Forestales	Rivera et al. (2011)

(continued)

Table 1 (continued)

Description	Country coverage	Sensors	Acquisition date	Method	Institution	References
Application of low resolution satellite data for the detection and monitoring of fire in Nicaragua	NI	AVHRR	1996–2003	AVHRR and MODIS fire detection algorithms	Ministerio de Agricultura y Forestal, Ministerio de Ambiente y Recursos Naturales	Mejía et al. (2004)
Classification of coastal marine ecosystems in Costa Rica	CR	RapidEye, Worldview 2	2009–2011	Digital classification of remote sensing data	Coastal Marine Biodiversity Project, Sistema Nacional de Áreas de Conservación, Deutsche Gesellschaft für Internationale Zusammenarbeit	BIOMARCC, SINAC, GIZ (2012)
Consistent historical time series of activity data from land use change in Costa Rica	CR	Landsat, RapidEye	2000–2012	Digital classification of remote sensing data	Agresta Sociedad Cooperativa, Digital Image Processing, Universidad de Costa Rica, Universidad Politécnica de Madrid	Agresta, Dimap, Universidad de Costa Rica, Universidad Politécnica de Madrid (2015)
High-fidelity national carbon mapping for resource management and REDD+	PA	Landsat, MODIS, airborne LiDAR	2008–2012	Modeling of the relationship between LiDAR, field above ground biomass measurements and optical remote sensing data	Department of Global Ecology, Carnegie Institution for Science	Asner et al. (2013)
High resolution mapping of forest and forest types in Central America	GT, BZ, HN, SV, NI, CR, PA	Rapideye	2011–2013	Digital object oriented classification of remote sensing data	REDD-CCAD/GIZ Program and national counterpart agencies	Jimenez (2013)

Country codes: MX = Mexico, GT = Guatemala, BZ = Belize, SV = El Salvador, HN = Honduras, NI = Nicaragua, CR = Costa Rica, PA = Panamá instrument (42 %), RapidEye and Advanced Very High Resolution Radiometer (AVHRR) (both with 16 %). Other data used included SPOT Vegetation, Medium Resolution Imaging Spectrometer (MERIS), Phased Array type L-band Synthetic Aperture Radar (PALSAR), Shuttle Radar Topography (SRTM), Tropical Rainfall Measuring Mission (TRMM), aerial photography, SPOT, Advanced Spaceborne Thermal Emission and Reflection Radiometer (ASTER), Worldview and Airborne Light detection and ranging (LiDAR). More than half of the applications (53 %) used multi-sensor, multi-resolution data, with two cases in which at least four sensors were integrated to obtain a final product (Cherrington et al. 2011; Blackman 2013).

The dominant application found was land use and land-use change mapping (53 %), often explicitly coupled with REDD+ support (42 %), followed by fire management mapping and support (21 %). Other applications included ecosystem mapping (16 %), biomass mapping (10 %), ocean monitoring (10 %) and only in one case (Anderson et al. 2008), biodiversity modeling was a primary objective of the application (4 %).

Applications working in an operational mode (21 %) include the production of fire monitoring outputs (Ressl et al. 2009), ocean monitoring (Graves et al. 2007; Cerdeira López 2011), and a REDD decision support system (Blackman 2013). Applications that do not produce outputs uninterruptedly, but that will likely do so periodically on an annual or multi-annual basis (37 %) are mostly related to REDD + support, fire management, mapping and land use and land-use change mapping (CONAP, INAB, CONRED, MARN 2010; Cherrington et al. 2010; INAB, CONAP, UVG, URL 2012; Rodríguez-Zúñiga et al. 2013; Jimenez 2013; Gebhardt et al. 2014 and Agresta, Dimap, Universidad de Costa Rica, Universidad Politécnica de Madrid 2015).

Government institutions participated in the majority of the applications (58 %), followed by regional or multilateral institutions (42 %) and academic institutions (26 %). CATHALAC, SERVIR and CONABIO were involved in almost half (47 %) of the applications and maintain all the ones considered operational.

With this general overview obtained from the selected applications reviewed, we can highlight several findings:

(a) Landsat and MODIS are the most used sensors in the region and are critical for several current and future applications, including those related to REDD+ and fire monitoring. It appears that the use of high resolution data, in particular RapidEye, is gaining traction in the region, in part because the need of more detailed products (for example to map forest degradation) and also because the REDD-CCAD/GIZ Program provided all the Central America countries with a wall to wall coverage year? Is notable the limited use of SAR data in a region even though there are zones (the Caribbean coasts of Honduras, Nicaragua, Costa Rica, and Panama) where cloud coverage makes the acquisition of useful optical data very difficult.

(b) With only one example of an application explicitly aimed to biodiversity issues, it appears to be necessary to promote a more active involvement of the conservation and protected area communities in the systematic use of EO data in the region. Several frameworks that propose specific ways in which EO can help biodiversity and conservation have emerged recently (Secades et al. 2014; Rose et al. 2015; Skidmore et al. 2015) and they should be examined to give direction to regional and national efforts.

(c) REDD+ is an emerging driving factor in the development of EO applications. All or almost all the countries in Mesoamerica have joined the Forest Carbon Partnership Facility (FCPF) and the UN-REDD Program. That implies that, to go forward with REDD, there is an obligation to build robust and transparent national forest monitoring systems that will, at least, produce baseline and consistent frequent information related to land use, land-use change, forest biomass, and forest degradation. That represents clear opportunities to develop EO applications, that will need to persist in the long term, and that may have additional uses and benefit fields like ecosystem and biodiversity monitoring.

(d) The role of SERVIR and CATHALAC has been crucial in the advances in the region and both deserve the credit for substantial progress in the transfer of knowledge, the training of human resources in the region and the development of relevant applications. However, the dependence on projects with a limited life-span that, with a few exemptions, wrap up after the project has finished and the funding has depleted is an issue that has to be addressed. If long-term applications are necessary and desirable, a bigger share of the costs of implementation have to be absorbed by the local governments and other local stakeholders with a clear long-term vision. Public institutions involvement in the selected examples is relatively high, but, they often act as cooperation recipients, and that needs to transition to full adoption of applications, included the costs that may entail.

(e) Mexico is, clearly, the regional leader in terms of systematic use and high-level of sophistication of EO applications used in land-use monitoring, biodiversity, and ecosystems. The experience and capabilities of Mexican institutions are already being used to transfer technology and knowledge to the other countries in the region trough the *"South-South cooperation to exchange experiences and capacities on MRV systems and REDD+"* project (http://mrv.cnf.gob.mx/index.php/en/) and also through the provision of EO-derived data to users in the region (Ressl et al. 2009).

2.2 Solid-Earth Hazards, Resources, and Dynamics

The unique location of Mesoamerica as an isthmus between North and South America and in the Inter-tropical Convergence Zone positions this region as one the most biodiverse territories on Earth. On the other hand, this unique location also makes the region more prone to a range of natural hazards, such as volcanic eruptions, earthquakes, hurricanes, heavy rains, floods, landslides, droughts, and storm surges, to mention some (REDLAC 2008). The high presence of hazards in combination with socioeconomical and political issues creates high vulnerability conditions for the population. More than half of the population in Central America and the Caribbean are at risk from multiple hazards (Dilley 2005).

This region has a high mortality risk from geophysical (earthquakes and volcanoes) and hydrological (floods, cyclone, and landslides) hazards (Dilley 2005). El Salvador, Costa Rica, Guatemala, Dominican Republic, Nicaragua, and Honduras rank among the top 15 countries at relatively high mortality risk from multiple hazards (Dilley 2005).

Further, according to the Climate Risk Index, the top ten countries more affected by disasters between 1997 and 2006 were led by Honduras and Nicaragua at a global level (Harmeling 2007; REDLAC 2008).

According to the EM-DAT International Disaster Database (Guha-Sapir et al. 2009) from 1960 to 2015 there have been 497 natural disasters in Mesoamerica and Dominican Republic that include drought, earthquake, extreme temperature, flood, landslide, mass movement (dry), storms, volcanic activity, and wildfire. Two hundred of these disasters occurred between 2001 and 2015. This means that about 40 % of the disasters reported in the past half century have occurred in the past 15 years. The accuracy of these disaster estimations may very depending upon the methodology used to collect the data, but it shows a general trend of disasters on the rise in Mesoamerica, which aligns with the global trend. Among the most frequent disasters in Mesoamerica and Dominican Republic are floods and storms accounting for 34 and 27 %, respectively, of the total disasters reported between 1960 and 2015 (Fig. 1). The third most frequent disaster in the region is earthquakes. Accounting for 13 %. In the 2001–2015 period, the trend of disaster affecting the region is the same.

The use of EO to provide actionable information in the case of disasters has been widely used in Mesoamerica and the Caribbean. There are several examples of the

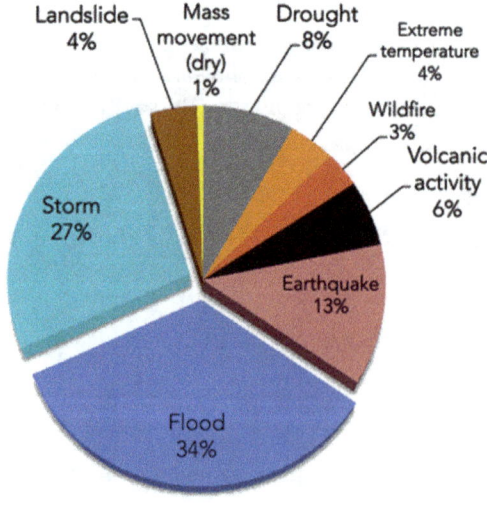

Fig. 1 Types of natural disasters which affected Mesoamerica and Dominican Republic from 1960 to 2015. *Data source* EM-DAT The International Disaster Database (Guha-Sapir et al. 2009)

use of satellite remote sensing to assess the damages inflicted by disasters, to monitor weather events, and to provide valuable information for early warning (REDLAC 2008). This section will focus on examples in which EO data have been used primarily in post-disaster assessment, particularly for response and recovery purposes.

An example of the use of EO in disaster response is the International Charter Space and Major Disasters, (Charter for short). The Charter is "an international collaboration among EO mission owners/operators to provide space-based data and information in support of relief efforts during emergencies caused by major disasters" (Gevorgyan and Briggs 2014).

The Charter only supports the phase of immediate response to a disaster (Courteille 2015; Gevorgyan and Briggs 2014). It was created in 1999, after Hurricane Mitch had struck Mesoamerica, at the Third United Nations Conference on the Exploration and Peaceful Use of Outer Space (UNISPACE) in Vienna, where the European Space Agency (ESA) and the French space agency (CNES) proposed to supply satellite imagery to emergency responders (Rocchio 2014).

Since its activation in 2000 the Charter has covered more than 400 disasters in over 110 countries worldwide (Courteille 2015). The charter is now composed of 15 members from different space agencies and/or mission operators around the world. From the U.S., the National Oceanographic and Atmospheric Administration (NOAA), with its fleet of meteorological satellites, and the U.S. Geological Survey are authorized members of the Charter.

The Charter first provided satellite data to Mesoamerica in 2001 for an earthquake in El Salvador. Since then it has provided data for 27 different disasters covering Mesoamerica and the Hispaniola (Haiti and Dominican Republic). See Table 2.

The Charter has been activated mainly for hydrological-related disasters in Mesoamerica (Fig. 2). Hurricanes, tropical storms, heavy rains, and the resulting flooding and landslides account for more than 70 % of the major disasters in Mesoamerica and the Hispaniola that have triggered the Charter activation (Fig. 2).

Once the Charter has been activated, a project manager (PM) takes charge to quickly analyze the collected data and create value-added products for first responders (Rocchio 2014). Different types of satellite data are acquired through the Charter such as passive and active satellite remote sensing data, and from medium to high spatial resolution.

The Water Center for the Humid Tropics of Latin America and the Caribbean (CATHALAC), in Panama, has acted as the project manager of 9 out of the 27 Charter activations in Mesoamerica and Hispaniola. Refer to Table 2. CATHALAC first hosted SERVIR in 2005. SERVIR links EO and geospatial technologies with developmental decision-making in SERVIR regions. Proven success of SERVIR in Mesoamerica has resulted in expansion to other regions. Currently SERVIR actively works in Eastern and Southern Africa, the Hindu Kush-Himalaya region, and the Lower Mekong region. CATHALAC—in the context of SERVIR and acting as the Mesoamerican hub- has served to connect the scientific, analytical, and decision-making needs of its users in the region. In Mesoamerica, SERVIR provided rapid response support during disasters by acquiring expedited satellite images, and

Table 2 International charter space and major disaster activations for Central America and Hispaniola (2000–Oct 2015), organized by activation date

No.	Call ID	Type of disaster	Country	Title	Date charter activation	Charter requestor	PM	Additional information
1	1	Earthquake	SV	Earthquake and Landslide in El Salvador	1/15/01	French Civil Protection Agency	CNES	
2	29	Cyclone	MX	Hurricane in Yucatan, Mexico	10/5/02	National Forestry Agency of Mexico (CONAFOR) via the French Ministry of the Interior (COGIC)	ESA	Hurricane Isidore
3	51	Flood and landslide	DO	Flooding in the Dominican Republic	11/27/03	UNOOSA	UNOSAT	
4	64	Flood and landslide	HT and DO	Floods in Haiti and in the Dominican Republic	5/26/04	UNOOSA on behalf of UNOCHA /Red Cross	UNOSAT	
5	72	Cyclone	HT	Hurricane in Haiti	9/21/04	United Nations Office for Outer Space Affairs (UNOOSA)	CNES	Hurricane Jeanne
6	106	Cyclone	GT and SV	Hurricane, floods, landslides in Central America; Volcanic eruption in El Salvador	10/5/05	SIFEM-Argentina; UN-OCHA	Instituto Nacional del Agua	Hurricane Stan, Santa Ana volcano eruption

(continued)

Table 2 (continued)

No.	Call ID	Type of disaster	Country	Title	Date charter activation	Charter requestor	PM	Additional information
7	173	Cyclone	MX	Hurricane in Mexico	8/21/07	USGS	CATHALAC	Hurricane Dean
8	176	Cyclone	NI and HN	Hurricane in Nicaragua	9/4/07	SIFEM	CATHALAC	Hurricane Felix
9	183	Cyclone	DO	Flooding and hurricane in the Dominican Republic	10/30/07	Public Safety Canada	CATHALAC	Tropical storm Noel
10	184	Flood and landslide	MX	Flooding in Mexico	11/2/07	SIFEM for the National Center for Disaster Prevention of Mexico	CATHALAC	
11	220	Cyclone	HT	Hurricanes in Haiti	8/30/08	USGS on behalf of Haiti	Office of U.S. Foreign Disaster Assistance; CNES	Hurricane Hanna, Hurricane Ike, Hurricane Gustav
12	229	Flood and landslide	HN	Flood in Honduras	10/27/08	UNITAR/UNOSAT on behalf of UN OCHA	ESA	
13	237	Earthquake	CR	Earthquake and landslide in Costa Rica	1/9/09	SIFEM (Sistema Federal de Emergencias)	CONAE	Turrialba Volcano eruption
14	278	Cyclone	SV	Flood and ocean storm in El Salvador	11/10/09	UNITAR/UNOSAT on behalf of UN OCHA; USGS on behalf of Centro de Gobierno, El Salvador	CATHALAC	Hurricane Ida

(continued)

Table 2 (continued)

No.	Call ID	Type of disaster	Country	Title	Date charter activation	Charter requestor	PM	Additional information
15	287	Earthquake	HT	Earthquake in Haiti	1/13/10	French Civil Protection, UNOOSA on behalf of UN Peacekeeping Mission in Haiti (MINUSTAH), Public Safety of Canada, American Earthquake Hazards Programme of USGS	CNES in collaboration with the SAFER project in the framework of the GMES initiative.	
16	312	Cyclone	GT	Ocean Storm in Guatemala	5/30/10	USGS on behalf of CONRED	CATHALAC	Tropical storm Agatha
17	329	Cyclone	MX	Hurricane Karl, Mexico	9/22/10	USGS on behalf of Mexico National Center for Disaster Protection	USGS	
18	342	Cyclone	HT	Ocean Storm in Haiti	11/5/10	Public Safety Canada	CSA	Tropical storm Tomas
19	348	Flood and landslide	PA	Flood in Panama	12/9/10	USGS on behalf of SINAPROC/Panama	CATHALAC	
20	378	Cyclone	Central America, SV	Flood in El Salvador	10/19/11	UNITAR/UNOSAT on behalf of UN OCHA	UNITAR/UNOSAT on behalf of UN OCHA	Tropical depression
21	396	Volcanic eruption	GT	Eruption of lava from Fuego volcano in Guatemala	5/21/12	SIFEM on behalf of SECONRED	CONAE	
22	418	Cyclone	HT	Hurricane Sandy in Haiti	10/29/12	UNITAR/UNOSAT on behalf of UNOCHA	SERTIT	Hurricane Sandy

(continued)

Table 2 (continued)

No.	Call ID	Type of disaster	Country	Title	Date charter activation	Charter requestor	PM	Additional information
23	420	Earthquake	GT	Earthquake Guatemala	11/8/12	SIFEM on behalf of SE-CONRED	CONRED	
24	496	Fire	DO	Fire in La Vega, Dominican Republic	7/30/14	USGS on behalf of Centro de Operaciones de Emergencias of the Dominican Republic (COE)	DLR	
25	501	Flood and landslide	PA	Flood and landslide in the Republic of Panama	8/18/14	USGS on behalf of SINAPROC /Chiriqui Province /Municipality of Tierra Alta	CATHALAC	
26	525	Volcanic eruption	CR	Turrialba Volcano in Costa Rica	3/14/15	USGS	OFDA-USGS Volcano Disaster Assistance Program	
27	545	Flood and landslide	GT	Landslide in Guatemala	10/7/15	USGS on behalf of National Emergency Operations Center of Guatemala	CATHALAC	

Source compilation of data from: https://www.disasterscharter.org/ Visited on October 2015. PM = Project managers
Call ID = Charter ID # per disaster. *CONAE* Comision Nacional de Actividades Espaciales de Argentina; *CONRED* National Coordination for Disaster Reduction of Guatemala, *CSA* Canadian Space Agency; *DLR* German Aerospace Center; *OFDA* Office of US. Foreign Disaster Assistance; *SERTIT* Service Regional de Traitement d'Image et de Teledetection; *SIFEM* Sistema Federal de Emergencias; *SINAPROC* Sistema Nacional de Proteccion Civil de Panama; *UNITAR* United Nations Institute for Training and Research; *UN OCHA* United Nations Office for the Coordination of Humanitarian Affairs; *UNOSAT* Operational Satellite Applications Programme

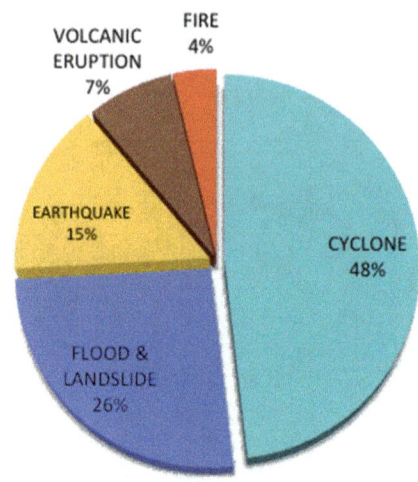

Fig. 2 Charter Activations by Disaster Type in Central America, South of Mexico, and Hispaniola from January 2001 to October 2015. *Source* compilation of data from: https://www.disasterscharter.org/, Visited on October 2015. *Note Cyclone disasters include hurricanes, tropical storms, and tropical depressions*

creating and sharing value-added products with end users (Flores et al. 2012). While it is no longer an active SERVIR hub, CATHALAC continues to meet the needs of its network of users by applying space data to disaster response support.

Flores et al. (2012) describes the type of EO data and information used in SERVIR-Mesoamerica from 2004 to 2011 to create value added products provided for monitoring, response, recovery and general disaster awareness.

Table 3 lists the EO-derived measurements and/or products used by SERVIR-Mesoamerica from 2004 to 2011 to generate added-value products for a variety of disasters.

Then, Table 4 lists the respective satellite sensors from which the derived EO measurement and/or product were generated, respectively.

Some Level 3 satellite-derived products such as land cover, and model products such as gridded population distribution were used to link the physical hazard (cyclone, heavy rain, flood, landslide, etc.) with the human or ecological factor at risk (Flores et al. 2012).

The majority of the satellite sensors used by SERVIR-Mesoamerica to provide added value products were from passive remote sensors that collect electromagnetic energy reflected or emitted by Earth's surface. Using this type of sensors proves difficult when estimating flooded areas during heavy rain and high cloud coverage conditions. Active remote sensors, such as C-band SAR and L-band SAR on board of Radarsat 1 and 2 and ALOS, respectively, were used heavily in cases with high cloud coverage particularly to map flooded areas during disasters. Hyper spectral satellite remote sensing was used primarily for water quality assessment during cases of algal blooms and was also used for monitoring volcano activity.

Some of the satellite data used by SERVIR-Mesoamerica to address disasters was acquired through the Charter, but not all. SERVIR-Mesoamerica also provided value-added products to smaller disaster events that did not trigger the Charter activation.

Table 3 EO-derived measurement and/or product used by SERVIR-Mesoamerica from 2004 to 2011 to analyze different type of disasters

Disaster type	Elevation data	Rainfall	Multispectral VNIR reflectance/radiance	Hyperspectral VNIR reflectance/radiance	Fire and thermal anomalies	Burned area	Surface temperature	Land cover	Relative Humidity	Synthetic Aperture Radar (SAR) images	SRTM-water body	Gridded population distribution	Near-real time cloud movement
Flood	✓		✓							✓	✓	✓	
Tropical cyclone/heavy rain		✓										✓	✓
Landslide	✓	✓	✓									✓	
Volcanic eruption	✓		✓	✓	✓		✓						
Earthquake								✓					
Fire					✓	✓		✓					
Drought	✓	✓					✓	✓	✓				
Algal bloom	✓	✓	✓	✓			✓	✓					

Source Summarized from (Flores et al. 2012)

Table 4 List of satellite sensors used to generate each derived measurement and/or product by SERVIR-Mesoamerica from 2004 to 2011

EO-derived measurement and/or product	Sensor, satellite
Elevation data	SRTM ASTER, Terra
Rainfall	Imager, GOES series MVIRI, Meteosat-7/5 VISSR, GMS TMI, TRMM SSM/I, DMSP series AMSR-E, Aqua AMSU-B, NOAA series
Multispectral VNIR reflectance/radiance	AVHRR, POES series MODIS, Aqua/Terra ASTER, Terra ALI, EO-1 Landsat series SPOT GeoEye-1 Ikonos QuickBird Worldview-1 & 2 Formosat-2
Hyperspectral VNIR reflectance/radiance	Hyperion, EO-1
Fire and thermal anomalies	MODIS, Aqua/Terra
Burned area	MODIS, Aqua/Terra
Surface temperature	Landsat Thermal Infrared (TIR)
Land cover, tree cover maps	MODIS, Aqua/Terra MERIS, Envisat Landsat series ASTER, Terra SPOT L-Band SAR (PALSAR), ALOS
Relative Humidity	AIRS and AMSU, Aqua
Synthetic Aperture Radar (SAR) images	C-band SAR, Radarsat-1 & Radarsat-2 C-band SAR, ERS-2 L-Band SAR (PALSAR), ALOS
SRTM-water body	SRTM
Gridded population distribution	DMS series
Near real-time cloud movement	Imager, GOES series

Source Summarized from (Flores et al. 2012)
ASTER Advanced Spaceborne Thermal Emission and Reflection Radiometer, *AVHRR* Advanced Very High Resolution Radiometer; *GMS* Geostationary Meteorological Satellite; *GOES* Geostationary Operational Environmental Satellite; *MERIS* Medium Resolution Imaging Spectrometer; *MVRI* Meteosat Visible Infra red Imager; *TRMM* Tropical Rainfall Measuring Mission; *PALSAR* Phased Array type L-band Synthetic Aperture Radar; *POES* Polar Operational Environmental Satellite; *SSM/I* Special Sensor Microwave Imager; *VISSR* Visible/Infra Spin Scan-Radiometer

SERVIR started in Mesoamerica when Landsat data was still not freely available, however, through SERVIR the region started getting access to such datasets and used for environmental monitoring and disaster response. As SERVIR-Mesoamerica started building up during its 5 years it also increased the generation of valued-added products for disasters to attend user needs in the region.

Today disaster response agencies across the region are heavily using EO data to address disaster response. As example national disaster institutions in Mesoamerica have become more involved in the Charter process to obtain satellite data, such is the case of SINAPROC in Panama, and CONRED in Guatemala. This last one has even acted as a project manager of the Charter.

3 Future Prospects

3.1 Land-Use Change, Ecosystem Dynamics, and Biodiversity

Despite the challenges, the future looks promising for EO in Mesoamerica. Access to free and open data is growing at a very fast pace. Beyond Landsat we can already use data of the Sentinel constellation that may solve problems related to cloud coverage (Sentinel-1) and provide a new and valuable resource for middle resolution optical data (Sentinel-2). Already free, or being put to public access are historical data of SPOT and ALOS PALSAR providing an additional opportunity to add more layers of EO data. Data continuity for middle resolution data appears to be solved in the mid-term, with for example, plans to build, and launch Landsat 9 underway.

We think that REDD+, fire and climate change monitoring are going to be pivotal for EO applications in the near future in Mesoamerica. REDD+ processes have started in all the countries in the region, and a great part of the success of the mechanism relies on the premise of being able to monitor forests continuously and with a level of detail that will need to resolve at a highest resolution possible deforestation and degradation of forests.

Fire, closely related to REDD+, because is a cause of forest degradation, is an increasing problem in the region and also a phenomenon of interest that can be monitored using EO. We will need to be able to go from just monitoring fire, to actually helping to modify significantly its management with early alert systems tied to regulations on the use of fire as an agricultural tool, for example.

REDD+ and fire applications have the potential to be designed in such a way, that products and outputs can be used in other fields, including biodiversity, ecosystem monitoring, climate change, water management, and agriculture. Because of that, communication and collaboration between remote sensing scientists, conservation practitioners and natural resource managers is fundamental.

3.2 Solid-Earth Hazards, Resources, and Dynamics

The current disaster trend indicates that disasters are increasing at the local and global scale (Guha-Sapir et al. 2009).[1] This is particularly dangerous for regions like Mesoamerica that have a high mortality risk from geophysical and hydrological hazards (Dilley 2005). The use of EO data has proven useful for disaster response and recovery and given the increasing positive trend of events it is expected that the role of EO in disaster management will intensify in years to come.

The current availability of new satellite data such as the Visible Infrared Imaging Radiometer Suite VIIRS on board of Suomi National Polar-orbiting Partnership (NPP), and Landsat 8 will allow continuity of products already used in Mesoamerica for disaster monitoring and response, such as the case of MODIS products used for fire monitoring, burned areas, and land cover to mention some. In addition, new missions such as the Soil Moisture Active Passive (SMAP) will generate new understanding of weather and climate patterns.

A pressing need of the technical community using satellite data for disaster assessment, particularly for response and recovery is the access to radar data to map floods under high cloud cover. The majority of the disasters in this region are related to hydrological hazards Figs. 1 and 2, which trigger other set of hazardous events, such as floods and landslides.

The new C-band SAR sensor, Sentinel 1 from the new European EO Programme, Copernicus, will be instrumental in providing information during high cloud coverage. The advantage of Sentinel 1 when compared to other radar data is that is freely available. This will open its use and application not only for disaster-related analyses but also for other applications.

NASA and Indian Space Research Organization (ISRO) SAR Mission (NISAR) also promises to become a source of valuable data for disaster risk management with the potential ability to measure deformation before and after geophysical hazard events.

4 Conclusion and Way Forward

There is no doubt about the fact that EO holds an immense potential in applications related to land use, ecosystems, biodiversity, and disaster risk management. We think that is also true that this potential has not been fully used in Mesoamerica (and elsewhere in the tropics). There are several challenges to improve on that:

---

[1] Disaster trends. Interactive graphs that show various trends and relationships within the EM-DAT data. EM-DAT: The OFDA/CRED International Disaster Database—http://www.emdat.be/disaster_trends/index.html.

(a) Governments in the region that, in most cases, are the potential users with the power to modify policies and effectively generate change, invest very little in agencies in charge of natural resources, ecosystems, and biodiversity management. Therefore, these agencies often lack the resources to maintain technical units in charge of applications of EO.
(b) EO data is often presented in formats that are not legible enough for the audiences that can influence and make decisions, including the general public. The message that EO delivers has to be clear and understandable to all.
(c) There is often a disconnection between user needs and products. We have to examine carefully if the processes we use to choose what to do is backed by enough feedback from the people and entities that will actually use EO products to make decisions.
(d) It is critical to understand that one of the main values of EO data relies on systematic, long-term observations that actually lead to information on changes and trends. Applications that emphasize research and demonstration obviously are critical for the development of EO, but the ones that can be used operationally and with a clear, practical purpose are those that function and provide societal benefits over a long time.
(e) We have to aim, when possible, at the use of EO application products to support compliance, regulations, and law enforcement. Doing that will represent a technical and perhaps legal challenge, but it would be a significant contribution to governance and environmental justice.
(f) At regional and local scale the use of EO data should be appropriately integrated into the operational processes of disaster risk management agencies to support all the phases of disaster risk management. The generation of geospatial-baseline information primarily in urban areas is critical to properly assess pre and post-event damage needed for immediate response and recovery. Such baseline information becomes more critical in this region that is undergoing population growth and is prone to natural hazards including earthquakes.

References

Agresta, DIMAP, Universidad de Costa Rica, and Universidad Politécnica de Madrid (2015). *Índice de cobertura como base para la estimación de la degradación y aumento de existencias de carbono*, Agresta, Dimap, Universidad de Costa Rica, Universidad Politécnica de Madrid, Costa Rica.

Anderson, E., Cherrington, E., Flores, A., Pérez, J., Carrillo, R., & Sempris, E. (2008). *Potential impacts of climate change on biodiversity in Central America, Mexico, Dominican Republic*. Panamá: Centro del Agua del Trópico Húmedo para América Latina y el Caribe (CATHALAC).

Asner, G. P., et al. (2013). High-fidelity national carbon mapping for resource management and REDD+. *Carbon Balance and Management, 8*(1), 7. doi:10.1186/1750-0680-8-7.

BIOMARCC, SINAC, & GIZ (2012). *Clasificación sistemas marino costeros costa Pacífica de Costa Rica*. Costa Rica: BIOMARCC, SINAC, GIZ
Blackman, A. (2013). Evaluating forest conservation policies in developing countries using remote sensing data: An introduction and practical guide. *Forest Policy and Economics, 34*, 1–16. doi:10.1016/j.forpol.2013.04.006.
Blackman, A., Pfaff, A., & Robalino, J. (2015). Paper park performance: Mexico's natural protected areas in the 1990s. *Global Environmental Change, 31*, 50–61. doi:10.1016/j.gloenvcha.2014.12.004.
Cartus, O., Kellndorfer, J., Walker, W., Franco, C., Bishop, J., Santos, L., & Fuentes, J. (2014). A national detailed map of forest aboveground carbon stocks in Mexico. *Remote Sensing, 6*(6), 5559–5588. doi:10.3390/rs6065559.
CELADE. (2004). *America Latina y el Caribe: Estimaciones y proyecciones de población 1950–2050*. Chile: CEPAL.
CELADE (2014). *Estimaciones y proyecciones de población a largo plazo* 1950–2100
Center for the Support of Native Lands, and National Geographic (2002). *Indigenous peoples and natural ecosystems in Central America and Southern Mexico*.
Cerdeira-Estrada, S., & López-Saldaña, G. (2011). Nuevo Sistema Satelital de Monitoreo Oceánico para México. *Ciencias marinas, 37*(2), 237–247.
Cherrington, E., Ek, E., Cho, P., Howell, B., Hernández, B., Anderson, E., Flores, A., García, B., Sempris, E., & Irwin D. (2010). *Forest cover and deforestation in Belize: 1980–2010*. Panamá: Centro del Agua del Trópico Húmedo para América Latina y el Caribe (CATHALAC)
Cherrington, E., Hernández, B., García, B., Oyuela, M. & Clemente, A. (2011). *Changes in land cover and deforestation in Central America 1990–2008*. Panamá: Centro del Agua del Trópico Húmedo para América Latina y el Caribe (CATHALAC).
CONAP, INAB, CONRED, & MARN (n.d.). *Sistema de Información Geoespacial para Manejo de Incendios en la República de Guatemala (SIGMA-I)*. Guatemala: CONAP, INAB, CONRED, MARN.
CONAPO (2014). Dinámica demográfica 1990–2010 y proyecciones de población 2010–2030
Courteille, J. -C. (2015). *International Charter "Space and Major Disasters" space-based information in support of relief efforts after major disasters*
Dilley, M. (2005). *Natural disaster hotspots: A global risk analysis*, disaster risk management (series no. 5). Washington, DC: World Bank.
Flasse, S., Ceccato, P. (1996). A contextual algorithm for AVHRR fire detection. *International Journal of Remote Sensing, 17*(2), 419–424.
Flores, A., Anderson, E., Irwin, D., & Cherrington, E. (2012). *Contributions of servir in promoting the use of space data in climate change and disaster management*. Naples, Italy: International Astronautical Federation.
Gebhardt, S., et al. (2014). MAD-MEX: Automatic wall-to-wall land cover monitoring for the Mexican REDD-MRV program using all landsat data. *Remote Sensing, 6*(5), 3923–3943. doi:10.3390/rs6053923.
Gevorgyan, Y., & Briggs S. (2014). International charter "Space and major disasters".
Giglio, L., Descloitres, J., Justice, C. O., & Kaufman, Y. J. (2003). An enhanced contextual fire detection algorithm for MODIS. *Remote Sensing of Environment, 87*(2–3), 273–282. doi:10.1016/S0034-4257(03)00184-6.
Graves, S., Yubin He M., & Hardin D. (2007). *SERVIR at the age of four: The development of an environmental monitoring and visualization system for Mesoamerica*. University of Maryland University College.
Grupo Regional Interagencial de Riesgo, Emergencia y Desastres de America Latina y el Caribe (REDLAC) (2008). *Diez años después del huracán Mitch: panorama de la tendencia de la gestión del riesgo de desastre en Centroamérica*.
Guha-Sapir, D., Below, R. & Hoyois, P. (2009). EM-DAT:International disaster database—www.emdat.be, http://www.emdat.be/advanced_search/index.html (Retrived from 2 November 2015)

Hansen, M. C., et al. (2013). High-resolution global maps of 21st-century forest cover change. *Science, 342*(6160), 850–853. doi:10.1126/science.1244693.

Harmeling, S. (2007). *Global climate risk index 2008*. Bonn, Berlin: Germanwatch.

INAB, CONAP, UVG, & URL (2012). *Mapa de Cobertura Forestal de Guatemala 2010 y Dinámica de la Cobertura Forestal 2005–2010*. Guatemala: INAB, CONAP, UVG, URL.

Jimenez, A. (2013). Metodología para el mapeo de tipos de bosques en el contexto de la cuantificación de la biomasa y el carbono forestal.

Mejía, S., Valerio, L., & Coronado, C. (2004). *Aplication of low resolution satellite data for the detection and monitoring of fire in Nicaragua* (pp. 413–417). Istanbul, Turkey: ISPRS.

Myers, N., Mittermeier, R. A., Mittermeier, C. G., da Fonseca, G. A. B., & Kent, J. (2000). Biodiversity hotspots for conservation priorities. *Nature, 403*(6772), 853–858. doi:10.1038/35002501.

Pierce, S. (2015). *Unaccompanied Child Migrants in U.S. Communities, Immigration Court and Schools*. Washington, DC: Migration Policy Institute.

Ressl, R., Lopez, G., Cruz, I., Colditz, R. R., Schmidt, M., Ressl, S., & Jiménez, R. (2009). Operational active fire mapping and burnt area identification applicable to Mexican Nature Protection Areas using MODIS and NOAA-AVHRR direct readout data. *Remote Sensing of Environment, 113*(6), 1113–1126. doi:10.1016/j.rse.2008.10.016.

Rival, A., & Levang, P. (2015). The oil palm (Elaeis guineensis): Research challenges beyond controversies. *Palms, 59*(1), 33–49.

Rivera, S., Lowry, J. L., Sánchez, A. J. H., Ramsey, R. D., Lezama, R., & Velázquez, M. (2011). A MODIS generated land cover mapping of Honduras: a base-line layout to create a national monitoring center. *Revista de teledetección: Revista de la Asociación Española de Teledetección, 35*, 94–108.

Rocchio, L. E. P. (2014). *Mapping disaster: A global community helps from space landsat science*

Rodríguez, M., et al. (2013). *Manglares de México: extensión, distribución y monitoreo*. México: CONABIO.

Rose, R. A., et al. (2015). Ten ways remote sensing can contribute to conservation: Conservation remote sensing questions. *Conservation Biology, 29*(2), 350–359. doi:10.1111/cobi.12397.

Secades, C., O´Connor, B., Brown, C., & Walpole M. (2014). *Earth observation for biodiversity monitoring: A review of current approaches and future opportunities for tracking progress towards the Aichi Biodiversity Targets*. Montréal, Canada: Technical Series, Secretariat of the Convention on Biological Diversity.

Skidmore, A. K., et al. (2015). Environmental science: Agree on biodiversity metrics to track from space. *Nature, 523*(7561), 403–405. doi:10.1038/523403a.

Turner, W., et al. (2015). Free and open-access satellite data are key to biodiversity conservation. *Biological Conservation, 182*, 173–176. doi:10.1016/j.biocon.2014.11.048.

UNEP-WCMC (2014). World database on protected areas (WDPA).

UNODC. (2013). *Global study on homicide*. Vienna, Austria: UNODC.

VAVILOV, N. I., & FREIER, F. (1951). *Studies on the origin of cultivated plants*.

Vreugdenhil, D., Meerman, J., Meyrat, A., Gomez, L. D., & Graham, D. (2002). *Map of the ecosystems of Central America : Main report*. Washington, DC.: World Bank.

Wang, J., Christopher, S. A., Nair, U. S., Reid, J. S., Prins, E. M., Szykman, J., & Hand, J. L. (2006). Mesoscale modeling of Central American smoke transport to the United States: 1. "Top-down" assessment of emission strength and diurnal variation impacts. *Journal of Geophysical Research, 111*(D5). doi:10.1029/2005JD006416.

World Bank. (2013). *Shifting gears to accelerate shared prosperity in Latin America and the Caribbean*. Washington, DC: World Bank.

Reform Earth Observation Science and Applications to Transform Hindu Kush Himalayan Livelihoods—Services-Based Vision 2030

M.S.R. Murthy, Deo Raj Gurung, Faisal Mueen Qamer, Sagar Bajracharya, Hammad Gilani, Kabir Uddin, Mir Matin, Birendra Bajracharya, Eric Anderson and Ashutosh Limaye

Abstract The Hindu Kush Himalayas (HKH) region with 210 million people living in the region poses significant scientific and technological challenges for livelihood improvement due to subsistence economy, livelihood insecurity, poverty, and climate change. The inaccessibility and complex mountain environmental settings carved special niche for Earth Observation (EO) science and significant contributions were made in the food security and disaster risk reduction sectors. The differentiated capacities of users to develop and use EO capabilities, challenges in outreaching the EO products to last mile users call for innovative ways of packaging EO products into actionable knowledge and services. This calls for great degree of reformation on EO community to tailor-made region specific EO sensors and models, mechanisms of synergizing EO knowledge with local traditional systems in addressing multiscale, and integrated end-to-end solutions. The paper addresses prospects and challenges of 2015–2030 to achieve success in three critical livelihood support themes viz food security, floods, and forest-based carbon mitigation. Different improvements in EO sensor and models to extend less than a day, all-weather imaging, improved hydro-meteorological forecasts, vegetation stress, and community carbon monitoring models are identified as priority areas of improvement. We envisage and propose mechanisms on how these EO advances could amalgamate into Essential HKH Variables (EHVs) on the lines of global Essential Climate Variables (ECVs) to provide turnkey-based actionable knowledge

---

M.S.R. Murthy (✉) · D.R. Gurung · F.M. Qamer · S. Bajracharya · H. Gilani · K. Uddin · M. Matin · B. Bajracharya
International Center for Integrated Mountain Development, Kathmandu, Nepal
e-mail: Manchiraju.Murthy@icimod.org

E. Anderson · A. Limaye
Earth Science Office, NASA Marshall Space Flight Center, Huntsville, AL 35808, USA

E. Anderson
Earth System Science Center, University of Alabama in Huntsville, Huntsville, AL 35899, USA

© Springer International Publishing Switzerland 2016
F. Hossain (ed.), *Earth Science Satellite Applications*,
Springer Remote Sensing/Photogrammetry, DOI 10.1007/978-3-319-33438-7_2

and services through global and regional cooperation. The complex web of users and orienting them toward adoption of EO services through multi-tier awareness, expertise development, policy advocacy, and institutionalization is also discussed. The paper concludes that the EO community needs to reform significantly in blending their science and applications with user-driven, need-based domains to provide better societal services and HKH livelihood transformation.

1 Introduction

The Hindu Kush Himalayan (HKH) region extends 3,500 km over all or part of eight countries from Afghanistan in the west to Myanmar in the east (Fig. 1). It is the source of 10 large Asian river systems and provides water, ecosystem services, and the basis for livelihoods to a population of around 210.53 million people in the region. The basins of these rivers provide water to 1.3 billion people, a fifth of the world's population. The human poverty, livelihood insecurity, food insecurity, gender, and social inequity continue to be the major detriments of livelihood quality across the region. The subsistence livelihood practices, globalization, liberalization, climate change, and vulnerability to disasters compounded with fragility, marginality, and inaccessibility are conceived as the critical factors regulating HKH region. Due to the very nature of mountain specificity, climate change is reported to largely govern the developmental trends of transitions from subsistence systems to commercial systems, increasing efficiency and productivity, high-value, niche-based and nonfarm products, and increased mobility–migration and remittances (Rasul 2014).

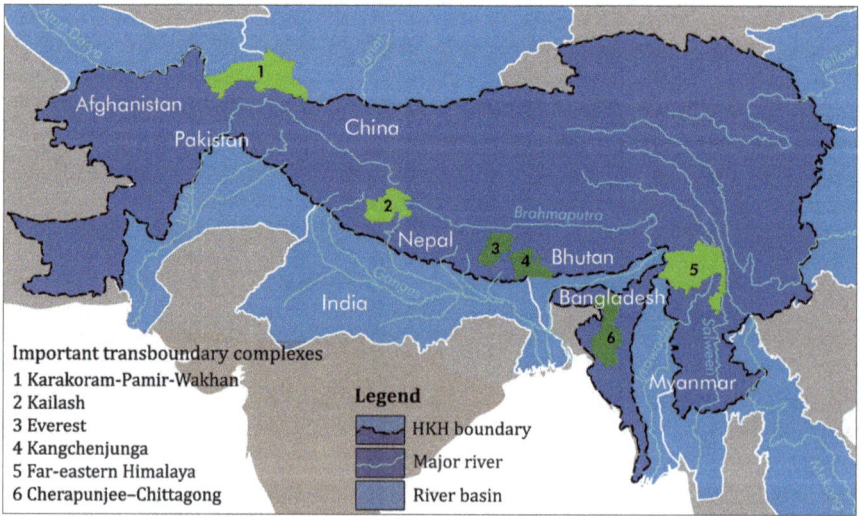

Fig. 1 The Hindu Kush Himalayan (HKH) region

The mean annual increase in temperatures by 2–3 °C are reported to have greater impacts on people and ecosystems of the HKH region (Anderson and Bows 2008; Hansen et al. 2013; Solomon et al. 2009). The high-resolution cryospheric–hydrological model projections have estimated increase in runoff levels due to increased precipitation in the upper Ganges, Brahmaputra, Salween, and Mekong basins, and from accelerated melt in the upper Indus (Lutz et al. 2014). The glaciers in the Himalaya are reported to be retreating faster than those in the Hindu Kush and Karakorum, with losses over 30 years of 5–55 % of area (Bajracharya et al. 2015). The impact of rising temperatures and loss of ice and snow in the region are affecting, for example, water availability (amounts, seasonality).

The ecosystem boundary shifts (tree line movements, high-elevation ecosystem changes), altered carbon fluxes, biodiversity change (endemic species loss, changes in predator–prey relations), and global feedbacks (monsoonal shifts, loss of soil carbon) are a few critical ecosystem impacts reported over HKH region due to climate change (Bates et al. 2008; Burkett et al. 2005; Holtmeier and Broll 2005; Parmesan 2006; Wang et al. 2000). The changes in rangeland calendars, rangeland degradation, and productivity losses across the entire HKH region due to unsustainable grazing practices and climate change are reported. The fragile mountain agriculture practices coupled with climate change impacts in terms of recurring droughts and floods have made mountain agriculture systems highly vulnerable to change and unsustainable production (Abbas et al. 2014; Sigdel and Ikeda 2010). The disaster frequency and associated societal and economic impact has increased by several folds (Etkin 1999; Kim and Marcouiller 2015).

While scientific data and information on the Himalayas are crucial to understand ongoing climatic and socioeconomic change processes in order to support policy planning and informed decision-making, the Himalayan region has a major data gap due to highly inaccessible terrain, inadequacy of resources and technologies, harsh climatic conditions, and lack of investments in long-term scientific research. Earth Observation (EO)-derived information bears special significance in the mountain areas, as it helps to access remote regions and to gain insights of regional scenarios, especially the climatic and broader environmental changes. The EO products today serve as major inputs to policy, planning and targeted interventions, contributing to building social capitals, natural resources assets, and environmental gains on the long-term basis (Jayaraman 2009).

The geospatial services help through cost and time savings, new/improved services, value of information, and extending strategic edge for planning. The in season crop production forecasts, disaster emergency and risk reduction, improved natural resources planning and nature conservation, building physical and social infrastructure, and natural resources assets are a few demonstrated examples, where geospatial systems contributed to effective economy and livelihood improvements. Sankar (2002) estimated that the value of Indian remote sensing products by providing value-added information and optimizing costs of conventional field surveys is around USD 427M. EO systems have contributed to disaster emergency planning and risk reduction processes, leading to avoided loss of life and property (Miller et al. 2011; Blaikie et al. 2014; Brown and Woolridge 2015).

Across the HKH region, there are good number of success stories in taking advantage of the science and observations afforded by satellites to make societal impacts and benefits (Gupta et al. 2015; Pilon and Asefa 2011). There are differentiated capacities among the HKH regional member countries with regard to the utilization of EO and geospatial technologies. The combined observational power of the multiple Earth observing satellites is currently not being harnessed holistically to produce more durable societal benefits. A fundamental challenge for the coming decade is to ensure that established societal needs help and guide scientific priorities more effectively, and that emerging scientific knowledge is actively applied to obtain societal benefits. New observations, analyses, better interpretive understanding, enhanced predictive models, broadened community participation, and improved means for information dissemination are all needed. Achieving these benefits further requires the observation and science program to be closely linked to decision support structures that translate knowledge into practical information matched to and cognizant of society's needs.

It is in this context, the present chapter reviews the current scenario and developments made during 2010–2015, and discusses prospects and challenges during 2015–2030 in evolving toward user-oriented EO applications for improving HKH livelihoods. Our analysis and suggestions are focused to address four essential elements: (1) new advances required in EO sensors to address increasing needs and challenges of natural resources and disaster sectors; (2) customization and new package of EO systems required to understand geophysical and biophysical complexities of mountain systems; (3) research and value-added systems to be improvised to move toward a *service provision* mode addressing societal benefits; and (4) gaps, challenges, and practices to be addressed for the better integration of stakeholder communities to achieve meaningful utilization of services. The data and information from the published literature, proceedings of stakeholder consultation meetings, current availability of datasets and infrastructure, and scenario of capacity building in the region formed the basis for our study. We present our analysis in three sections: (1) current scenario assessment, (2) prospects and challenges for the period 2015–2030 and (3) conclusions and recommendations. The study focused on three major livelihood critical areas viz food security, forest-based carbon mitigation and floods.

2 Current Scenario Assessment (2010–2015)

The current scenario assessment has drawn the findings and recommendations of needs and status assessment studies from SERVIR-Himalaya program. SERVIR (an acronym meaning "to serve" in Spanish) was initiated during 2005 as a joint venture between NASA and the U.S. Agency for International Development (USAID), to provide satellite-based EO data and science applications to help developing nations in Central America, Eastern and Southern Africa (in 2008), and the Hindu Kush Himalayas (in 2010) to improve their environmental decision-making. The requirements of the local and national institutions and diverse end users have been conceived as the basis for developing geospatial science applications. The key regional issues, national priorities, and capacities of the institutions of the regional member countries were assessed before initiating the design and development of information products and services within the SERVIR-Himalaya framework. A literature review was also carried out to analyze the existing documents, workshops, and study reports dealing with the issues of the HKH and the use of geospatial technologies to understand the past and ongoing initiatives, and identification of suitable institutions and people to participate in the needs and capacity assessment process.

For a more comprehensive need assessment and identification of priority applications, country-level stakeholder workshops were held consisting of focused group discussions and questionnaire under the themes of cryosphere, ecosystems and biodiversity, disaster risk reduction, and transboundary air pollution. An assessment of decision support tools and geospatial portals were also done in view of potential SERVIR products development. For each of these themes, SERVIR-Himalaya program assessed of major issues, identified and engaged key stakeholders in the countries, maintained awareness of ongoing initiatives, and synthesized gaps and opportunities. Further, a regional inception workshop was organized for more focused discussion on regional-level needs of SERVIR-Himalaya consisting of experts and professionals from all the eight countries of the HKH including the representatives from USAID and NASA.

The generic framework on current scenario assessment and user integration in application development was done in two phases: (1) product design, testing, and development (2) operational product delivery systems. The Phase-1 has essentially guided the strategic application development based on user diversity, needs, current ground support systems, technology potential facilitating the product design, testing, and development. The second phase addressed the ground operational segment, user diversity, and application use. Accordingly three critical application areas viz, Geo-societal applications using EO data for food security and disasters, assessment and monitoring support for climate change mitigation and adaptations, enabling systems/platforms for multithematic data and products organization, analysis, and dissemination.

Based on the information and knowledge gained through needs and capacity assessment studies undertaken as mentioned above, a table consisting of qualitative scoring against different parameters for six geospatial application products over the three thematic areas is developed (Table 1). The assessment included seven major variables with 22 subvariables addressing information needs, development, and uptake. The land cover change assessment monitoring and forest biomass quantification are the two applications considered as part of forest-based climate change mitigation programs of Ministries of Forests and Soil Conservation. The crop sown area and condition assessment are the two application considered to support Nepal Food Security program of Ministry of Agriculture Development. The multiscale disaster risk assessment and flood risk decision support system to strengthen flood risk reduction systems operated by Ministry of Home Affairs, Nepal. In this article,

Table 1 Needs and Status of Geospatial Applications over HKH Region-2010 (Based on Stakeholder discussions held as part of NASA-SERVIR-Himalaya program during 2010) (3 = High, 2 = Medium, 1 = Low)

	Forest carbon mitigation		Season crop monitoring		Flood DRR	
	Cover monitoring	Periodic carbon assessment	Area assessment	Condition assessment	Risk reduction	Early warning
1. Needs/application						
Direct/basic	3	3	3			3
Complimentary				3	3	
Periodicity	3	3	3	3	2	3
Long-term need	3	3	3	3	3	3
2. Databases/tools/systems available						
Historical/current databases	1	2	1	1	1	1
Standard procedure/methods followed	2	1	1	1	1	1
Operational systems	2	2	1	1	1	1
3. Awareness						
Strategic level	2	1	2	2	1	1
Operational level	2	2	3	3	3	2
4. Expertise available						
National	2	2	1	1	1	1
Local	1	1	1	1	1	1
5. Ground operational systems						
Infrastructure	3	2	1	1	2	1
Dedicated expertise	2	2	1	1	1	1
Information uptake systems	2	1	2	2	2	2
Recurring budgetary support	NIL	NIL	NIL	NIL	NIL	NIL
Inter institutional systems	1	1	3	3	3	3
6. Technology operational systems						
Open access satellite data continuity	3	3	3	3	2	2
Open access software	1	1	2	2	3	2
Reliable operational methods	2	1	3	3	2	1
Open access dissemination tools	3	3	3	3	3	2
7. Information use						
Government	2	2	1	1	1	1
NonGovernment	1	1	1	1	1	1

we focus on the experiences and challenges regarding geospatial data and information availability, user capacities on product development and use. A very limited discussion on scientific methods and accuracies followed during product development will be mentioned, whereas such details can be obtained from supporting references.

3 Agriculture and Food Security

Agriculture is the most important livelihood sector in the HKH region and provides a substantial proportion of rural income and employment opportunities to its estimated 210 million inhabitants. Typically, the very nature of subsistence-oriented farming is structured around a traditional weather calendar based on undisturbed climate (Vedwan and Rhoades 2001). The dramatic climatic and environmental changes that are taking place in the Himalaya are reported to have profound impact on the conditions for food production. Alongside socioeconomic changes including economic globalization, increasing accessibility, and land use conflicts, dynamic demography is affecting the subsistence agriculture systems across HKH region. Geographical constraints often restrict the flow of goods to and from isolated areas; landslides, mud falls, and soil degradation not only adversely affect crop yields and farming practices but also restrict transport and the advantages it brings.

Food security is a significant challenge in the HKH region, where the harsh biophysical conditions and short growing seasons constrain agricultural productivity and can create food deficits (Kurvits et al. 2014). As a so-called climate change hotspot, climate change, and extreme weather events like floods and droughts are projected to impact food security in mountain regions like the HKH particularly hard (Murthy et al. 2014). Recent vulnerability assessments show that over 40 % of households in the mountainous region of the HKH are facing decreasing yields in their five most important crops as a result of floods, droughts, frost, hail, and disease (Bhandari et al. 2015).

A few of the basic challenges addressing food security in the region lies on how to improve the underlying productivity of natural resources and cropping systems to meet increasing farmers demand and to produce food which is safe, wholesome and nutritious and promotes human well-being. Case studies on agricultural transformation of mountain areas show how farming of High-Value Crops (HVC) has increased food security and employment, thus improving living conditions for mountain people (Dadhani 1998; Partap 1999; Sharma and Sharma 1997; Tulachan 2001). The development of multifunctional agriculture systems, sustainable management of mountain agriculture practices such as terrace agriculture, shifting cultivation, developing narrow climate, and geographic niche-based local cropping systems, evolving resilient productive systems against climate disasters, improving water efficient and high crop intensity regimes in low elevations are a few priority areas which require scientific support and policy-level interventions on a continual basis. On the other hand, the development of effective communication systems and improved accessibility are also identified as potential windows through which useful and timely information and material would reach the farmers to face both environmental and social challenges.

Recent developments in use of Earth observation satellite data in combination with ground-based information, can reveal valuable information about environmental conditions that can subsequently impact the livelihoods of farmers. A Geographic Information System (GIS) and remote sensing technologies are helpful to identify regions experiencing unfavorable crop growing conditions and food supply shortfalls and to determine food insecure areas and/or populations (Minamiguchi 2004). Similarly, increase in use of Information Communications Technologies (ICTs) in facilitating rural development by providing agricultural extension services and real-time market price information, availability of rural payphone can play a significant role in enhancing the ability of rural families to continue, and perhaps enhance their contribution to national agricultural production and postharvest activities (Anderson and Bows 2008). These scientific and technological interventions involve in-depth operational research and application with a focus to leap forward into useful ground applications. In this context, an understanding of the current capacities on technological and user operational segments and the future scenario and needs over the decade in adoption of EO-based actionable information would stand as most supporting information.

3.1 Current Scenario Assessment

In the current national food security assessment system across the region, the District Agricultural Development Offices are responsible for the collection of crop situation data from field for national- and subnational-level analysis. These well-placed traditional systems need to be strengthened for better decision-making using substantial capacity developed for crop condition monitoring and production estimation using EO systems. Globally, satellite remote sensing has become an integral component of operational agricultural monitoring systems, enabling crop analysts to track development of growing season, and provide actionable information to decision-makers for developing effective agricultural policies and timely response to food shortfalls (Wu et al. 2014; Zhang et al. 2010). Yet, the lack of synergistic use of space and ground-based systems in the HKH region is limiting the potential of spatially explicit decision-making processes and translating the knowledge as crop advisory systems. The integration of satellite-based climate and soil moisture information with crop growth models is urgently need to allow for and improve assessments of drought scenarios and crop yield estimations and forecasts.

Across the HKH region, differentiated capacities and operational systems prevail, requiring different levels of technological intervention (Table 1). For example, the significant efforts are being made through FEWSNET program in Afghanistan and in Pakistan, The Pakistan Space and Upper Atmosphere Research Commission

(SUPARCO) in collaboration with provincial agriculture departments of Sindh and Punjab in Pakistan, has recently established satellite-based crop monitoring system, where high-resolution optical data acquired during peak growth seasons of February for rabi crops and September for kharif crops to assess seasonal yield forecast for major crops. However in Nepal, Bangladesh, and Myanmar, the adoption of remote sensing data is at very preliminary level. While there is a general agreement on significant need on such information, there are no dedicated institutional setup and expertise available. Practitioners in operational positions tend to possess moderate experience in remote sensing technology, but such awareness at higher strategic levels is very limited. On the other end of the spectrum, targeted end users of remote sensing products have very low technical expertise in interpreting technically complex products. This stresses the need for developing more customized and actionable information.

The recent efforts of NASA/SERVIR-Himalaya in developing national crop and drought monitoring systems over Nepal using satellite-based information and their integration with quarterly food security bulletins of Ministry of Agriculture and Development and World Food Program reveal possible leads could be achieved through effective user involvement. However, formidable challenges remain for the EO community in downscaling the national-level near-real-time crop monitoring systems into farm/site-level advisory systems, designing systems to effectively inform evolving operational crop insurance mechanisms, and improving Agro-Pastoral, Silvi-Agro-Pastoral flows and linkages to enhance livelihoods over high-altitude Himalayas.

3.2 Prospects and Challenges

The small farmer holdings, mixed crop conditions, persistent cloud cover, rain fed systems experiencing temperature stress, diseases, and land degradation are a few of the key elements within food security domain, where EO and improved sensors need to play vital role. The frontier technologies and modeling frameworks should help to integrate local knowledge with scientific data and technical skills to improve knowledge on mountain-specific traditional production systems, as well as local cropping systems within narrow climate and geographic niches. Given that unique climate and geographic niches often occur across boundaries, frameworks for regional and transboundary collaboration are important. Such frameworks would promote international benefit of technology interventions that would otherwise not be realized if only national perspectives are considered.

3.2.1 Crop Identification and Monitoring

A summary of the availability and continuity for remote sensing data products for the effective agriculture monitoring in the HKH region is give in Table 2.

Table 2 Needs and challenges of geospatial application—agriculture development in HKH region

Thrust areas/crop types	Current systems (2010–2015)	Technology continuity/new systems (2015–2030)	Ground segments (2015–2030)
Major cereal crops	• Medium resolution Optical + SWIR (MODIS 250 m, MERIS 300 m)	• Continuity of similar system SAR (Sentinel-1A, 1B)	• Improving accuracy of subnational agriculture statistics
High-altitude local crops	• Medium to high-resolution Optical + SWIR and Hyperspectral data (Landsat 30 m, CBERS, Hyperion) • No such data available	• High-resolution Optical + SWIR (Sentinel-2) • High-resolution multispectral data	• Introducing basic GIS/Mobile apps to agriculture extension staff • Significant technology effort to be made on methodology standardization to develop microwave-based crop area estimates
Crop sown area	• Medium resolution Optical + SWIR (MODIS 250 m, MERIS 300 m) • In season sown area assessment generated • Time lag in generation, reliability of estimates due to elimination of small crop holdings, cloud infestation effects the operational use	• High-resolution Optical + SWIR (Sentinel-2) with high temporal capability, better radiometric resolution • Availability of Microwave sensors to address could effects	• Introducing area-frame sampling approach and integration of UAV
Crop stress	• Medium resolution Optical + SWIR Landsat TM Water stress, disease stress-based operational methodologies lacking • Use of satellite/model-based soil moisture inputs lacking	• Continuity of similar systems Landsat TM type of sensor with additional bands of red edge and moisture • Experimental Hyperspectral data • SMOS-based/similar products assimilation to be achieved	• Improving infrastructure on automated ago-met stations • Research efforts/projects to be implanted over red edge and hyperspectral-based pilot studies • Pilot studies on soil moisture assessment and integration to be introduced
Crop advisories	• Reliable real time spatial climate data, stream flow, soil moisture products almost lacking • Crop production models integrating hydromet data, soil moisture, crop specific data as above are at research level	• Integrated regional specific models to developed to provide policy and operational level crop advisories	• Research studies on developing region specific spatial hydromet products and crop production models to be augmented

The current MODIS/SPOT-based temporal crop monitoring systems need continuity with more customization in data processing and delivery mechanisms linking diversity of users. The improved spatial resolution, as well as, enhanced radiometry are required for achieving better accuracy to assess small farmer holdings, and discrimination of mixed cropping systems. Most of the rice crop is grown in kharif season (monsoon), coinciding with the over cast cloudy conditions of the sky most of the time. The dedicated microwave missions, data access mechanisms, and R&D to identify observables needed in terms of polarizations, frequencies, and spatial and temporal resolutions are essential to study kharif season crop growth systems. High-resolution thermal infrared data along with shorter wavelengths (VIS, NIR and SWIR) would enhance the process modeling pertaining to energy and water balance and characterizing thermal environments of the agricultural crops. This would help toward estimating the water requirements/evapotranspiration which is useful in modeling crop stress and productivity in the context of rainfed and drought scenario.

Atmospheric effects are significant source of errors in multitemporal analysis of satellite data over the mountain regions. This calls for inclusion of specific atmospheric correction channels or independent atmospheric correction sensor to facilitate estimation of atmospheric parameters such as water vapor content and aerosol optical thickness in the upcoming sensors. Toward improving the accuracy of surface parameters, retrieval multiangular observation system is required for determining the Bidirectional Distribution Function (BRDF). This could be implemented with multiangle imaging sensor of five views having four bands in VNIR region. The improved information and knowledge from such new sensor systems could extend extensively to food production and availability components of food security in mountainous regions.

3.2.2 Supporting Food Security Through Agro Advisories—Moving Across Scales

Negative impacts of climate trends on crop production have been more common than positive impacts, as evident in some high-altitude regions. There have been several periods of rapid food and cereal price increases following climate extremes in key producing regions, indicating a sensitivity of current markets to climate extremes, among other factors affecting the food availability (Trostle 2010). In addition, the episodic response of food production to climate also impairs the supply chain systems of local people, threatening food availability. EO-based crop monitoring and geospatial models on crop productivity provide only a partial assessment of food security–climate change relationships in terms of food production and not specifically on food availability.

It is of particular concern how to link this multitude of information to policy-based issues involving sustainable development communities, assessing and ensuring food security prospects. Engagement of these policy communities requires much a broader and comprehensive scalable framework to translate productivity and production information from crop and farming systems respectively to food security systems. The geospatial technology plays a critical role in monitoring, assessing, and modeling food production through crop simulation models, continual monitoring of growth profiles over crop cycle, crop production estimations using spectral and ground-based yield data, and risk assessment, with the goal of minimizing losses and improving production stability against natural disasters (Belmonte et al. 2003; Grigera et al. 2007). The applicable satellite systems on precipitation (GPM) and soil moisture (SMAP) and regionalized climate forecast systems constitute primary information support needed for crop growth monitoring systems (El Sharif et al. 2015). In addition to biophysical monitoring, information on different farm specific local practices, agronomic challenges, and regional micro- and macroeconomics are drivers of food production and availability. This emphasizes that significant coupling of information across scales and disciplines is needed to understand linkages of food production and availability (Fig. 2; Table 2).

In order to move across sectoral scales, the EO community could play a critical role in coupling environmental models (e.g., climate forecasts, hydro-meteorological dynamics, crop growth, and productivity) with spatial econometric models. Such an effort is only possible with joint efforts of multiple expert communities, relevant subject domain experts, and policy-level decision-makers. Currently, such efforts are far from beginning in this region, and hence call for long-term multi-institutional initiatives to address the serious problems of food security exacerbated by climate change.

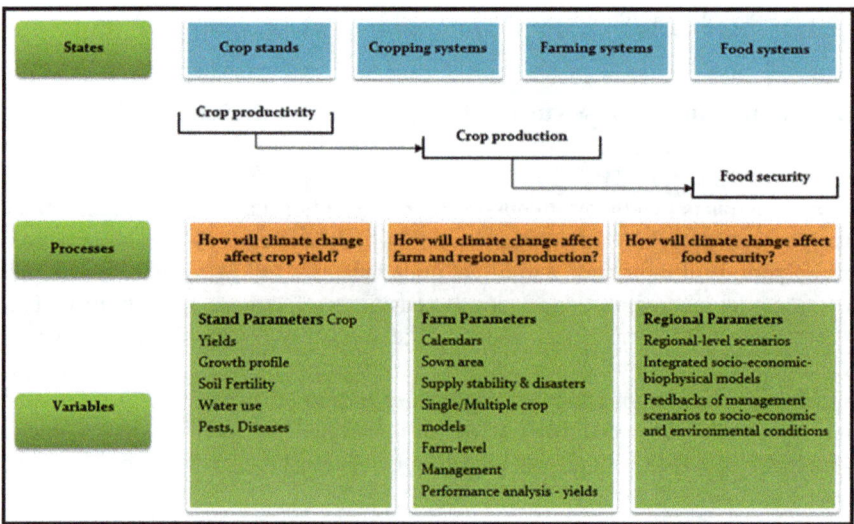

Fig. 2 Climate change impacts and food security moving across spatial scales: From crop production to food security

3.2.3 Alleviating Food Distress Through Bio-Resources Management

The high-altitude Himalayan livelihoods largely comprise of natural resource-based subsistence ecosystems. Apart from less agriculture land availability, the harsh climate, rough terrain, poor soils, and short growing seasons often lead to low agricultural productivity and food availability. The sustained food security hence is also determined by grasslands function and associated livestock patterns, availability of forest-based resources, and balanced inter flow of material regulating these three interconnected ecosystems. The mountain-specific practices like terrace cultivation, shifting cultivation, narrow niche-based cropping systems, multifunctional agriculture, and harnessing local water springs become pertinent in such areas. The Agro-Pastoral, Silvi-Pastoral, and Silvi-Agro-Pastoral systems also stand as great players in the region. By looking at HKH region mountain farming trends, Tulachan (2001) suggests focusing on crop–livestock farming systems, which will pose less pressure on common property resources such as forests and community lands, leading to positive impacts on the environment.

The effective functions of interactive ecosystems are largely governed by local innovations and practices of indigenous and local communities embodying traditional lifestyles. This is demonstrated in the wisdom developed over many generations through holistic, traditional, and scientific utilization of the lands, natural resources, and environment. It is generally passed down by word of mouth, from generation to generation and is, for the most part, undocumented. However climate change, globalization, and increased disasters are forcing many livelihoods to shift toward new practices like unsustainable production of cash crops, water harvesting, and use of pesticides in agricultural fields, leading to vanishing traditional cultural practices. While traditional knowledge is valid and necessary, however in the age, it is only beneficial when blended with enormous contemporary scientific knowledge. The adaptability to new practices requires appropriate infrastructure and scientific knowledge associated with vulnerability, disaster management, and unsustainable extraction of bioresources to ensure food security.

In this context, EO systems and associated geospatial technology play promising roles at multiple scales for understanding local practices, relevance, and changing internal and external drivers. Contemporary EO systems and models are natural frameworks to integrate downscaled high-resolution climate forecasts, high-resolution Digital Elevation Models (DEMs), large-scale land cover and bioresource inventories, crop and rangeland calendars and productivity, land degradation, and spatial explicit ecosystems services assessment and flows (Table 2). Such comprehensive biophysical information previously not available could be integrated with traditional knowledge and practices and social and economic drivers, through agent-based models, spatial econometric models, and other self-learning algorithms to delineate or stratify socioecological regimes across large bioclimatic regions of the high-altitude systems.

Locations and characteristics of such socioecological strata as defined by a certain degree of similarity in structure and function would constitute a cluster of food distress units. These clusters could be further studied using three models:

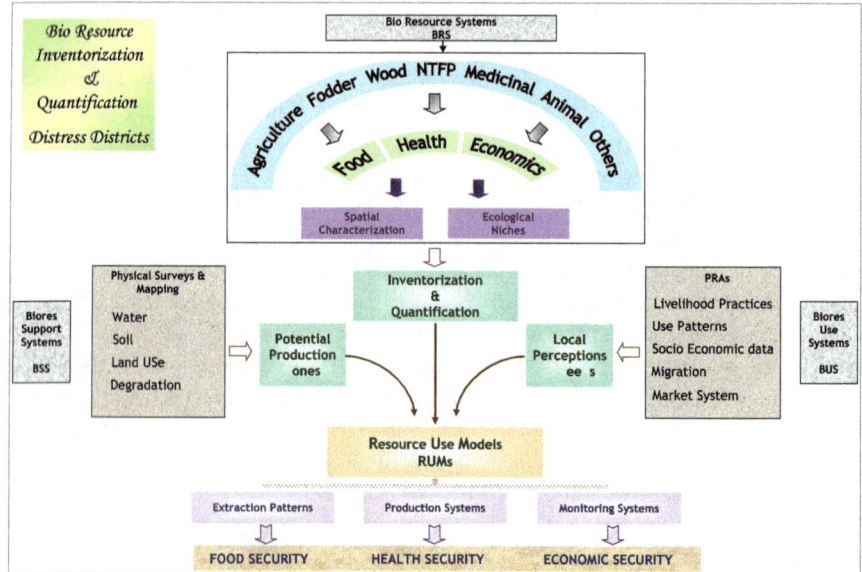

Fig. 3 Bioresource inventorization and quantification

(1) bioresources capital system, (2) bioresources support system, and (3) bio-resources use system. The integrated analysis of these different systems on spatial and temporal dimension would provide alternate scenarios of ensuring food security through sustainable management of bioresources. The different components of three subsystems and integration frameworks are given in Fig. 3.

The pilot testing and implementation of best possible scenarios across food distressed villages provide model concepts of bridging traditional and scientific knowledge for ensuring food security. Such an approach would also provide a framework for upscaling over the similar socioecological strata identified elsewhere. In this light, it necessary that the EO community reaches the last mile to stakeholders to implant technological bottom-up simple innovations, developing coupled traditional and scientific knowledge bases, cost optimization, and scalability. The strengthening of local participatory and governance systems, identifying appropriate capacity building needs and institutionalization, are also very critical to effective application of EO to addressing food security challenges.

4 Forest-Based Climate Change Mitigation

Carbon sequestration, carbon substitution, and carbon conservation are the processes through which forests can contribute to the mitigation of climate change. Contrasting a typical Land cover, Land-use and Forestry (LCLUF) approach, an

Agriculture, Forestry and Other Land-Use (AFOLU) approach greatly expands the potential for carbon sequestration in the land-use sector with greater benefits for the countries of the HKH. The forests of HKH region are estimated to mitigate 3500 megatons (MT) of carbon through avoided deforestation and degradation and 47 MT of carbon through improved forest management over next 20 years. Therefore, flexible, robust, credible, and well-tested Sustainable Forest Management (SFM) and integrated land-use management frameworks are needed to support the simultaneous reduction of carbon emissions, sequestration of carbon, and ensured provision of ecosystem goods and services.

The Collaborative Partnership on Forests (CPF) consisting of 14 major forest-related international organizations, institutions, and secretariats, developed the strategic framework on forest-based carbon mitigation and climate change. Among the six critical messages of the framework, one of the messages stresses the need for reliable, consistent, and continual information support to develop effective assessment and diagnosis, planning, implementation, and monitoring. In addition to the strong local context-based knowledge and action, the use of geospatial information emanating from multisensor remote sensing information is strongly advocated both for mitigation and adaptation strategies (IPCC 2006).

Globally, the practice of using geospatial science-based evidential information has significantly increased with the advent of open access satellite data, information products, mobile, and web-based data collection and dissemination platforms (Elwood 2010). However, due to differentiated capacities of HKH countries, there is a wide gap in terms of: (1) application and use of geospatial systems for forest cover change assessment and monitoring, (2) cost effective multiscale carbon monitoring to support the Reducing emissions from Deforestation and Forest Degradation (REDD) strategies, and (3) evolving planning and monitoring strategies for forest-based vulnerability and adaptation planning (Boyd and Foody 2011). Periodic land cover change assessments, forest degradation assessments, and biomass quantification, stand as three fundamentals for forest-based carbon mitigation strategies.

4.1 Current Scenario Assessment

A significant effort has been made over the region in developing periodic national land cover change data bases in the region as part of ICIMOD regional collaboration with regional member countries through the NASA-SERVIR-Himalaya initiative (Gilani et al. 2015a, b; Uddin et al. 2015). The details of current status on data availability are given in Table 3. However, dedicated operational systems and framework are lacking to produce periodic change assessment. Globally available annual tree cover and global change monitoring systems using Landsat TM data need significant improvement for regional-level applications. Satellite-based community-level forest cover monitoring and degradation assessment tools using high-resolution data are at research stage to bring out cost effective, user friendly open source methodologies (Gilani et al. 2015a, b).

Currently, assessment of forest growing stock is done at national and subnational levels. National growing stock estimates are developed using national-level multisource forest inventories. In order to address sustainable forest management, the national forest inventories are made exhaustive in terms of parameters and data collected across the entire country. Most of the national inventories follow systematic sampling with fixed grids and proportional temporary and permanent sample points chosen for data collection. In view of this complexity, national inventories involve intensive field sampling and are thus time and cost intensive. The estimates are generally planned over a 5-year time interval. However, due to time and cost constraints, biomass assessments in most HKH countries, apart from China and India, have not been carried out at regular intervals to support the need for forest carbon monitoring (Du et al. 2014; FSI 2011).

Equally, national-level growing stock estimates over large countries do not provide realistic subnational scenarios, while the next lower level assessments done at the district level are designed with district-specific requirements in terms of sampling design and time. The application of satellite remote sensing in optimization of cost and time to produce periodic biomass assessment is at early stages of application. Significant emphasis has been placed on developing such information, awareness, and expertise at both operational and at strategic levels, and institutional frameworks exist across the different countries of the region. However, improvements are needed on scientific methods being followed as well as on capacity building efforts on developing operational frameworks and assimilating new technologies to address periodicity, reliability, cost, and turnaround time.

4.2 Prospects and Challenges (up to 2030)

4.2.1 Land Cover Change Assessments

In recent years, at the global level, numerous efforts have been made to provide satellite-based medium-resolution land cover and forest cover information. Most commonly used are Landsat images, which are of long standing and widely used for continuous land cover monitoring (Gong et al. 2012; Hansen et al. 2013; Kim et al. 2014). At the global to regional level, these data products are developed through supercomputing techniques to provide reasonable results. At the national levels, more coordination, collaboration and ground validation are very much needed. Further synergies between national and global and regional partners need to be established. At this stage, significant effort needs to be made, to improve operational frameworks on Monitoring, Reporting, and Validation (MRV), ranging from ground data collection techniques to institutionalized delivery and use of products. The land cover monitoring needs in HKH over the next decade definitely require data continuity from Landsat TM type of high-resolution satellite systems. The local scale, very-high-resolution-based land cover change assessment is in its infancy in HKH due to several operational constraints. The likely challenges for the next decade are likely to be open source object-based algorithms, cost effective very-high-resolution data, scientific and technical expertise, and necessary infrastructure development (Table 3).

4.2.2 National Biomass Estimations and Mapping

In the current forest carbon estimation systems, excessive cost related to extensive field measurements and spatially inconsistent estimations along with limited control on uncertainty due to errors from allometry and sampling design are major challenges to addressed through new tools and approaches (Lei et al. 2009). Also for conservation management strategies, there is a need to determine the essential measurable properties such as species distribution, stand density, basal area, and canopy density that describe the forest vegetation and also influence the forest biomass/carbon stocks (Kim et al. 2010; Kwak et al. 2007; Lovell and Graetz 2001; Yang et al. 2013). According to the Intergovernmental Panel on Climate Change Good Practice Guidance (IPCC GPG) (IPCC 2003), remote sensing methods are especially suitable for independent verification of national LULUCF carbon pool estimates, especially for the aboveground biomass. Remote sensing instruments and techniques (space borne/airborne) are widely in use for forest cover monitoring, biomass estimation, mapping, and accuracy assessments at local, national or regional scales (Baccini et al. 2004; DeFries et al. 2007). Most of the developing countries do not have forest inventory data to get accurate biomass figures. Furthermore, several methods have been proposed for estimating forest biomass using remote sensing techniques that make use of a combination of regression models, vegetation indices, and canopy reflectance models (Cho et al. 2012; Gonzalez et al. 2010; Huang et al. 2013; Kajisa et al. 2009). Remote sensing-based biomass estimation, mapping, and accuracy have increasing attraction for scientists (Ahmed et al. 2013; Foody et al. 2003; Franklin and Hiernaux 1991; Lu 2006; Nelson et al. 1988; Steininger 2000; Zheng et al. 2008). Gibbs et al. (2007) reviewed different remote sensing data sets by discussing the befits, limitation, and uncertainty.

Approaches that make full use of remote sensing techniques to estimate Above Ground Biomass (AGB) are therefore needed (Le Toan et al. 2011). To overcome this issue, NASA's Global Ecosystem Dynamics Investigation Lidar (GEDI) is planned to launch in 2019. GEDI will use a laser-based system to study a range of climates, including the observation of the forest canopy structure over the tropics, and the tundra in high northern latitudes. This data will help scientists better understand the changes in natural carbon storage within the carbon cycle from both human-influenced activities and natural climate variations (Stysley et al. 2015). The European Space Agency (ESA) the BIOMASS Mission is also planned to launch in 2020 from 70° N to 56° S with accuracy not exceeding ±20 % (or ±10 ton/ha in forest regrowth) with forest height with accuracy of ±4 m at spatial scale of

Table 3 Needs and challenges of geospatial spatial applications—land cover and forest degradation at national and local level

Thrust areas	Current system (2010–2015)	Technology continuity/new systems (2015–2030)	Ground segments (2015–2030)
National level			
Land cover	• Consistent periodical land cover change data available except Myanmar • Dedicated operational systems and framework lacking to produce periodic change assessment • Global annual Tree cover monitoring system available using LANDSAT TM and MODIS data needs improvement in terms of product type and quality	• Continuity of Landsat TM type of sensors • Regionalized automated algorithms and provision of computing platforms attracts users to adopt space-based monitoring systems	• Significant effort need to be made to evolve operational framework on monitoring, awareness expertise and institutionalizing • Synergies with global and regional partners to be established • Research studies on regional algorithms to take place
Forest degradation	• Landsat-based periodic crown closure changes used as one of the measures. Need to be made operational	• Operational algorithms to be evolved on crown closure change assessments as terrain effects and shadows limits reliable classification • Adoption of red edge and moisture-based remote sensing studies shown significant promise on degradation assessment. • Availability of red edge, moisture and thermal sensors on similar lines of ASTER is very important.	• Significant research studies dovetailed with user needs to take place
Local level			
Land cover	• Satellite-based community forest monitoring tools using high-resolution data are at research stage • Cost and technology optimization and framework for operational use lacking	• Multispectral high-resolution sensors systems are limited, need for developing countries in spatial planning process is critical, open source object-based algorithms and S/W are lacking to analyses high-resolution images	• Enhanced localized EO planning acquisitions helpful (e.g., from International Space Station (ISS) platforms such as ISS-SERVIR Environmental Research Visualization System (ISERV) or other) • Interventions of GEO or Systems like International Disaster charter, Subsidy provisions from commercial systems are a few cost optimizations to be attempted • Research focus to go on object extraction algorithms

(continued)

Table 3 (continued)

Thrust areas	Current system (2010–2015)	Technology continuity/new systems (2015–2030)	Ground segments (2015–2030)
Degradation assessment	• Critical need exists to support community forestry programs promising research and operational aspects to be tested • Very limited studies were available	• Future satellite systems should consider this global need of local degradation assessment, large proportion of stakeholders for space data could belong to this community, sensors on the similar lines RapidEye supported with 1 m BW will be in demand	• Research work on piloting, demonstration, capacity building is needed. Involvement of global expertise is important • Studies on linking to national scales and packaging for use at local level as simple monitoring tools is challenge

100–200 m (Le Toan et al. 2011). Another NASA-ISRO Synthetic Aperture Radar, Mission (NISAR) is also in pipeline to be launched in 2020. According to Alvarez-Salazar et al. (2014) NISAR's unprecedented coverage in space and time will reveal biomass variability far more comprehensively than any other measurement method.

In the HKH region, where steep terrain and complex forest characteristics can play problem for accurate inversion of the satellite data, robust validation efforts based on field measurements are needed. Although, sporadic efforts in the form of filed campaign are being made to produce national and subnational measurements. A more structure effort and network of institutions are needed to bring the validated satellite products and to make them available for national usages.

Capacity building and long-term technical back stopping are also essential to strengthen the local partners who can take advantages of global products. Indigenous human resource adequately trained with new technology will ensure continuity of efforts as well as influence the decision-making process.

4.2.3 Forest Degradation Assessment—REDD MRV Systems

A few of the critical challenges of REDD MRV systems are reliable estimation of deforestation, afforestation, forest enhancement, and degradation. Currently, fairly good satellite-based global to sub national monitoring systems are available to monitor forest cover gain and loss (Gibbs et al. 2007). However, quantification of changes in carbon fluxes due to forest cover change dynamics is still a complicated task (Sasaki and Putz 2009) and stands important as it contributes to larger proportion of carbon dynamics. The national-scale degradation assessments to quantify carbon fluxes are related to timber and fuelwood extractions, grazing, and fire impacts (Miettinen et al. 2014). Significant efforts need to be made to provide crown closure changes-based degradation assessment to support carbon assessment using remote sensing data. At the national level, operational algorithms to be evolved on crown closure change-based forest degradation assessment. Availability of red edge, moisture and thermal sensors, and application are important for detection of changes and mapping degraded areas. However, such national approaches are not suitable at local scale community forestry systems due to lack of driver data and microlevel canopy changes. In this context, adoption of community-based forest monitoring tools involving local stakeholders is being tested as one of the alternate systems. Danielsen et al. (2011) have explained the role, accuracies, and economics of community-based, forest-based monitoring systems. The community-based monitoring systems involve scientifically identified permanent sample plots over individual community forest areas, developing protocols and training and periodic measurements of different parameters at specified intervals (Poudel et al. 2014). One of the critical challenges of these community-based monitoring systems is efficacy of low-intensity permanent sample points representing the large area under study, lack of spatial extrapolation power to address changes beyond the sample plots such as forest cover losses and degradation over the entire study area (Härkönen 2002). In such a case, difficulties also arise in the inability of point-specific measurements to reliably quantify carbon dynamics due to leakage, additionality, and persistence. In an effort to handle these challenges, integration of remote sensing systems with field-based measurements are being evaluated to assess potential application and operational feasibility for community-level MRV systems (Palmer Fry 2011) addressing challenges on detecting local-level changes and integration with local filed monitoring systems. Remote sensing-based spatial explicit estimates provide wall to wall coverage, enabling decision-makers, and managers to understand the dynamics beyond point-based estimates (Baccini et al. 2012; De Sy et al. 2012). Integration of forest measured and monitoring data collected by the local communities with satellite systems could provide precise and accurate reporting of degraded hotspots (Fig. 4). Affordable, very-high-resolution multispectral data is going to be a huge need in the context of upcoming REDD initiatives to enable assessments of local species and stand-level degradation and biomass change assessment (Table 4). The adoption of multiresolution satellite systems also helps to address biomass estimations at species-, stand-, and forest-type levels, enabling a spatial linkage of different scales of information to reach from community level to national estimates.

5 Flood

The HKH region encompasses high-mountain landscapes and is prone to natural disasters of myriad types due to active geotectonic setting, sensitivity to climate extremes, fragile ecology, large degree of poverty, lack of infrastructure, and poor accessibility to services. An increasing trend in the recurrence of natural disasters and associated impacts due to floods, glacial lake out bursts, landslides, and forest fire is reported over HKH region. The total recorded major disasters events during 1900–1910 to 2000–2010 have increased from 7 to 617 involving millions of human and economic losses, even when not considering local-level events. The

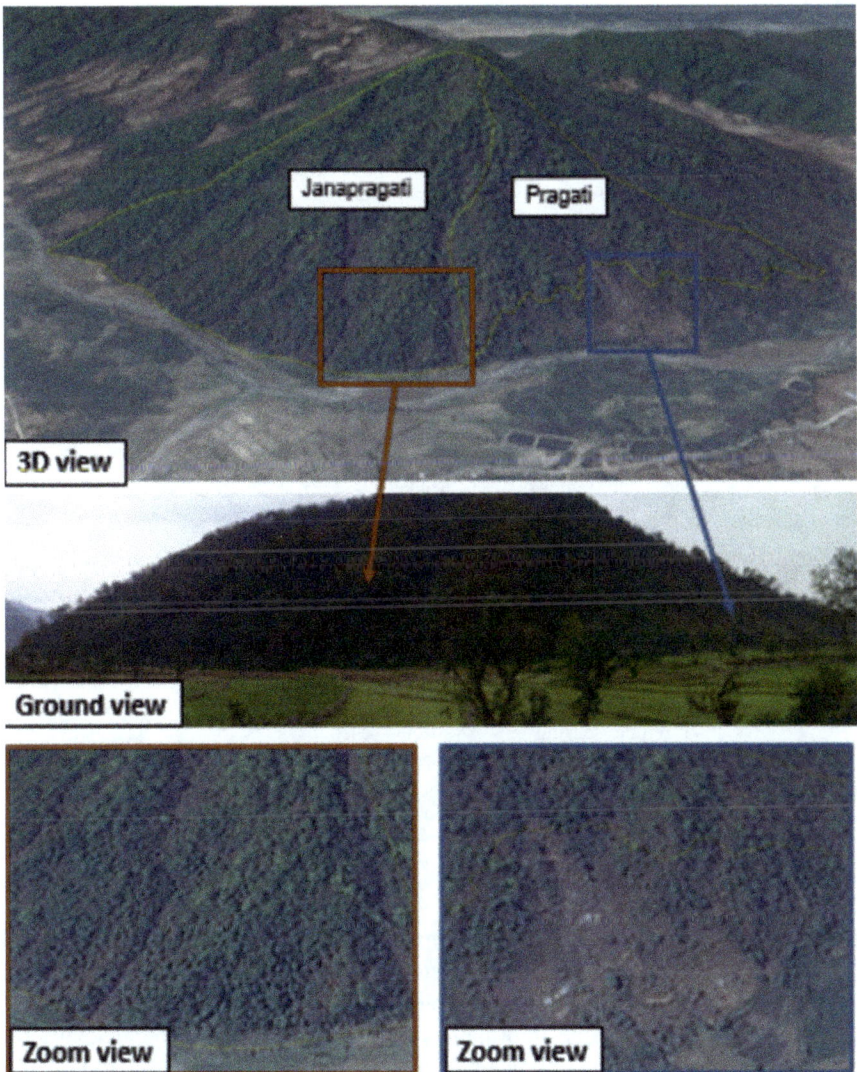

Fig. 4 Synchronization of satellite images and ground photographs

unprecedented scale of disaster events like flood (Fig. 5), landslides, and forest fire in recent years are found to have close links with increasing incidents of extreme climate events. In addition, rapid but unsustainable development, population explosion, and economic development in the HKH region are expected to inflate the cost of flood impacts.

The uncertainties involved in understanding high-altitude land surface and atmospheric processes and extreme climate dynamics are identified as primary

Table 4 Needs and challenges of geospatial spatial applications—forest carbon monitoring at the national and local level

Thrust areas	Current system (2010–2015)	Technology continuity/new systems (2015–2030)	Ground segments (2015–2030)
National carbon accounting	• National growing stock assessment designs and measurements in place. • Limited interventions from remote sensing takes place to improve sampling designs and estimations • Periodic assessments, needs and framework lacking to support monitoring carbon fluxes	• Sampling designs and demonstrations involving optical, microwave and LiDAR data depicting cost and data optimization required • Regional microwave-based height retrieval algorithms are needed	• Capacity building on microwave-based biomass assessments and spatial modelling at user end is an urgent need • Ground campaigns through mobilization of local forest communities under the guidance of forest officers
Community carbon monitoring	• Community-level field-based monitoring tested in a few countries • Complimentary use of remote sensing and field technology to enhance thematic content and cost optimization to be implanted	• Species- and canopy-level proxy indicators and stratification, canopy geometry and species-based parametric and nonparametric models to be developed	• Research collaboration, technical training and infrastructure development needed

factors for such increased risk. To overcome these challenges, we need to invest on innovative and low-cost systems to develop reliable predictive models and tools with monitoring capabilities that integrate local knowledge and ensure timely communication of actionable information. A significant need to build long-term resilience against disasters, acting upon hazard and risk by building robust and planned infrastructure and mitigation systems is also advocated.

5.1 Current Scenario Assessment

The role of geospatial tools which includes both remote sensing and GIS in spatial planning in the context of different thematic focus, including disaster has long been recognized in flood management (Abdalla and Li 2010; Goodchild 2006; Tahir 2007; Uddin et al. 2013). The improved availability of operational all-weather capability satellites, constellation of satellite systems, satellite-based rainfall mapping and estimation systems, Unmanned Aerial Vehicles (UAV), low-cost

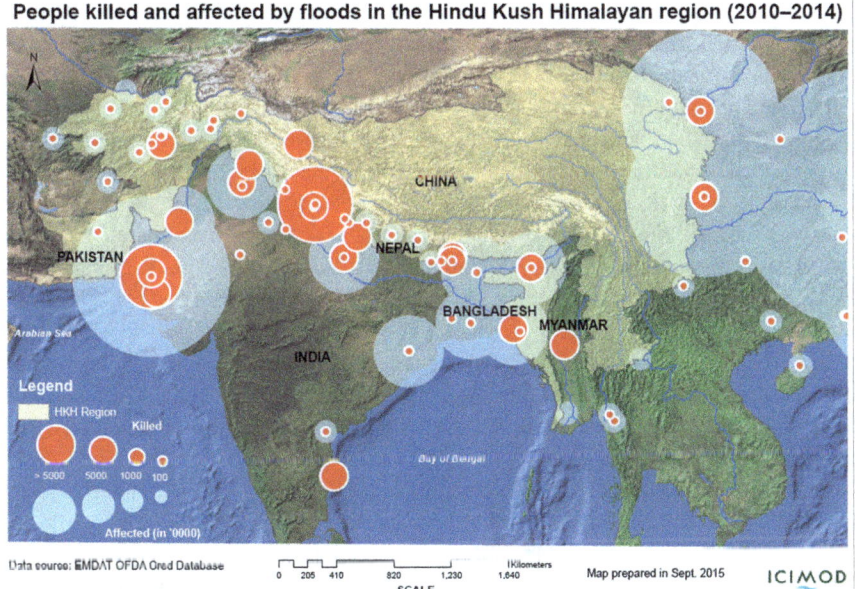

Fig. 5 HKH causality by flood

ground-based sensors, mobile-based communications, spatially explicit simulation and statistical models, and open access high-resolution data and information products have improved globally the efficiency of disaster early warning and emergency response systems.

The cooperation among regional and international agencies in terms of data sharing, response synergies, and risk reduction platforms are also contributing to the disaster management. During the last decade, HKH countries stand high in invoking and obtaining support from the International Charter for Space and Major Disaster for disaster response. The current level of adoption of these systems across the region is presented in Table 1. These seed attempts need be to further strengthened as several technological and institutional gaps and challenges do exists which needs to be addressed over the coming decade. The major thrust need to be drawn on flood risk reduction systems consisting reliable flood risk reduction information and early warning systems which are described as below.

5.2 Prospects and Challenges

5.2.1 Flood Risk Reduction Information and Planning

The Sendai Framework articulates amongst others, the need for improved understanding of disaster risk in all its dimensions of exposure, vulnerability and hazard characteristics to mainstream Disaster Risk Reduction (DRR) in development through risk informed policy and investment (Sendai Framework for DRR 2015–2030). Risk informed policy and investment has become imperative in view of rising trend of extreme events due to climate change, and rapid economic development on the other hand compromising safety concerns. Tracking loss and damage associated with natural hazards at national and subnational levels help us monitor the impacts of disaster, thus forming the basis for risk informed policy and investment, and a means to evaluate success of DRR investment. Pre- and Post-2015 framework for DRR puts emphasis on "Record, analyse, summarize and disseminate statistical information on disaster occurrence, impacts and losses," (Sendai Framework for Disaster Risk Reduction 2015–2030).

An overview of the status, gaps, and challenges on evolving geospatial support for disaster risk reduction systems is given in Table 5. The spatial delineation of flood risk zones using historical flood events, inundation maps, land cover, settlements, topographic variability, drainage patterns, and river network have been prepared as a case example for a watershed in Nepal to support planning and reduce associated life and property losses (http://apps.geoportal.icimod.org/raptiflood). However, the currently available risk reduction databases over highly critical areas are not appropriate available and if available are not for operational working at local level. The availability of local scale high-resolution DEM, infrastructure and settlements databases, and inundation scenario for various return periods and incremental rainfall situation has been a serious gap of information and is of urgent need in the HKH region.

The adoption of very-high-resolution data due to cost constraints and lack of reliable feature extraction algorithms in developing such information also needs to be addressed. The calibration and development of open source physical models to incorporate run off and inundation scenarios to support operational planning is of urgent need. The development of Decision Support Systems (DSS) to integrate and analyze such diverse thematic datasets to support informed decision-making for pre-flood risk reduction planning at local levels is also lacking. The technology demonstration with cost effective and user friendly DSS has still to go a long way in the region. These database and information systems are also most needed during flood emergencies for damage assessment, relief, and recovery operations. The cross-sectoral institutional understanding and institutional strengthening toward evolving national disaster emergency data infrastructure, associated policy support systems, integration with traditional knowledge systems, and adoption of DSS on the similar lines of China and India is of immediate need.

In this regards, the recent collaborative effort of Ministry of Home Affairs (MoHA), Nepal and ICIMOD, Kathmandu under SERVIR-Himalaya program on establishing online multiscale disaster information and risk assessment is worth mentioning (http://apps.geoportal.icimod.org/disaster/). It is essentially based on a decade-long disaster loss database from MoHA. The online product is directly relevant to policy makers and planners to make risk informed policy decisions and investments through better comprehension of disaster and impact in the temporal and spatial perspectives. The system provides temporal trends of disaster impacts

Table 5 Needs and challenges of geospatial spatial applications—flood risk reduction planning

Thrust area/response, recovery	Current systems (2010–2015)	Technology continuity/new systems (2015–2030)	Ground segment (2015–2030)
High-resolution terrain	• 2–5 m DEM available over limited area • 90 and 30 m DEM in public domain	• Ortho rectified DEM needs to be prepared • Higher resolution DEM needs to be generated over critical flood plains	• Ground-level local organization to be involved and strengthened
Infrastructure/settlement inputs	• Less than 1 m data-based mapping of vulnerable areas lacking	• Feature extraction algorithms, periodic data updating • Data cost optimization mechanisms to be addressed	• Significant research on feature extraction • Local institutional involvement to be addressed
Inundation inputs	• Inundation maps produced based on optical data • Optical data limited by cloud cover • Microwave-based maps not in operational use • Processing of microwave data in steep terrain still a big challenge	• Operational data availability mechanisms and algorithms • Cost optimization methods to be optimized • Collaboration with Global Flood Monitoring System (GFMS)	• Dedicated SAR mission less than a day revisit, all-weather capability, satellite constellation
Real time damage assessment	• MODIS and Landsat TM used but do not meet all needs due to resolution limitations • Significant limitation in providing damage related inputs in real time • Lack of framework to integrate data sources from multisensors and systems	• Strengthen hydrological modelling approaches and undertake research to prepare inundation maps • Large swath, high-resolution, high-temporal optical data to provide coverage within less than 1 day • Continued availability of SAR missions on open access basis • Possible synergies between low Earth orbit (LEO) and available geostationary orbit (GEO) systems in the region	• Ground segments capability to process high temporal and spatial data products • Satellite constellation, formation flying concepts • Microwave-based algorithms and processing capacities at ground • Research studies on LEO-GEO integration studies

represented by loss of lives and economic loss, a basis for evaluating success of DRR interventions at national, district, and subdistrict level. The local scale DSS helping flood risk planning developed over local watershed of Nepal is an another good example. Similarly, great relevance exists to develop such systems over Bangladesh, Bhutan, Myanmar, and Pakistan.

5.2.2 Flood Risk Reduction and Flood Early Warning

Flood management necessitates the real-time event capture both spatially and temporally. Disaster events like floods, for example, are relatively faster in nature, having different temporal and spatial dimensions than drought and crop pests/diseases, which are slow onset types. This necessarily calls for critical inputs from EO systems. The significant cloud cover problems during flood seasons in the HKH region necessitate all-weather capability EO systems. The quantification of catchment physical characteristics, such as topography and land use, and catchment variables such as soil moisture and snow cover supporting flood risk reduction are well supported using high and low-resolution polar orbital earth resource satellites. However, there has been a significant weakness in turnaround time in providing reliable information during flood events in the region. The reliable medium-term climate forecasts, real-time inundation levels and associated damages have been always a limiting element from EO systems (Table 6). It is important to progress toward all-weather capability systems to: (i) increase time and frequency of coverage, (ii) improve coverage access and delivery, (iii) increase resolution of Digital Terrain Model (DTM) for local applications—this includes very local (1 m) information (town, small watershed, etc.), and (iv) achieve the ability to access EO products and services on real and near-real-time basis.

The satellite constellation with elements of autonomy and 'formation flying' mode on a long-term basis should aim to provide less than a day revisit and all-weather capability. The suite of satellite systems of LEO, GEO satellites, and a dedicated SAR mission would add substantial strength to real-time monitoring of the floods and also capturing the moving targets like cyclones (Table 6). There are several global and regional LEO systems and GEO systems of China and India operating over the region. However, there is no significant effort made to assess the realistic quantitative assessment on synergetic benefit of all these systems, associated gaps, and challenges. Most of the collaborations and responses are limited to the emergency response during flood events thorough international space charter. These aspects should be discussed at international GEO summits or evolving regional GEOs to evolve more synergistic suite of satellite systems. The disaster monitoring satellite if launched is perhaps one of the first attempts in this direction. The regional partners need to be encouraged to participate in such global missions before launch for fruitful product development and use.

One of the major challenges in the context of floods is on the understanding of climate and its variability. It necessitates studies on various atmospheric parameters and processes leading to operational applications and predicting major climatic events with high accuracy. The aim would be to deliver reliable climate information in useful ways to maximize opportunities for decisions and help to minimize risks due to floods. In this context, the efforts from SERVIR on providing climate data and forecasts through ClimateSERV are worth mentioning (http://climateserv.nsstc.nasa.gov). The real-time and effective use of GPM and SMAP missions to provide regional products is one of the important elements for the coming decade, especially in terms of validating and customizing products. Such efforts could lead to reduced

Table 6 Needs and challenges of geospatial spatial applications—flood early warning systems of HKH region

Thrust areas	Current systems (2010–2015)	Technology continuity/new systems (2015–2030)	Ground segment (2015–2030)
Real time satellite and station blended rainfall data and delivery	• Limited use of TRMM 3B42V7, TMPA, CMORPH, GSMaP, CPC RFE • Limited efforts on blended data, limited efforts to adjustment or modify in precipitation retrieval algorithm in mountainous region in HKH region • Limited tools: Geo-WRSI	• Adoption of GPM data and blending • Undertake research to fill critical gaps in those fields, modify precipitation retrieval algorithms in mountainous region in HKH region • Integration with new satellite missions • Integration with Geosynchronous systems of China and India • Adoption of high-resolution Integrated Multisatellite Retrievals for the Global Precipitation Measurement Mission (IMERG) precipitation product	• Significant research on GPM validation, awareness to take place • Pilots on customized products • Partnerships with national systems and agencies and global teams and efforts
Communication system	• Dependent on local network, not dependable during disaster event • Satellite systems are very expensive	• Service from communication satellite systems to be made affordable	• Active participation of national organizations
High-altitude observatory	• Very scattered, lack of resilient and dedicated systems • Difficulties in ground maintenance • No opportunities for mountain (HKH) specific mission	• New Innovations in instrumentation, recording and survey against all-weather capabilities • Installation of AWS to validate the space products	• Significant research and funds should be engaged to observe, measure, assimilate into scientific and decision-making systems
Soil moisture	• Very limited awareness and use of Satellite-based Soil Moisture Products	• Adoption of SMAP data, blending and integration into hydrological and landslide models	• Significant awareness building and research on validation and application of SMAP data and new sensors
Rainfall forecasts	• No dedicated national forecasts 3–4 h of lag (latency) time—critical for flash flood • Limited use of Global forecast system (GFS) data or Global Data Assimilation System (GDAS)	• Integration with global efforts on short and long-term forecast • Improve global forecasts with local surface and ground data	• Global collaboration efforts in evolving regional systems need to be focused

(continued)

Table 6 (continued)

Thrust areas	Current systems (2010–2015)	Technology continuity/new systems (2015–2030)	Ground segment (2015–2030)
Run off/flow levels	• Hydrological models used • Satellite Altimetry models over plains are being attempted	• Adoption of GFS data generated by IMD, and Weather and Research Forecast (WRF) data generated by ICIMOD • Improved hydrological forecasts using better satellite-based input information as above • Altimetry models to be further tested • Continuity of Altimetry-based sensors • Dedicated SAR Mission (regional hydrology) • Satellite Constellation-Less than a day revisit, large swath, high-resolution optical sensors • Small satellite missions process understanding • Improve regional coordination and sharing of data	• Significant research on hydrological modelling models to be strengthen Institutional linkages and traditional knowledge systems to be integrated to make actionable products

uncertainty in climate projections by providing timely information on the forcing and feedbacks contributing to changes in the Earth's climate. The blending of national hydromet measurements systems and regional hydromet systems like HYCOS with GPM missions would enable the provision of real-time climate information for both flood forecasts and flood damage assessments. In addition, real-time applications of satellite altimetry-based flow information in conjunction with regional specific in situ measurements and hydrological models on evolving flood warning systems as developed in Bangladesh should be further tested. A summary of progressive events leading to development of effective flood early warning and risk management systems is presented in Fig. 6.

5.2.3 Flood Emergency Response

Emergency responders and managers need near-real-time situation reports as flooding unfolds to be able to appropriately deploy response interventions. Traditional data collection means are limited in timely and updated information gathering at wider geographical scope. Satellite data with wide viewing scope and high temporal frequency when considered in unison with multiple sensors bridges this critical gap, effectively feeding in near-real-time information into

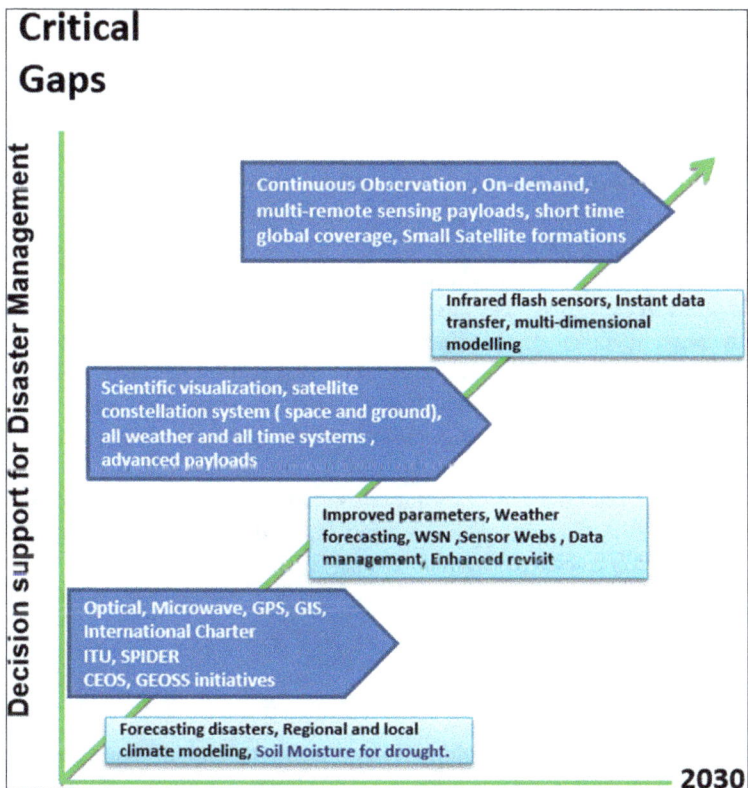

Fig. 6 Summary of progressive events leading to development of effective flood risk management systems

decision-making processes. Advances of space science are now able to provide multispectral satellite images with submeter spatial resolution of any corner of the Earth. Under ideal environmental situations, satellite-based mapping can provide information on progression of flood hazards, pockets of vulnerable exposures, and safe sites for evacuation, minimizing impact from the event. On the software side, there are algorithms to automatically delineate inundation layers to map flood extent and assessment of situation when event happens. Cloud cover precludes ground observations and poses a major challenge, but microwave EO satellite systems have shown a lot of promise in overcoming this challenge. Operational challenges with microwave systems are affordability, inadequate temporal frequency, and need for rigorous processing techniques to adequately address terrain influence, especially in the HKH region.

The role of models in flood scenario development, though a proven concept, is challenged by unavailability of exposure data and necessary input data such as rainfall and topography. Rainfall pattern in mountainous terrain are affected by orographic influence and current satellite-based rainfall product (GPM with 0.1°

spatial resolution) do not capture rainfall distribution and spatial variability adequately. There is need for satellite-based rainfall products with higher spatial resolution to capture spatial variability of rainfall for improving discharge estimates and inundation scenario.

The internet has revolutionized the way information sharing and delivery happens. Internet-based delivery systems reliant on elaborate infrastructure are often absent in disaster prone far-flung communities, where decisions must be made that are often not within national delivery mechanisms. Mobiles services, despite remarkable penetration, have not reached many poor communities living in disaster prone area. Moreover, SMS-based delivery mechanisms, despite being a preferred communication mechanism, have not made the difference as one would expect (http://sm4good.com/2013/05/02/code-conduct-sms-natural-disasters/). A new form of delivery mechanism needs to evolve such that it operates in conjunction with community-based DRR apparatus. Satellite-based communication systems such as VSAT are still limited by high cost, and unless high-value infrastructure are involved it is often not the choice. Solutions targeting developing countries need to be simple and cost effective so that scaling up of the system is not hindered by management and financial constraints.

6 Conclusions

The HKH mountains serve as globally significant natural resource regimes and climate regulation systems. In the context of climate change and transboundary issues, the understanding of mountain ecosystems needs utmost priority and importance. There is an urgent need to orient EO efforts to characterize, monitor, and model the mountain systems in the way designed to study oceans and atmosphere. Current satellite-based EO systems and associated products play a limited role in understanding high mountain cryopshere dynamics, associated hydrology regimes, and the erosion of ecosystem services. The ability of current ground sensor networks is also limited in monitoring high-altitude climatology, atmospheric chemistry, and dispersion, vegetation stress degradation of biodiversity and associated services.

Primary efforts for the region with respect to EO include the validation and customization of satellite-based rainfall and soil moisture products, their integration into climate and hydrological forecasts, and building end-user capacity and strengthening institutions that can benefit from EO. The future endeavors should also focus toward better quantification of snow/ice depth to understand precursors of hydrological dynamics, disaster vulnerability, and upstream/downstream dynamics across nations (water availability, life security). This should facilitate packaging EO observations into more reliable, readily usable and customized region specific climate services related to natural resource and disaster sectors. Essential EO-informed products needed to support mountain livelihoods will be easy-to-access and interpret, and be of high-spatial and temporal resolution. It will cover vegetation composition, stress, and productivity for food security and ecosystem services assessment, disaster support information for response and recovery, and vegetation degradation and biomass maps to address carbon monitoring. The combined use of satellite, ground and model-based studies to monitor and assess land degradation and air quality also contributes to ecosystem health.

All these applications needed to be packaged into Essential EHVs in line with ECVs and provided on turnkey basis through global and regional cooperation on the similar lines MODIS value-added products. The EHVs underlines the concept of converting information into *actionable products and services*. In this context, various global efforts like WRI, Silva carbon, WMC, GEO, NASA, MAIRS and SAARC, and regional/national efforts would be closely networked through a dedicated mountain centers and platforms such as ICIMOD. An improvement on the currently disaggregated nature of these projects would be to work on a more programmatic basis.

The different dimensions of enhancements in EO capabilities are required to develop above-mentioned products. A few of the critical to address include:

- Improved satellite-based stereogramatic capabilities to understand and quantify snow and ice depth
- Oriented and customized sensors and products to better monitor and forecast mountain climate
- Satellite constellation systems to enhance repetitive cover to address disaster and resources monitoring-related aspects
- Strengthening advanced planning and data acquisition systems for disasters
- Active sensor (microwave)-based turnkey products for disasters
- Short-term experimental small satellites to study geophysical and biophysical mountain complexities
- Satellite—Sensor webs to address problems like flash floods, glacier melts, and glacier lake outbursts
- High radiometric and spectral channels-based systems to monitor ecosystem processes, invasion, stress, productivity, and water quality

These developments could only be realized through the integration of appropriate ground measurement and validation systems. The lack of reliability or nonusage of ground-based information/data while developing sophisticated models and algorithms are found to ultimately minimize the accuracy, dependability, and use of global turnkey products. The following would strengthen the region:

- Provision of common pool data
- Development and provision of algorithms and products to encourage replication, improvements, and utilization
- Strengthening of forecasting abilities and related systems
- Cloud computing platforms

With the advent of developing and availability of tailor-made services, the user capacity building efforts should focus more on services specifications, potential application, need and method of integration with local datasets for value-added services, sustainability, and institutionalization approaches.

It is also essential to reorient and enable user segments so as to achieve enhanced uptake of actionable products. These efforts could include:

- Popularizing and bringing larger awareness among users
- Infusing local flavor to dissemination systems
- Empowering simple local mechanisms, institutions, and uptake systems
- Developing and building dedicated a larger tier and network of policy, scientific, and local user communities
- Introducing key facilitators and practitioners who can take actionable products to a diversity of users
- Enhancing the layperson's understanding of the use of online web applications and ability to contribute more accurate data through crowd source systems.

The weak bridge between researchers/products and policy makers on the utilization of available satellite resources should be addressed as one of the major capacity building initiatives in the HKH countries. Integration of global expertise and experience in blending innovation mechanisms with local knowledge both from top-down and bottom-up approaches facilitates for policy advocacy at local scales. Regular interaction between researchers and application developers with policy makers for better understanding and ensured uptake of EO-informed systems is also necessary. While these approaches and mechanisms are standard, the major focus has traditionally been oriented toward service development with minimal policy support. Focusing on drawing policy support for EO service use is the formidable challenge to overcome in the coming decade to ensure EO systems advance the last mile toward sustained use.

Acknowledgement The authors thank ICIMOD and NASA for all the support in developing this manuscript. We also thank NASA- and USAID-supported SERVIR-Himalaya program for the necessary support to undertake the current scenario assessment. The views expressed by ICIMOD authors are at individual level and do not necessarily represent the institution.

References

Abbas, S., Nichol, J. E., Qamer, F. M., & Xu, J. (2014). Characterization of drought development through remote sensing: a case study in Central Yunnan, China. *Remote Sensing, 6*(6), 4998–5018.

Abdalla, R., & Li, J. (2010). Towards effective application of geospatial technologies for disaster management. *International Journal of Applied Earth Observation and Geoinformation, 12*(6), 405–407.

Ahmed, R., Siqueira, P., & Hensley, S. (2013). A study of forest biomass estimates from LiDAR in the northern temperate forests of New England. *Remote Sensing of Environment, 130*, 121–135.

Alvarez-Salazar, O., Hatch, S., Rocca, J., Rosen, P., Shaffer, S., Shen, Y., Sweetser, T., & Xaypraseuth, P. (2014). Mission design for NISAR repeat-pass Interferometric SAR. Pages 92410C-92410C-92410. SPIE Remote Sensing. International Society for Optics and Photonics.

Anderson, K., & Bows, A. (2008). Reframing the climate change challenge in light of post-2000 emission trends. *Philosophical Transactions of the Royal Society of London A: Mathematical, Physical and Engineering Sciences, 366*(1882), 3863–3882.

Baccini, A., Friedl, M., Woodcock, C., & Warbington, R. (2004). Forest biomass estimation over regional scales using multisource data. *Geophysical Research Letters, 31*, 1–4.

Baccini, A., et al. (2012). Estimated carbon dioxide emissions from tropical deforestation improved by carbon-density maps. *Nature Climate Change, 2*(3), 182–185.

Badhani, K. (1998). Enterprise-based transformation of hill agriculture: a case study of vegetable growing farmers in Garampani area, Nainital district, India.

Bhandari, T. S., Shrestha, H. M., & Dhital, K. R. (2015). Soil Study in Mustang. http://www.icimod.org/?q=20204.

Bajracharya, S. R., Maharjan, S. B., Shrestha, F., Guo, W., Liu, S., Immerzeel, W., & Shrestha, B. (2015). The glaciers of the Hindu Kush Himalayas: Current status and observed changes from the 1980s to 2010. *International Journal of Water Resources Development* (ahead-of-print), pp. 1–13.

Bates, B., Kundzewicz, Z. W., Wu, S., & Palutikof, J. (2008). *Climate change and Water: Technical Paper vi*, Intergovernmental Panel on Climate Change (IPCC).

Belmonte, A. C., Jochum, A. M., & Garcia, A. C. (2003). Space-assisted irrigation management: Towards user-friendly products, in *paper presented at ICID Workshop on Remote Sensing of Crop Evapotranspiration,* Montpellier.

Blaikie, P., Davis, C. I., & Wisner, B. (2014). *At Risk: Natural Hazards, People's Vulnerability and Disasters*. Routledge.

Boyd, D. S., & Foody, G. M. (2011). An overview of recent remote sensing and GIS based research in ecological informatics. *Ecological Informatics, 6*(1), 25–36.

Brown, M., & Wooldridge, C. (2015). Identifying and quantifying benefits of meteorological satellites. *Bulletin of the American Meteorological Society* doi:10.1175/BAMS-D-14-00224.1.

Burkett, V. R., Wilcox, D. A., Stottlemyer, R., Barrow, W., Fagre, D., Baron, J., et al. (2005). Nonlinear dynamics in ecosystem response to climatic change: Case studies and policy implications. *Ecological complexity, 2*(4), 357–394.

Cho, M. A., et al. (2012). Mapping tree species composition in South African savannas using an integrated airborne spectral and LiDAR system. *Remote Sensing of Environment, 125*, 214–226.

Danielsen, F., et al. (2011). At the heart of REDD+: A role for local people in monitoring forests? *Conservation Letters, 4*(2), 158–167.

DeFries, R., Achard, F., Brown, S., Herold, M., Murdiyarso, D., Schlamadinger, B., & de Souza, C. (2007). Earth observations for estimating greenhouse gas emissions from deforestation in developing countries. *Environmental Science & Policy, 10*(4), 385–394.

De Sy, V., Herold, M., Achard, F., Asner, G. P., Held, A., Kellndorfer, J., & Verbesselt, J. (2012). Synergies of multiple remote sensing data sources for REDD+ monitoring. *Current Opinion in Environmental Sustainability, 4*(6), 696–706.

Du, L., Zhou, T., Zou, Z., Zhao, X., Huang, K., & Wu, H. (2014). Mapping forest biomass using remote sensing and national forest inventory in china. *Forests, 5*(6), 1267–1283.

El Sharif, H., Wang, J., & Georgakakos, A. P. (2015). Modeling regional crop yield and irrigation demand using SMAP type of soil moisture data. *Journal of Hydrometeorology, 16*(2), 904–916.

Elwood, S. (2010). Geographic information science: Emerging research on the societal implications of the geospatial web. *Progress in Human Geography, 34*(3), 349–357.

Etkin, D. (1999). Risk transference and related trends: Driving forces towards more mega-disasters. *Global Environmental Change Part B: Environmental Hazards, 1*(2), 69–75.

Foody, G. M., Boyd, D. S., & Cutler, M. E. J. (2003). Predictive relations of tropical forest biomass from Landsat TM data and their transferability between regions. *Remote Sensing of Environment, 85*, 463–474.

Franklin, J., & Hiernaux, P. Y. H. (1991). Estimating foliage and woody biomass in Sahelian and Sudanian woodlands using a remote sensing model. *International Journal of Remote Sensing, 12*, 1387–1404.
FSI. (2011). Carbon stock in India's forests reports. Forest Survey of India (FSI), Ministry of Environment and Forest Dehradun.
Gibbs, H. K., Brown, S., Niles, J. O., & Foley, J. A. (2007). Monitoring and estimating tropical forest carbon stocks: Making REDD a reality. *Environmental Research Letters, 2*(4), 045023.
Gilani, H., Shrestha, H. L., Murthy, M. S., Phuntso, P., Pradhan, S., Bajracharya, B., & Shrestha, B. (2015a). Decadal land cover change dynamics in Bhutan. *Journal of Environmental Management, 148*, 91–100.
Gilani, H., Murthy, M. S. R., Bajracharya, B., Karky, B. S., Koju, U. A., Joshi, G., Karki, S., Sohail, M. (2015b). Assessment of change in forest cover and biomass using geospatial techniques to support REDD+ activities in Nepal, 60.
Gong, P., et al. (2012). Finer resolution observation and monitoring of global land cover: first mapping results with Landsat TM and ETM+ data. *International Journal of Remote Sensing, 34*(7), 2607–2654.
Gonzalez, P., Asner, G. P., Battles, J. J., Lefsky, M. A., Waring, K. M., & Palace, M. (2010). Forest carbon densities and uncertainties from LiDAR, QuickBird, and field measurements in California. *Remote Sensing of Environment, 114*(7), 1561–1575.
Goodchild, M. F. (2006). GIS and disasters: Planning for catastrophe. *Computers, Environment and Urban Systems, 30*(3), 227–229.
Grigera, G., Oesterheld, M., & Pacín, F. (2007). Monitoring forage production for farmers' decision making. *Agricultural Systems, 94*(3), 637–648.
Gupta, A. S., Tarboton, D., Hummel, P., Brown, M., & Habib, S. (2015). Integration of an energy balance snowmelt model into an open source modeling framework. *Environmental Modelling and Software, 68*, 205–218.
Hansen, M. C., et al. (2013). High-resolution global maps of 21st-century forest cover change. *Science, 342*(6160), 850–853.
Härkönen, E. (2002). 1999 inventory of Nepal's forest resources. *Mountain Research and Development, 22*(1), 85–87.
Huang, W., Sun, G., Dubayah, R., Cook, B., Montesano, P., Ni, W., & Zhang, Z. (2013) Mapping biomass change after forest disturbance: Applying LiDAR footprint-derived models at key map scales. *Remote Sensing of Environment, 134*(0), 319–332.
Holtmeier, F. K., & Broll, G. (2005). Sensitivity and response of northern hemisphere altitudinal and polar treelines to environmental change at landscape and local scales. *Global Ecology and Biogeography, 14*(5), 395–410.
IPCC. (2003). Good Practice Guidance for Land Use, Land-Use Change and Forestry, IPCCC National Greenhouse Gas Inventories Programme, Institute for Global Environment Strategies, Kanagawa, Japan.
IPCC. (2006). Guidelines for National Greenhouse Gas Inventories, IPCCC National Greenhouse Gas Inventories Programme, Institute for Global Environment Strategies, Kanagawa, Japan.
Jayaraman, V. (2009). Status and trends of space applications for achieving internationally agreed development goals.
Kajisa, T., Murakami, T., Mizoue, N., Top, N., & Yoshida, S. (2009) Object-based forest biomass estimation using Landsat ETM+ in Kampong Thom Province, Cambodia. *Journal of Forest Research, 14*(4), 203–211.
Kwak, D.-A., Lee, W.-K., Lee, J.-H., Biging, G. S., & Gong, P. (2007). Detection of individual trees and estimation of tree height using LiDAR data. *Journal of Forest Research, 12*, 425–434.
Kim, S. R., Kwak, D. A., Lee, W. K., Son, Y., Bae, S. W., Kim, C., & Yoo, S. (2010). Estimation of carbon storage based on individual tree detection in Pinus densiflora stands using a fusion of aerial photography and LiDAR data. *Science China Life Science, 53*, 885–897.

Kim, D.-H., Sexton, J. O., Noojipady, P., Huang, C., Anand, A., Channan, S., et al. (2014). Global, Landsat-based forest-cover change from 1990 to 2000. *Remote Sensing of Environment, 155*, 178–193.

Kim, H., & Marcouiller, D. W. (2015). Natural disaster response, community resilience, and economic capacity: A case study of coastal Florida. *Society & Natural Resources*, 1–17.

Kurvitz, T., Kaltenborn, B., Nischalke, S., Karky, B., Jurek, M., & Aase, R.H. (Eds.). (2014). The Last Straw: Food security in the Hindu Kush Himalayas and the additional burden of climate change. ICIMOD and GRID-Arendal, 60 p.

Le Toan, T., et al. (2011). The BIOMASS mission: Mapping global forest biomass to better understand the terrestrial carbon cycle. *Remote Sensing of Environment, 115*(11), 2850–2860.

Lei, X. D., Tang, M. P., Lu, Y. C., Hong, L. X., & Tian, D. L. (2009). Forest inventory in China: Status and challenges. *International Forestry Review, 11*(1), 52–63.

Lovell, J. L., & Graetz, R. D. (2001). Filtering pathfinder AVHRR land NDVI data for Australia. *International Journal of Remote Sensing, 22*(13), 2649–2654.

Lu, D. (2006). The potential and challenge of remote sensing based biomass estimation. *International Journal of Remote Sensing, 27*, 1297–1328.

Lutz, A., Immerzeel, W., Shrestha, A., & Bierkens, M. (2014). Consistent increase in High Asia's runoff due to increasing glacier melt and precipitation. *Nature Climate Change, 4*(7), 587–592.

Miettinen, J., Stibig, H.-J., & Achard, F. (2014). Remote sensing of forest degradation in Southeast Asia—Aiming for a regional view through 5–30 m satellite data. *Global Ecology and Conservation, 2*, 24–36.

Miller, H. M., Sexton, N. R., Koontz, L., Loomis, J., Koontz, S. R., & Hermans, C. (2011). The users, uses, and value of Landsat and other moderate-resolution satellite imagery in the United States—Executive report: U.S. Geological Survey Open-File Report 2011-1031, 42 p.

Minamiguchi, N. (2004). Drought and Food Insecurity Monitoring with the Use of Geospatial Information by the UN FAO, paper presented at Regional Workshop on Agricultural Drought Monitoring and Assessment Using Space Technology, Hyderabad, India.

Murthy, M., Bajracharya, B., Pradhan, S., Shrestha, B., Bajracharya, R., Shakya, K., Wesselman, S., Ali, M., Bajracharya, S., & Pradhan, S. (2014). Adoption of Geospatial Systems towards evolving Sustainable Himalayan Mountain Development. *ISPRS-International Archives of the Photogrammetry, Remote Sensing and Spatial Information Sciences, 1*, 1319–1324.

Nelson, R., Krabill, W., & Tonelli, J. (1988). Estimating forest biomass and volume using airborne laser data. *Remote Sensing of Environment, 24*, 247–267.

Palmer Fry, B. (2011). Community forest monitoring in REDD+: The 'M' in MRV? *Environmental Science & Policy, 14*, 181–187

Parmesan, C. (2006). Ecological and evolutionary responses to recent climate change. *Annual Review of Ecology, Evolution, and Systematics*, 637–669.

Partap, T. (1999). Sustainable land management in marginal mountain areas of the Himalayan region. *Mountain Research and Development*, 251–260.

Pilon, P. J., & Asefa, M. K. (2011). Comprehensive Review of the World Hydrological Cycle Observing System.

Poudel, M., Thwaites, R., Race, D., & Dahal, G. R. (2014). REDD + and community forestry: Implications for local communities and forest management: A case study from Nepal. *International Forestry Review, 16*(1), 39–54.

Rasul, G. (2014). Food, water, and energy security in South Asia: A nexus perspective from the Hindu Kush Himalayan region. *Environmental Science & Policy, 39*, 35–48.

Sankar, K. (2002). Evaluation of groundwater potential zones using remote sensing data in Upper Vaigai river basin, Tamil Nadu, India. *Journal of the Indian Society of Remote Sensing, 30*(3), 119–129.

Sasaki, N., & Putz, F. E. (2009). Critical need for new definitions of "forest" and "forest degradation" in global climate change agreements. *Conservation Letters, 2*(5), 226–232.

Sharma, H. R., & Sharma, E. (1997). Mountain agricultural transformation processes and sustainability in the Sikkim Himalayas, India.

Sigdel, M., & Ikeda, M. (2010). Spatial and temporal analysis of drought in Nepal using standardized precipitation index and its relationship with climate indices. *Journal of Hydrology and Meteorology, 7*(1), 59–74.

Solomon, S., Plattner, G.-K., Knutti, R., & Friedlingstein, P. (2009). Irreversible climate change due to carbon dioxide emissions. *Proceedings of the National Academy of Sciences, 106*(6), 1704–1709.

Steininger, M. K. (2000). Satellite estimation of tropical secondary forest aboveground biomass data from Brazil and Bolivia. *International Journal of Remote Sensing, 21*, 1139–1157.

Stysley, P., Coyle, B., Clarke, G., Poulios, D., & Kay, R. (2015). High Output Maximum Efficiency Resonator (HOMER) Laser for NASA's Global Ecosystem Dynamics Investigation (GEDI) Lidar Mission.

Tahir, M. A. (2007). The needs and geospatial technologies available for disaster management in urban environment. In *IEEE 3rd International Conference on Paper Presented at Recent Advances in Space Technologies, 2007. RAST'07*.

Trostle, R. (2010), *Global Agricultural Supply and Demand: Factors Contributing to the Recent Increase in Food Commodity Prices (rev)*. DIANE Publishing.

Tulachan, P. M. (2001). Mountain agriculture in the Hindu Kush-Himalaya: Aregional comparative analysis. *Mountain Research and Development, 21*(3), 260–267.

Uddin, K., Gurung, D. R., Giriraj, A., & Shrestha, B. (2013). Application of remote sensing and GIS for flood hazard management: A Case Study from Sindh Province, Pakistan. *American Journal of Geographic Information System, 2*(1), 1–5.

Uddin, K., Shrestha, H. L., Murthy, M., Bajracharya, B., Shrestha, B., Gilani, H., et al. (2015). Development of 2010 national land cover database for the Nepal. *Journal of Environmental Management, 148*, 82–90.

Wang, G., Gertner, G., Parysow, P., & Anderson, A. (2000). Spatial prediction and uncertainty analysis of topographic factors for the revised universal soil loss equation (RUSLE). *Journal of Soil and Water Conservation, 55*(3), 374–384.

Wu, B., Meng, J., Li, Q., Yan, N., Du, X., & Zhang, M. (2014). Remote sensing-based global crop monitoring: experiences with China's CropWatch system. *International Journal of Digital Earth, 7*(2), 113–137.

Yang, X., Strahler, A. H., Schaaf, C. B., Jupp, D. L. B., Yao, T., Zhao, F., et al. (2013). Three-dimensional forest reconstruction and structural parameter retrievals using a terrestrial full-waveform lidar instrument (Echidna®). *Remote Sensing of Environment, 135*, 36–51.

Zhang, X., Wu, B., Ling, F., Zeng, Y., Yan, N., & Yuan, C. (2010). Identification of priority areas for controlling soil erosion. *Catena, 83*(1), 76–86.

Zheng, D., Heath, L. S., & Ducey, M. J. (2008). Spatial distribution of forest aboveground biomass estimated from remote sensing and forest inventory data in New England, USA. *Journal of Applied Remote Sensing, 2*, 021502.

Integrating Earth Observation Systems and Data into Disaster Preparedness in the Lower Mekong: Experiences from the Asian Disaster Preparedness Center

David J. Ganz, Peeranan Towashiraporn, N.M.S.I. Arambepola, Anisur Rahman, Aslam Perwaiz, Senaka Basnayake, Rishiraj Dutta and Anggraini Dewi

Abstract The Lower Mekong Region, which comprises Cambodia, Lao PDR, Myanmar, Thailand and Vietnam, faces significant challenges associated with rapid economic, social, and environmental changes. Many of these challenges are exacerbated by climate change and natural disasters afflicting the region. The regions' sustainable development challenges will worsen as people recognize that they are exposed to more and more dangerous and unacceptable levels of risk from disasters. This chapter addresses the pressing challenges for sustainability and disaster preparedness and explores the role of earth observation systems.

1 Overview

The Lower Mekong Region, which comprises Cambodia, Lao PDR, Myanmar, Thailand, and Vietnam, faces significant challenges associated with rapid economic, social, and environmental changes. Many of these challenges are exacerbated by climate change and natural disasters afflicting the region. The regions' sustainable development challenges will worsen as people recognize that they are exposed to more and more dangerous and unacceptable levels of risk from disasters. In addition, major environmental changes such as alterations to water regimes and the loss or degradation of natural forests are linked to challenges in maintaining ecosystem services, conserving biodiversity, and maintaining soil and water quality. All of these challenges highlight the need for improved governance and decision making to ensure better outcomes across all three aspects of triple bottom line

---

D.J. Ganz (✉) · P. Towashiraporn · N.M.S.I. Arambepola · A. Rahman
A. Perwaiz · S. Basnayake · R. Dutta · A. Dewi
Asian Disaster Preparedness Center (ADPC), 979/69 Paholyothin Rd.,
Sam Sen Nai, Phayathai, Bangkok 10400, Thailand
e-mail: david.ganz@adpc.net

© Springer International Publishing Switzerland 2016
F. Hossain (ed.), *Earth Science Satellite Applications*,
Springer Remote Sensing/Photogrammetry, DOI 10.1007/978-3-319-33438-7_3

accounting—economic, social, and environmental. Accounting or auditing systems are being integrated into a risk assessment framework so that decisions and well-designed Disaster Risk Management (DRM) lead to outcomes which maintain economic, social, and environmental services (i.e., maintaining livelihoods) as well as the DRM fundamentals of protecting lives and traditional assets.

Geospatial data and Earth observation data collected by a growing number of satellite-borne instruments and new techniques for robust geospatial analyses contribute significantly to more timely and informed decision making for societal benefits, especially in the context of climate change and increasing exposure to disasters. Despite an increase in the amount of data and the frequency of data acquisition, in order for it to be deemed useful, such information and technologies must reach the right people and institutions at the right time and in the right form for a given planning, policy, or other decision-making context. Disaster risk management benefits greatly from the use of geospatial and Earth observation technologies because spatial methodologies can be fully explored for serving different purposes from assessment/analysis to application in decision-making process. The use of geospatial and Earth observation data has become an integral part of the DRM community of practice. Some of the technologically advanced tools in different application areas have been pilot tested by the Asian Disaster Preparedness Center (ADPC), both in the past and now as it is applied within sound project management for DRM principles.

ADPC has found this integration of geospatial and Earth observation data as critical for successful DRM as well as for widely accepted societal applications. These types of data have demonstrated a wide diversity of applications, from addressing the types of hazards to covering all aspects of the risk management spectrum—starting with prevention and mitigation, through preparedness and response, to rehabilitation and reconstruction. In recent years, there is a growing international recognition that governments must take the full lead in assessment and recovery planning, with support from various development partners. Therefore, ADPC has been working with these development partners to build up the country's institutional arrangements and capacities, using internationally accepted methodologies which need to adapted to specific country requirements in order to get appropriate buy-in. A similar approach is going to be needed for integrating geospatial and Earth observation data into the full risk management spectrum. In this paper, the authors highlight nearly 30 years of experience from one of the capacity development leaders in the broader Asian region. Two case studies present the lessons learned for the broader geospatial and Earth observation community, especially perspectives from the DRM subsector that are engaged for better uptake of geospatial information. It should be noted that the geospatial community has been quick to expand into the disaster relief and response arenas; however, in this review of its experience, the ADPC highlights a few applications of geospatial data under two disaster risk reduction (DRR) contexts where the focus is on better decision making utilizing the best of what is available and the ability of these countries to properly integrate this kind of information into better disaster preparedness planning, safer land-use planning, regulatory zoning for disaster-resilient development, and appropriate and timely weather forecasting and early warning systems.

These two case studies document the existing use of geospatial information through its projects: what type of data was developed, how it was integrated, and then served into successful decision-support tools for making different decisions in a disaster risk management context. As in other fields of application, the most effective use of remote sensing and GIS is when it builds local capacity to use geospatial data freely for specific intended purposes and outcomes. DRM is no different in that regard and consideration must be given to the types of software packages for modeling, scenario development, information storage, and sustainability of these approaches when we make them available to technical specialists/practitioners. But the benefits of these geospatial approaches and techniques also should be enjoyed not only by technical specialists but also the other stakeholders including the general public. All of them to a different degree should be able to use them for analysis of the hazard environment, to assess the vulnerabilities and risk, as well as to plan for responding to early warning systems, prepare better and to mitigate the impacts of disasters. For that more investments need to be made in the area of data generation, development of key information, facilitate sharing, capacity building, and to promote wide utilization of related geo-information in disaster-resilient development.

1.1 Current State of Play and ADPC's Regional Role

Over the last 10 years, significant progress has been made in developing institutions, policies, and legislation for disaster risk management, especially in the Lower Mekong region. Further, capacities for risk assessment and identification, disaster preparedness, response, and early warning capacities and in reducing specific risk have been significantly strengthened through robust institutional development and governance structures. Despite these advancements, the DRM community in the Lower Mekong will still need to learn how to use and integrate the best available spatial technologies and Earth observation data for creating disaster-resilient development. The region continues to rapidly develop and take advantage of its natural resources, cheap labor, and newly founded economic community.

ADPC's programs are being developed in support of the above needs and meting priorities of the national and local Governments of the region. ADPC has developed and delivered the Strategy 2020 in year 2011 during its 25th anniversary celebrations to present its approach for attending to regional needs and addressing gaps. The ADPC strategy 2020 has been constituted under three core programmatic areas of focus (Fig. 1), founded upon ADPC's comparative technical advantage, as well as the results and outcomes that could be expected and would be needed for disaster risk reduction (DRR) and climate change adaptation (CCA), over the next decade. The ongoing three core programs are

SCIENCE—Enhanced capacity of countries in utilization of science-based information to understand risk.

SYSTEMS—Strengthened systems for effective management of risk at all levels in countries, especially at subnational and local levels.

Fig. 1 ADPC core programmes

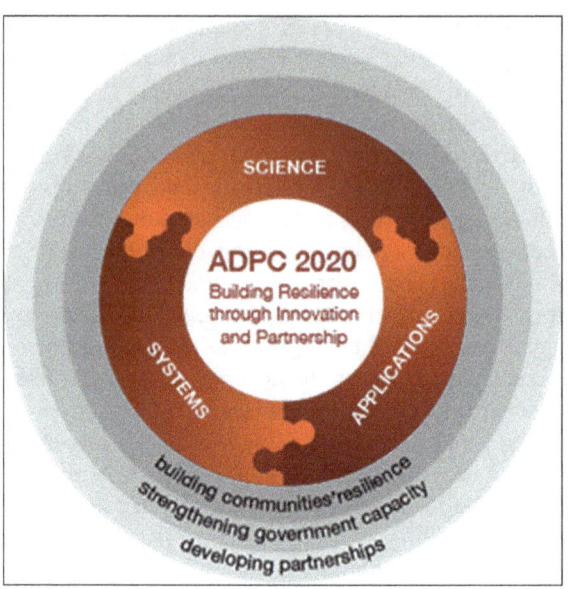

APPLICATIONS—Improved and grounded application of risk reduction measures in development.

ADPC programs are designed to encompass a wide diversity in application, adopt a multi-hazard approach, and cover all aspects of the risk management spectrum. The core programs that will be implemented under the above strategy (during 2012–2020) are focused on "Investing in People" and pursue a broad scientific approach to risk reduction, engrossed with unmet needs of sustainable development and poverty reduction, driven with an overall aim of safer communities and sustainable development through DRR in accordance with the United Nations Millennium Declaration and the Millennium Development Goals (MDG) and the Sendai Framework for Action (2015–2030). It encompasses "Gender Equality and Women's Empowerment" by addressing inequalities and promoting women's rights as well as inclusiveness of all vulnerable groups in society.

The objectives of ADPC's core programs to provide targeted, sustained, technical support, and services to the Asia Pacific region in DRR, CCA and development are echoed in many of the regional support strategies and priorities of development partners.

1.2 Key Challenges

Geo-information, especially those related to hazards, vulnerability, and risk, are not well-integrated or mainstreamed into the spatial planning process. Those professionals and policy makers mandated by Government institutions to take on land-use planning and implement sustainable development will need to develop a harmonized approach to apply hazard and risk knowledge for risk mitigation through integrated spatial planning, especially in a highly risk-prone region like the Lower Mekong. This has developed in a number of DRR subdisciplines or fields, such as Integrated Fire Management, but more emphasis is needed to apply geospatial and Earth observation data when designing multi-hazard resiliency strategies and adopt more holistic and integrated land-use planning. Similarly, very limited research has been carried out to assess the relationship between the cascading effects of land cover changes with hydro-meteorological hazard occurrence and the subsequent long-term impacts on the environment, especially through the loss of ecosystem services. These are significant challenges facing the integration of geo-information.

Recent scientific advances in satellite technologies and the availability of spatial and Earth observation data and products are still not well understood by the vast array of stakeholders in the Lower Mekong region. There is significant demand from these stakeholders/end users for this kind of information, especially in the form of decision-support tools for applying new science to a fairly well-disciplined field such as DRR. There is an understanding that the science exists but varying levels of awareness on how it can be applied to different decision contexts. There is also a varying level of understanding of technical matters within the governing institutions of the Lower Mekong, often prompting a response by donors and NGOS to develop capacity building programs and follow-up trainings. More emphasis needs to be given on capacity building and transfer of knowledge from the national level to subnational levels as well as down to the community levels. In order to get this kind of information to end users at the subnational, district, commune, and village levels, much of the capacity building materials needs to be translated in the national languages and sometimes even in the local dialects. Training of trainers is needed to help disseminate these capacity building programs to the appropriate end users.

The implementation of risk management measures such as disaster preparedness planning, safer land-use planning, regulatory zoning, weather forecasting, and early warning systems are needed to be considered essential components in sustainable development in the Lower Mekong subregion. Related vulnerabilities and risks, whether it be to dam infrastructure, irrigation, and transportation infrastructure, have not been integrated or mainstreamed into the investment planning process by those mandated to review cost benefits of these projects within Lower Mekong government institutions. Subsequently, there is no legal basis or institutional process to review the added costs or avoided costs of building infrastructure in areas of high vulnerability and risk. If hidden costs or benefits are not fully accounted for, people will tend to make uninformed choices leading to inefficient outcomes and unsustainable development. The geospatial community must continue to advocate the use of the best available information in investment planning and build in resiliency measures with a triple bottom line approach, or just avoid the hidden costs of disaster relief and recovery, and be explicit to these Lower Mekong Government institutions how these benefits from sustainable development are in the best interest of all involved.

1.3 Other Potential Challenges

- Weak capacities in government institutions that inhibit them from taking advantage of the opportunities offered by the space community, including lack of skills of staff, inadequate hardware, and software to process satellite imagery, as well as lack of access to high-resolution satellite imagery due to its high cost;
- Poor coordination among stakeholders at the national level to make use of Earth observations (EO);
- Weak cooperation and coordination among international and regional organizations involved in Earth observation when it comes to assisting countries to make use of Earth observation in efforts focusing on disaster risk reduction.

2 Future Directions

2.1 Measuring and Accounting for Ecosystem Services

Analyzing extensive risk highlights a key development challenge of our time: how to strengthen risk governance capacities fast enough to address the rapidly increasing exposure of population and assets that accompanies economic growth and recognizes the values associated with functioning ecological systems. Extensive risk exists wherever development occurs, and risk drivers such as badly planned and managed urbanization, environmental degradation, and poverty directly construct it. Extensive disaster losses and their impacts on health, education, structural poverty, and displacement go unaccounted for in the Lower Mekong subregion, hiding the real cost of disasters. The actors involved in Lower Mekong development remain unconcerned about ecosystems due to inadequate understanding of the connection between ecosystems and life-threatening disasters. The Lower Mekong region also harbors some of the world's most diverse yet threatened natural systems in the world. Rapid and unsustainable economic development is threatening the integrity and continuity of high priority conservation landscapes in the region. A challenge facing the region's governments therefore is how to develop their economies sustainably and equitably while at the same time conserving ecosystems and their services upon which our societies and economies are built. To do this effectively and to enable countries to make the transition toward a 'green economy,' one significant piece of missing information is the value of natural capital—the stock of natural assets (land, water, biodiversity) that supports the provision of ecosystem services (WWF Living Planet Report 2014)—and an understanding of how these values are to be mainstreamed into development planning, DRM, and other public policy as well as local business decisions that are often driving land-use changes.

The undervaluation of ecosystem services poses a major problem to effective, equitable, and sustainable conservation and development decision making, especially in the Lower Mekong subregion. Because economists and the DRM community have traditionally focused on the value of commercial land and resource uses as well as traditional damage/loss assessments, some of the most valuable ecosystem services and the true costs of disasters have remained hidden, unaccounted for in infrastructure and post-disaster assessments. As a result, decisions have often been made on the basis of incomplete and flawed information—and, unsurprisingly, ecosystem resilience and biodiversity conservation have typically not emerged as a profitable land, resource, or DRM investment option. This has, in turn, resulted in missed economic opportunities, and at worst has incurred substantial costs and losses.

ADPC has recognized the need to incorporate ecosystem resilience in its DRM strategies. It has begun to support officials at the national and subnational levels to enhance their existing knowledge on ecosystem management and restoration of degraded ecosystems. Enhancing weather and climate data utilization and introducing new technological interventions are critical in order to mitigate undesirable impacts within the ecosystem, especially the flow of ecosystem services for DRM approaches like maintaining natural floodplain and watercourses that feed into deltas and mangrove ecosystems. Space-based information not only provides information on extent of ecosystems, but also provides much insight into the dynamics and health of these ecosystems. Geospatial information can be analyzed in conjunction with the field-based information at all functional levels including biospheres, biomes, landscape, ecosystems, habitats, and communities. As a tool, it may be used to monitor remote and difficult areas such as highlands and steep slopes, where the role of ecosystem is often critical for downstream impacts. There are also a number of advanced sensors which now provide the capacity to monitor ecosystems on a more frequent basis (i.e., return intervals and higher resolution).

The Lower Mekong countries—along with the international community, regional civil societies, the DRM community, and private sector actors—recognize that the invisibility of natural capital risks unsustainable economic development in the long term because of the high likelihood of loss of critical benefits that nature provides and the lack of linkages between these critical benefits and disaster preparedness. To ensure more informed decisions for sustainable development are made, an important next step is to help Lower Mekong government institutions, the DRM community, and private sector actors to collect, map, and generate information on natural capital stocks and flows as well as their impact on climate change resilience and disaster mitigation. In this context, there is a need for increased and improved use of decision-support tools and capacity that

- Integrates ecosystem protection with developmental planning to ensure that ecosystems remain balanced and robust, and development is resilient to disasters and climate change;
- Uses the best available Earth observation data to identify and monitor what ecosystem services are generated in key ecosystems in the Lower Mekong;

- Provides Lower Mekong government institutions, the DRM community, and private sector actors with the socioeconomic value of these services to local populations, national, and regional economies;
- Uses the best available Earth observation data to provide insights on how ecosystem services may be affected, including trade-offs between ecosystem services and economic return, by different policy interventions (e.g., land-use planning, improved forest management regimes), risk events like climate change, flooding, drought, and/or the immediate business decisions that need to be taken to balance these trade-offs using a comprehensive accounting framework (such a triple bottom line accounting).

It is this data generation and subsequent mainstreaming which will help facilitate a transition within Lower Mekong to a green economy and support the restoration and maintenance of the region's natural capital.

2.2 Measuring Urban Resiliency

Healthy communities are the basis of our physical, mental, and social well-being. And the core foundation of healthy communities is a healthy environment and lifestyle security. For hundreds of millions of people whose livelihoods depend directly on the resources and services that nature provides, the link is obvious. Southeast Asia's economic rise is being fueled by the booming growth of its cities. Today, just over one-third of ASEAN's population lives in cities, and these urban areas account for two-thirds of the region's GDP. More than 90 million people are expected to move to cities by 2030, bringing the urban share to almost 45 % of the population and 76 % of GDP (ASEAN Community 2015). For the ever-growing number of people who live in cities, increasingly detached from the natural world, the importance of healthy ecosystems may not be so immediately apparent—and yet the effects of environmental problems can be just as striking, from air and water pollution to extreme weather events.

The global population landscape has changed in the past decade. For the first time in history, the majority of the world's population lives in cities, with urbanization growing fastest in the developing world. Statistically, this is the result of natural growth, rural–urban migration and reclassification of rural to urban land (Buhaug and Urdal 2013). The urbanization trend in the Lower Mekong has a lot to do with environmental security, resource scarcity (farming and grazing lands, forests, and water), environmental degradation, frustration with reliance on increasingly unpredictable natural systems, and natural hazards all pushing people to leave behind their rural settings in search of greater livelihood and lifestyle security. This in turn has significant consequences for the health of the cities that absorb them.

The majority of the world's population lives in cities with urbanization growing fastest in the developing world. This trend has a lot to do with environmental security: resource scarcity (farming and grazing lands, forests, and water), environmental degradation, frustration with reliance on increasingly unpredictable natural systems, and natural hazards all pushing people to leave behind their rural settings in search of greater livelihood and lifestyle security. This in turn has significant consequences for the health of the cities that absorb them.

Since 1995, ADPC has been working in the field of Urban Disaster Preparedness across Asia and the Pacific. While the urban areas in Asia are rapidly expanding so has the associated urban disaster risk. With a goal to address challenges associated with increased urban risk and sustainable development, ADPC is working with the respective cities to develop future resilience by assisting cities and its communities to quickly identify challenges and gaps needed to ensure greater resilience is built up at the local level. ADPC has also developed a robust set of specific indicators to measure the resilience at this local level. Measures of urban resiliency and the stock and flow of ecosystem services in a city environment may be built into a city geospatial database if city planners and regulators can see the power and application of these types of datasets. With the development of these indicators and geospatial datasets, ADPC also utilizes a methodology to measure the effectiveness of past projects, namely in Mandalay City, Myanmar and Dhaka City, Bangladesh (presented in the case study below), building upon risk-sensitive, urban land-use planning and contingency planning. Following this analysis, ADPC and City authorities are able to integrate the lessons learned from this methodology into any upcoming and future DRM project design allowing for flexibility in how urban DRM is applied dependent on available financial resources, available data relating to hazards and its population, and institutional capacity of a given city. Based on their identified needs and priorities, City governments may select the most appropriate level of risk assessment as well as design DRM strategies that are cost-effective, building upon the strengths of the existing institutional capacity. The integration of geospatial data will certainly assist in developing a risk-informed urban society and build awareness to the prevailing risk environment. However, more investments will be needed in the area of urban resiliency data generation, indicator tracking, capacity building, and data sharing to promote wide utilization of related geo-information in urban disaster-resilient development. In this context, there is a need for increased and improved use of decision-support tools and capacity that

- Uses the best available Earth observation data to identify and monitor urban risk through better urban resiliency data generation;
- Defines growth patterns and reviews existing land-use and zoning plans based upon new information, in particular areas with high risk;
- Builds upon existing metrics and urban parameters such as impervious surfaces, land surface albedo, land surface emissivity, land surface temperature, and land cover/land use;
- Contributes toward the establishment of a global data infrastructure for urban data and products. GEO is developing the architecture, data management policy, data portals, software, and tools needed to improve global coverage and data

accuracy of urban observing systems through integrating satellite data observed from multiple platforms/sensors and different resolutions, with in situ data; and
- Provides well-designed scenarios for urban growth where outputs can improve the adaptation and resiliency of urban society even in a context where weather-related disasters may increase in frequency and impacts due to climate change.

It is this data generation and subsequent mainstreaming which will help facilitate a transition for Lower Mekong cities to move toward urban disaster-resilient development.

3 Case Studies

3.1 Natural Hazard Risk Mapping and Improved Post-disaster Recovery in Lao PDR

In September 2009, Typhoon Ketsana ravaged Lao PDR's southern provinces causing widespread destruction, affecting more than 180,000 people. Twenty-eight storm-related deaths were reported. It was the first time Lao PDR had experienced disasters of this magnitude with losses to life and livelihoods.

Two years later, Typhoons Haima and Nokten (June 2011) hit the northern and central parts of Lao PDR causing serious erosion and landslides from heavy rain and flooding. Some of the most important infrastructures were severely damaged including the irrigation system, roads, bridges, hospitals, and schools. Again, these typhoons had a direct impact on local people's lives and livelihoods, and significantly disrupted the delivery of public services.

Because of these typhoons as well as other natural hazards in Lao PDR, UNDP —with technical support of ADPC—initiated a project to assess and map out various hazards and their corresponding risks including those from floods and landslides for the entire country. Through a series of geospatial analyses, the result of that initiative was presented as maps where the areas of varying levels of hazards (e.g., flood depth and the area's susceptibility to landslides) and risk (the consequences on the people and assets) were demarcated. These maps helped the government of Lao PDR in identifying risk-prone provinces where risk mitigation activities should be prioritized. Figure 2 illustrates the national landslide susceptibility map for Lao PDR.

Learning that certain provinces are more prone to floods and landslides than the others, the Ministry of Planning and Investment (MPI) of the Government of Lao PDR, with support from the World Bank and technical assistance of ADPC, took an initiative to further the study to detailing out possible impacts of those natural phenomena on important infrastructure sectors such as irrigation and roads in selected pilot provinces. The outcomes of this initiative were aimed at mainstreaming disaster risk information into public infrastructure investments by the

Integrating Earth Observation Systems and Data ... 73

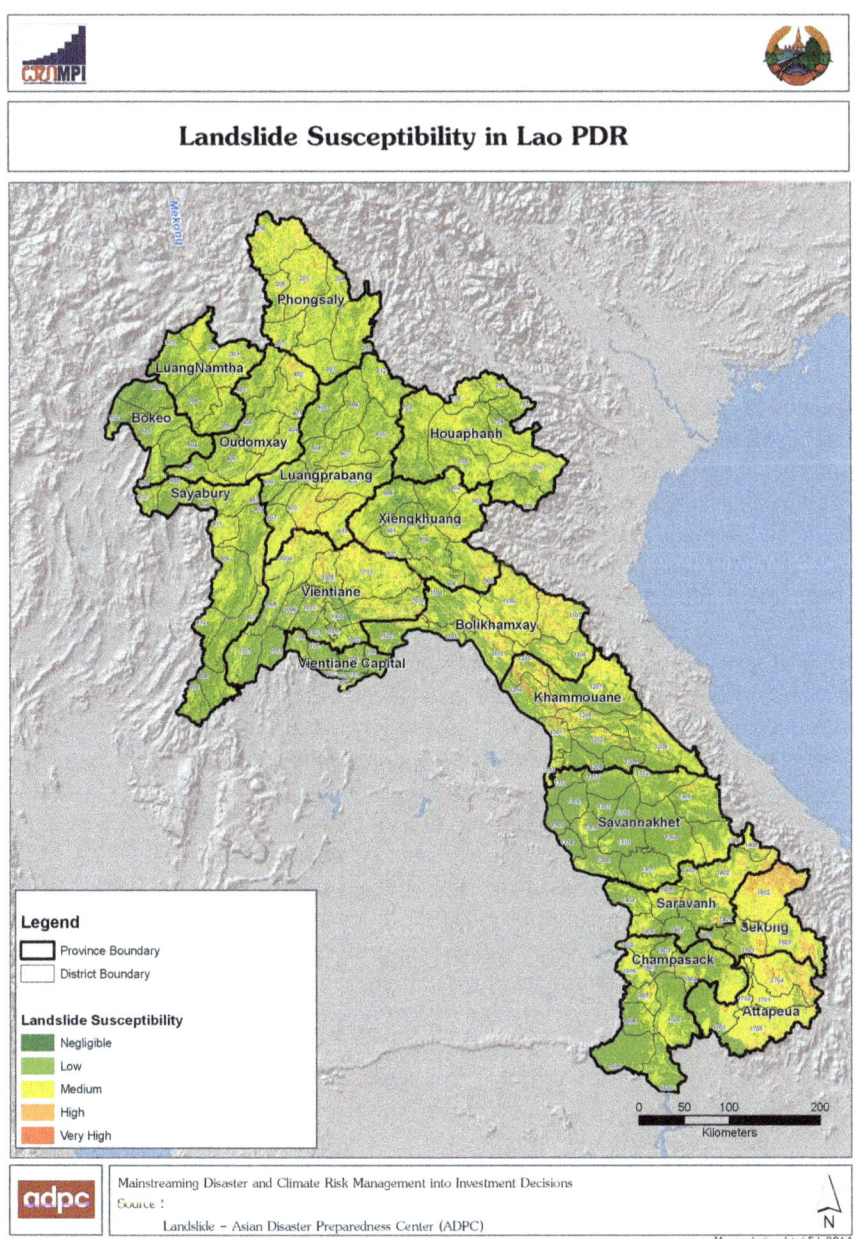

Fig. 2 Example national-level landslide susceptibility map for Lao PDR

Ministry of Agriculture and Forestry (MAF) and the Ministry of Public Works and Transport (MPWT)—the project's governmental partners, thereby reducing the vulnerability of the provincial and national economy to natural hazards.

The project determined, based on the national mapping initiative, that Bolikhan district, Bolikhamxay Province, and Beng district, Oudomxay Province, are the pilot districts for detailed evaluation of the risk and its applications for informed investment decision. Assessments of the frequency, severity, and geographical location/coverage of floods and landslides were carried out for the two pilot districts, the results of which were displayed as maps in a GIS environment (Fig. 2).

Assets that would be at risk from future floods and landslides such as roads, bridges, irrigation head works, and rural houses in the two districts were added to a geospatial database, containing information not only on the location, but also the attributes of each asset type. By doing so, it allowed straightforward geospatial analyses with the hazard intensity layers. Moreover, it will allow the provincial/district-level agencies to maintain and update the data with ease in the future.

Before this initiative by MPI, the World Bank, and ADPC, the occurrences of landslides were not recorded in a systematic manner in Lao PDR. Whether it is major or minor road systems, there were rarely any proper assessments or subsequent slope stabilization measures taken to improve the stability of the slope. One of the project objectives was to address this issue. ADPC worked closely with the Lao PDR government to develop an inventory system using GIS that could easily record landslide data and demonstrate the level of vulnerability for decision makers to take appropriate decisions on road improvements, reconstruction, and rehabilitation of damaged areas. It was estimated that landslide and road repairs may reach as high as 50 and 80 % of annual emergency maintenance costs (Hearn et al. 2008a, b). During some years, this can amount to direct cost of $5 million USD annually from emergency road and irrigation maintenance budgets (SEACAP 2008).

The developed geospatial framework for documenting these landslides and slope failures also went so far as to document the roles and responsibilities of different levels of the Department of Roads (DoR) in inventory preparation, data storage, and sharing as well as in follow-up interventions, utilization of inventory data in managing a reliable road transportation system in Lao. The geospatial inventory has provided the Government of Lao PDR to analyze the vulnerability and risk of its transportation and irrigation systems in a systematic way so that they may take appropriate decisions in a timely manner. This has in turn facilitated risk-averse decision making that takes on suitable proactive measures that are taken in a timely and cost-effective manner to reduce disruptions to the transportation network due to landslide occurrences.

Inevitably, tropical storms and flooding will continue to affect various government sectors in Lao PDR and especially susceptible areas like Khammouane Province. Many people, especially the poorest, their livelihoods, properties, and assets are directly affected by these natural and man-made events. The Provincial Government in Khammouane Province has identified the need to boost its capacities for post-disaster recovery and reconstruction relating to damage and loss assessment, institutional and financial arrangements, disaster-resilient investments, and project management. The Khammouane Department of Planning and Investment, through the Khammouane Development Project, was supported by the World Bank to provide technical services for institutional development and capacity building activities for post-disaster reconstruction in the province.

ADPC has been working closely with the World Bank and the Government of Lao PDR to enhance existing provincial and district-level institutional arrangements for post-disaster reconstruction. This has been accomplished by developing clear guidelines and procedures that estimate the post-disaster damages, losses, and corresponding needs of the various sectors affected. Ultimately, the Provincial Government must have the capacity to respond swiftly to the impacts of a disaster so the community and local economy can recover quickly. This includes the ability to use geospatial information in successful post-disaster planning and reconstruction.

APDC's work in Khammouane Province is distinctive from other projects in which it provides support specifically to the subnational levels while fully aligned with national-level arrangements. It is ADPC's experience that much of the geospatial information used for post-disaster reconstruction is held primarily at the national level. This geospatial information is critical for the post-disaster reconstruction, especially for performing damage and loss assessments, institutional and financial arrangements, disaster-resilient investments, and project management at the Provincial and District levels. In this Laotian context, ADPC has found that the relationship between the Provincial and District levels is really the critical catalyst for post-disaster reconstruction to be successful. The investments can only be sustainable if the responsibilities, functions, and procedures are well understood by the relevant authorities. ADPC's technical assistance in Khammouane Province forms a key component of its mission to reduce disaster and climate risk impacts on communities and countries in the Lower Mekong subregion by working with governments, development partners, and key stakeholders at multiple levels. Key lessons learned from ADPC's interventions in Lao PDR include

- Geo-information related to flood and landslide hazards, related vulnerabilities and, risk to irrigation and road sector infrastructure have not been integrated or mainstreamed into investment planning process by professionals attached to the mandated Government institutions. Hence, there is no legal basis or institutional process to continue utilization of such information in investment planning.
- The road sector engineers find it difficult to follow the methodology provided for inventorying landslides in the road network although several training and capacity building programs have been organized to transfer the knowledge. A similar situation has been observed in generating flood return period maps for some of the minor catchments. Further interventions are needed for better knowledge transfer to these sectors.
- Certain efforts are essential to be undertaken for developing data sharing protocols, data standardization, and also to develop a common data sharing platform for DRR-related information in future. One mandated institution should be given the responsibility for creating enabling environment for data collection, storage, and sharing with other agencies.

3.2 Improved Earthquake Risk Assessment and Risk Reduction Strategies in Bangladesh and Mandalay City, Myanmar

The use of GIS and Remote Sensing (RS) technology in DRM is extremely beneficial because spatial methodologies can be fully explored throughout a risk assessment process. The risk assessments for DRM to be designed and function properly must be carried out at different scales of analysis, ranging from regional/national scale to the community level. Each of these levels of analysis has its own objectives and spatial data requirements for hazard inventories, exposure coverage of elements, environmental data, and dependent and independent factors to consider that will trigger additional downstream and/or post-disaster effects. Cost effectiveness, speedy delivery, and data accuracy are some of the key advantages of using GIS-/RS-based tools and geospatial products for DRR decision making utilizing integrated risk information. Due to the possibility of generating alternative scenarios in a spatial context, the Government authorities will have the advantage for taking best and cost-effective solutions in reducing the risk while maximizing their investments in DRM strategies.

Unfortunately, the development and utilization of risk information for disaster-resilient urban development is not widely practiced in Asia, especially the Lower Mekong region. The use of GIS and RS technology is especially applicable when conducting risk assessment in large cities, where the level of detail is inherently needed to capture the heterogeneity. Comprehensive building and infrastructure databases are needed for exposure surveys and well-integrated, vulnerability assessment surveys. If city-level risk information can be provided in open-access formats which can used and understood not only by professionals but also the general public, then urban development and risk reduction decision making would become an open and transparent process. Therefore, there is tremendous value in providing risk information using a geospatial platform that people can use and understand easily. Spatial visualization of risk information within the city will help authorities to identify geo-political areas, population, infrastructure, sectors of the economy, and vulnerable members of civil society (elderly, handicapped, youth, etc.) that will be impacted by a pending earthquake event. This also allows for estimating the value of physical damages, economic losses, and reductions in ecosystem services arising after a potential earthquake event.

A widely used application that has been tested extensively by ADPC is the integration of GIS and RS Earth observations into earthquake hazard and risk assessments. This was conducted for nine large cities including Dhaka, Bangladesh, and Mandalay City in Myanmar. In this case study, we describe how this Earth observation data can be used to

- help city authorities to define the growth patterns in the city and reviewing existing land-use plans based upon new information, in particular areas with high potential risk;

- analyze the factors which enhance the futuristic risk and to take appropriate measures to introduce risk-sensitive land-use planning regulations, including but not limited to;
  - siting, design, and construction of public and private structures considering earthquake and other associated hazard risk (such as landslides, liquefaction, subsidence, etc.),
  - guidelines for earthquake resistant buildings suggesting improvements to the quality of buildings via building codes and best practices, and
  - retrofitting solutions and safer construction practices in high-risk zones;
- assist city authorities to monitor and enforce various land-use policies, especially building standards and codes. This may include enhancing policies and systems for quality assurance;
- provide a baseline to monitor trends in changes of environmental conditions, climate change impacts, and impact from climate sensitive hazards; and
- analyze vulnerability considering the aspects of gender, culture, livelihoods, etc. and take appropriate measures in reducing the vulnerability of those traditionally vulnerable groups.

Spatial databases were developed for nine Asian cities including the base maps for conducting the hazard and vulnerability assessments, comprehensive risk assessment, and scenario-based contingency planning in electoral wards. These spatial databases for each city have first considered which physical features are most important to collect. Based upon the availability of existing data and baseline information on these respective cities, ADPC designed an appropriate and cost-effective methodology to acquire missing information by conducting physical feature surveys and collecting and verifying necessary attribute information.

In Myanmar, ADPC carried out a seismic hazard and risk assessment for Mandalay City, the country's second largest city. The assessment involves extensive use of geospatial information that includes the high-resolution satellite data such as Corona, Geo-Eye, IKONOS, and World View 2 which helped experts to extract information on building footprints, infrastructure, geo-morphological features, etc. for earthquake hazard and risk assessment. Space-based geodetic measurements using the Global Positioning System in synergy with ground-based seismological measurements, interferometric data, and high-resolution digital elevation models as well as imaging spectroscopy were all used effectively to develop seismic hazard and risk assessments. The same approach for this kind of seismic hazard and risk assessment in Mandalay City was used for cities of Bangladesh with the Comprehensive Disaster Management Program (CDMP), under the Ministry of Disaster Management and Relief. Earthquake indicators that may be detected from space were also observed using ground-based monitoring stations. These include factors such as surface deformation, land surface changes, and geo-morphological features such as land form changes, electromagnetic and ionospheric anomalies in the fault-line areas, as well as seismicity. Multispectral and multi-temporal satellite images were analyzed for building continuity and regional relationships of active faults as well as for geologic and seismic hazard mapping.

QuickBird images of the cities were used to demarcate physical features such as the road alignment, building outline, water body boundary, and river boundary. However, the maps generated here were still missing information that would be critical for the earthquake risk assessment process, but could not have been gathered through remotely sensed techniques. As a result, a field data collection and verification process was introduced. Each of the digital features would need a set of attribute data that would dictate its risk to earthquakes. Those information include the number of floors in a building, construction materials, structural weak points, the number of bridge spans, bridge type, and road surface materials, for examples. Base maps for those cities can be considered complete for use by the risk assessment specialists only after these attribute information could be added to the geospatial database. These city-level base maps are useful not only for risk assessment purposes, but city officials can also use them for non-disaster related purposes such as monitoring the progress of collection of taxes and revenues, and designing public service facilities (such as water and electricity distribution, garbage disposal, impervious surfaces, drainage/storm drain runoff and releases, transportation, and school design). As previously discussed with issues that will need to be addressed in future Earth observations, measures of urban resiliency and the stock and flow of ecosystem services in a city environment may also be built into a city geospatial database if city planners and regulators can see the power and application of these types of datasets.

The steps followed during base map preparation are given in detail within Fig. 3.

The final outputs of the initiative for assessing earthquake risk in these cities are hazard and risk maps that would inform city officials and DRR practitioners on the

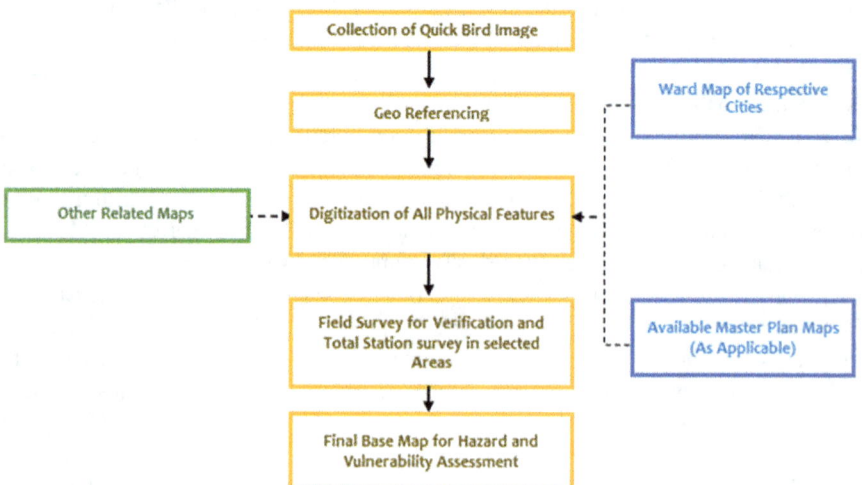

Fig. 3 Base map development process using QuickBird images

areas in the city that is prone to damage caused by earthquakes. Estimated damages to buildings, roads, bridges, pipelines, as well as the number of casualties and its geographical distribution across the cities were illustrated. Figure 4 shows an example of the risk maps showing damage expected on concrete buildings in Mandalay City of Myanmar.

The main beneficiaries of geospatial information related to earthquake risk are the city authorities who are supposed to make major decisions on sustainable urban development investments in order to develop safer and resilient urban habitat and infrastructure. The city residents and other migrant populations to the city also benefit directly and indirectly from this type of baseline assessment. In theory, they should have access to better information and potentially relocate to live within safer environments (either outside of a high-risk zone or within buildings that are built to appropriate building standards). Therefore, it is critically important for city authorities to take steps to communicate this geospatial risk information and all associated issues to the communities, disaster management practitioners, private sector, and all vested stakeholders, whether they are part of the tax base or just vulnerable populations living within high-risk areas.

Typically, when city authorities introduce revisions to their legal and institutional mandates and/or jurisdictions, they will include developing risk management schemes and plans as a part of safer urban development planning process. During this planning process, Government institutions may also give priority to the high-risk areas with the greatest potential for loss of lives and/or economic damages. Quantitative inputs from risk assessment utilizing Earth observation data provide city authorities with critical information to define changes or modifications to public policies to minimize disaster impacts. Prime examples can be drawn from the case of the earthquake risk assessment initiatives in Myanmar and Bangladesh. The city of Mymensingh, Bangladesh, benefited from being one of the cities that is equipped with geospatial earthquake risk information, had taken another important step to apply such information into its formal urban planning procedure. This was done through careful engagement of stakeholders to review the hazard and risk information and existing policies before making a proposal to demarcate earthquake-prone areas into the Strategic Development of the city. Similarly, in Myanmar, the outcomes of the earthquake risk assessment of Mandalay City was incorporated as examples in the Guidelines for Integrating Disaster Risk Information into Urban Land Use Planning in Myanmar, which was accepted by the Ministry of Construction, Myanmar, as a framework for future planning decisions (Fig. 5).

The geospatial risk information developed in these nine cities will serve as the baseline for monitoring the progress of risk reduction interventions undertaken. Risk information also serves to define the city government's capacity to meet its own post-disaster needs and address these by identifying the external assistance needed to respond in the case of devastating earthquake event. This would include defining the financial needs and priorities for economic recovery and reconstruction

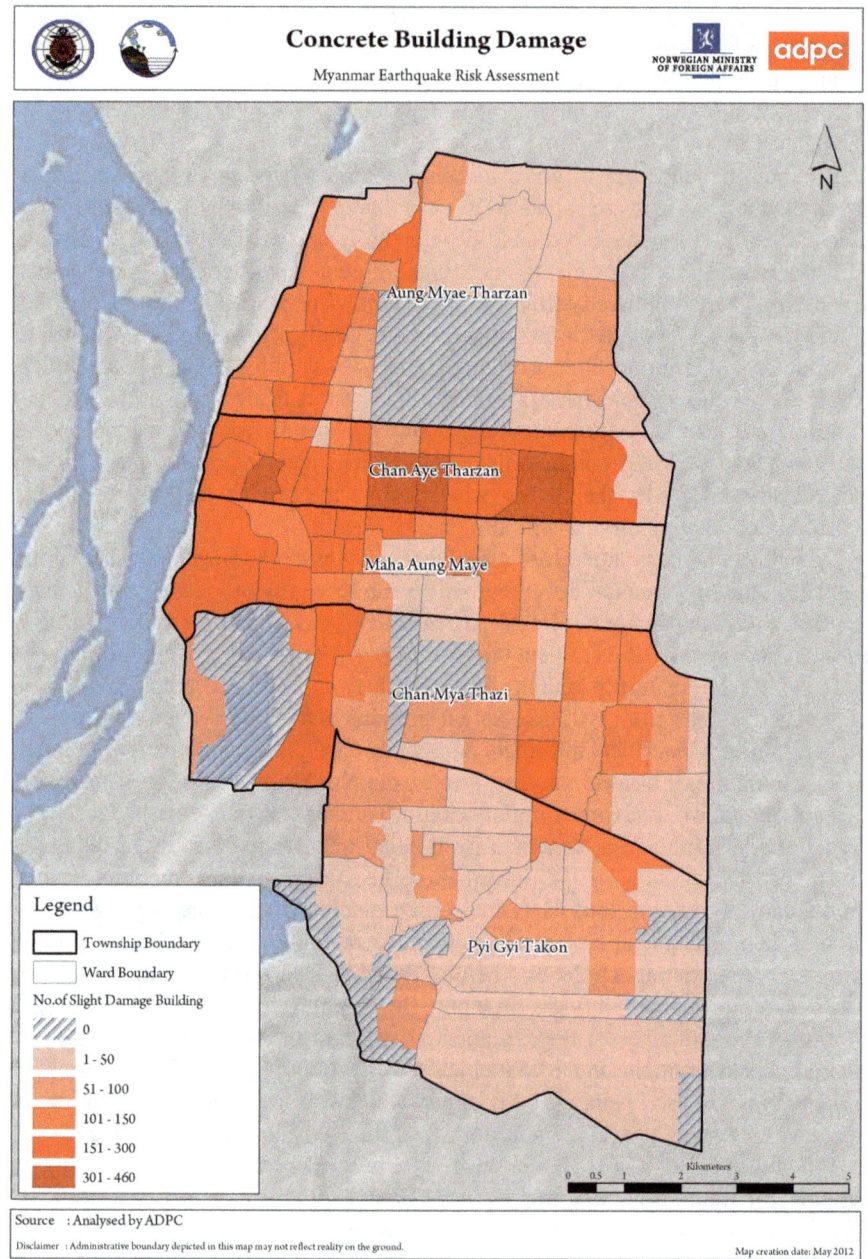

Fig. 4 Earthquake risk map for Mandalay City: distribution of damage expected on concrete buildings in the city

Fig. 5 Vulnerable housing in Mandalay City

in case of such an event. High-impact earthquake events usually create emergency needs within the city to treat the injured, dispose of the dead, provide temporary shelter to the affected population, and organize the provision of food, water and other important logistics for immediate relief as well as economic recovery and reconstruction. Risk information will help city authorities to identify these needs and develop a well-integrated earthquake response mechanism that includes the proper coordination among all city jurisdictional levels (city, wards, individual institutions, community institutions, institutions addressing vulnerable populations such as the elderly, etc.).

Therefore, given the spatial implications of the impact of earthquakes, incorporating spatial analysis based on hazard and vulnerability mapping into the scenario-based contingency planning process is also essential. It allows planners and emergency managers to look at risk and potential impact as per the emergency clusters and identify cluster needs/gaps/challenges. The inclusion of spatial planning during the contingency planning process also promotes an integration of disaster preparedness into long-term physical planning to make sure the cluster level requirements are met during the long-term city planning.

4 Conclusions

Developing a geodatabase inventory structure ensured the integration of Earth observation solutions into traditional approaches for assessing seismic hazard and landslide occurrences. Effective utilization of GIS and remote sensing data for earthquake disaster risk management-related decision making and scenario-based contingency planning was carried out for the first time in numerous cities. Developing an accurate database consisting of large number of building footprints, infrastructure, lifeline, and service facilities for cities like Chittagong, Dhaka, Sylhet, and Mandalay required research and development in many areas, including data collection, structuring, and field verification of comprehensive set of spatial data. Similarly, database development for landslide occurrences in Lao PDR has been transformative for local governments that needed institutional strengthening to respond swiftly to the impacts of a disaster so that community and local economies can recover quickly. This includes the ability to use geospatial information in successful post-disaster planning and reconstruction. In both contexts, incorporating geospatial and Earth observation information into hazard and vulnerability mapping allowed Governments to consider scenario-based contingency planning. In the Lower Mekong, which has seen its share of weather-related disasters, this kind of integration has allowed planners and emergency managers to look at risks, potential impacts, and identify needs and gaps in all aspects of the risk management spectrum —starting with prevention and mitigation, through preparedness and response, to rehabilitation and reconstruction.

There are many different actors involved in the DRM process at local, national, regional, and international levels. Their capacity to integrate technology and information changes on a geographic and development basis, and their different mandates determine the nature of their varying needs for data and information.

No single organization or government has a responsibility to meet international requirements for satellite-based DRM data. Working with user organizations to identify resources to leverage initial 'in-kind' contributions of data, the international community can make meaningful progress toward increasing the use of Earth observation data in DRM. This may enable the purchase of commercial imagery and ultimately increase the number of satellites designed and built to support DRM. It is also clear from the texts of agreements such as the UNFCCC or other related conventions on biodiversity and desertification that satellite data can help to develop a wealth of information to address monitoring and verification needs, not just for disasters but the underlying conditions that lead to decisions that impact the way disasters are addressed. The United Nations agencies are planning for the post-Sendai framework (United Nations 2015) with renewed emphasis on concrete actions to implement the recommendations. In this new framework, satellites and Earth observation data will play a critical role, particularly in reducing the underlying risk factors and strengthening disaster preparedness for effective response. Satellite data can supply regular, detailed updates on the status of hazards on a global, regional, or national basis. The Lower Mekong is an important place for such regional data to be consolidated to assist the transition to a green economy and support the restoration and maintenance of the region's natural capital.

Reducing the severity of disasters requires the integration of observations, exploiting predictive modeling, and disseminating timely and accurate information needed by all actors involved in response and risk mitigation. The implementation of risk management measures such as disaster preparedness planning, safer land-use planning, regulatory zoning, weather forecasting, and early warning systems are needed and should be considered essential components for sustainable development of the Lower Mekong subregion.

Earth observation's contribution to the provision of refined risk assessment includes up to date localization and characterization of the asset at risk; information to support prevention plan elaboration; supporting anticipation (such as weather forecasting and early warning systems); as well as crisis management operations (rescue, recovery) and to help understand the resulting environmental damages, ecosystem services, and natural recovery mechanisms better.

It is true that many of the current generation of satellite systems were not designed specifically for DRM activities; however, the rapid pace of technology advance has led to the launch of increasingly flexible and powerful systems. These missions are now multipurpose and offer exceptional coverage and scope. Taken collectively, the world's satellite systems offer a unique tool for DRM. For example, the requirements of disaster management centers concerning plain flood hazard have been gathered in consultation with national civil protection authorities: reference mappings are needed within the day of the hazard impact, while rapid mapping of the flood extent is needed within a few hours or within a day, every day. Similar analysis has produced a characterization of requirements for other hazard types and other phases of DRM such as early warning—for a range of different hydro-meteorological or geophysical hazards—situational awareness during and after disasters, precise damage assessment, and support to recovery and reconstruction.

DRR depends on concrete measures put in place at the community level by informed national and local authorities (ministries of interior, environment, development planning, land use, agriculture, river basin administrations, mayors, civil protection, etc.). DRM work relates to different aspects of development and territorial management that concern a variety of domains such as water, sanitation, habitat, transportation, and energy. They are managed by different types of organizations with mandates attached to a specific domain, like basin authorities looking at hydrological risks, and geologists and road engineers looking at geohazards such as landslides. Beyond local and national actors, international development organizations are increasingly more involved in development programs with risk management components. As DRR activities are increasingly undertaken with national and regional users (such as the Mekong River Commission), through the sponsorship and guidance of international players (international financial institutions such as the World Bank or development agencies such as the UNDP and USAID), the uptake of new systems and satellite-based data in operational support to disaster management will certainly increase. There is nothing planning can do to reduce the number and severity of hazards. Nonetheless, the integration of relevant, timely, and comprehensive EO datasets into disaster mitigation activities will be the largest factor contributing to reducing the loss of life and damage to property in the Lower Mekong region. ADPC will continue to and further advance the use of EO for improved disaster risk management both pre- and post-events, specifically on disaster risk assessment, climate-related early warning, emergency response planning, and post-disaster recovery, as well as an integration of EO and disaster risk reduction into future development planning. A better preparedness and systematic efforts toward risk reduction based on long-term development and environmental concerns will achieve the ultimate goal of achieving a green economy while reducing the number and scale of actual disasters caused by hazards.

References

Asian Development Bank and the International Labour Organisation. (2014). ASEAN Community 2015: Managing integration for better jobs and shared prosperity.
Buhaug, H., & Urdal, H. (2013). An urbanization bomb? Population growth and social disorder in cities. *Global Environmental Change 23*(1).
Hearn, G. J., Hunt, T., Aubert, J., & Howell, J. (2008a). Landslide impacts on the road network of Lao PDR and the feasibility of implementing a slope management programme. In *The First World Landslide Forum. Satellite Conference, Sendai, Japan,* 11–12 Nov 2008.
Hearn, G. J., Hunt, T., Howell, J., & Carruthers, N. (2008b). Mainstreaming stabilisation trials to strengthen slope management practices in Lao PDR. *South East Asia Community Access Programme*. Department for International Development (DFID), UK. Report SEACAP21-004 2008 (p. 9).
SEACAP. (2008). Feasibility study for a national programme to manage slope stability. *South East Asia Community Access Programme*. Department for International Development (DFID), UK. Final Report SEACAP21/002 (p. 41).
United Nations. (2015). *Sendai Framework for Disaster Risk Reduction: 2015–2030* (p. 35).
WWF 2014. Living Planet Report. (2014). *Species and spaces, people and places*. WWF International, Gland, Switzerland (p. 180).

Land Cover Mapping for Green House Gas Inventories in Eastern and Southern Africa Using Landsat and High Resolution Imagery: Approach and Lessons Learnt

Phoebe Oduor, Jaffer Ababu, Robinson Mugo, Hussein Farah, Africa Flores, Ashutosh Limaye, Dan Irwin and Gwen Artis

Abstract Africa is rich in natural resources including minerals, natural habitats and vegetation (African Bank 2007). However, these resources are under sustained pressure from a population that is growing at a faster rate than anywhere else on the globe (African Institute for Policy Development 2012). Substantial changes in land cover have been observed at national and local scales on the continent by various studies in the recent past. The rapidly growing population increases demand for food and land for settlement, a challenge that is further compounded by climate change. The effects of climate change make the African populace vulnerable to unpredictable weather patterns, floods, droughts and rising sea levels that threaten low coastal towns. This chapter addresses the pressing challenges of environmental management and sustainability in Eastern Africa.

1 Introduction

1.1 Drivers of Land Cover Change in Africa

Africa is rich in natural resources including minerals, natural habitats, and vegetation (African Bank 2007). However, these resources are under sustained pressure from a population that is growing at a faster rate than anywhere else on the globe (African Institute for Policy Development 2012). Substantial changes in land cover have been observed at national and local scales on the continent by various studies in the recent past. The rapidly growing population increases demand for food and land for settlement, a challenge that is further compounded by climate change.

P. Oduor (✉) · J. Ababu · R. Mugo (✉) · H. Farah
Regional Centre for Mapping of Resources for Development, Nairobi, Kenya
e-mail: poduor@rcmrd.org

R. Mugo
e-mail: rmugo@rcmrd.org

A. Flores · A. Limaye · D. Irwin · G. Artis
SERVIR Science Coordination, NASA Marshall Space Flight Centre, Huntsville, USA

The effects of climate change make the African populace vulnerable to unpredictable weather patterns, floods, droughts, and rising sea levels that threaten low coastal towns. One of the important characteristics of land cover changes in Sub-Saharan Africa is the increasing areas under cropland and settlement, as well as decreasing natural vegetation including forestland and bushland.

A number of studies on the continent's ecosystems has shown similar land cover conversion patterns as well as driving forces. Mertz et al. (2007) identified woodlands and dry forests as the least protected and thus most threatened ecosystems on the continent, findings they attributed to population growth, climate change, as well as weak environmental management and policy frameworks. Tropical deforestation is also caused by multiple factors and not just single variables (Helmut and Lambin 2002). The wide spectrum of the key drivers is the combination of settlement expansion, agricultural expansion, and wood extraction with distinct regional variations. Again, most households in Africa remain unconnected to power grids, and therefore the majority of them still rely on firewood and charcoal for energy, making forest and grazing land more vulnerable across the continent (EIA 2009).

Research on the interactive effects of poverty and rural population indicates that faster cropland expansion is attributed to poverty in densely populated rural areas (Nkonya et al. 2013). On the other hand, the rapid rate of urbanization in Africa and Sub-Saharan Africa in particular is explained by the rate of urban growth on the continent, which far overwhelms the capacity and efforts by governments to deliver essential services such as housing, water, education, electricity, health services, and waste disposal (African Institute for Policy Development 2012). The result is often overcrowded informal settlements. This is an area that requires extensive study, because the understanding of change trajectories and potential future scenarios is essential for a sustainable development strategy by any geographic region. While Africa has witnessed the largest forest loss over the past few decades than ever before (Nkonya et al. 2013), afforestation and other conservation measures in some countries have significantly reduced the net loss of forest area, and set the area under forest cover on the increase (FAO 2010).

Since the 1970s, satellite technology has enabled the monitoring of the earth from space. The continent of Africa has been receptive and supportive, studying otherwise expensive or inaccessible areas often repeatedly. However, low to medium resolution satellite images are often used, providing limited quantitative data on land cover change (Bégué et al. 2011; Cabral et al. 2011). The continent struggles to create a balance between simply sustaining a growing population, and investing in economic development.

1.2 The Challenge of an Increasing Carbon Footprint for Africa

The total carbon emissions for Africa are still less than that of some developed and middle level economy countries like the United States, China, India, and Russia. Since the 1950s, Carbon emissions for Africa have increased 12fold, measuring 311 million metric tons of Carbon in the year 2008 (Boden et al. 2011). Africa's biggest emitter is South Africa, accounting for some 40 % of the continent's CO_2 emissions. South Africa derives most of her electricity from coal, and is the 12th biggest emitter in the world (EIA 2009). Egypt is the second largest emitter contributing 17 % of the continent's emissions. This data were based on annual 2009 CO_2 emissions from energy consumption alone excluding other GHG sectors. Other GHG sectors such forestry, agriculture, land use, and waste, are rather complex and require investment, which is a major challenge for many countries. The most consistent, transparent, and accurate ways of handling information related to climate change have been proposed, but the rate of uptake in Africa is disturbingly low. African countries need to put in place national GHG inventory systems that will simplify their climate change reporting obligations to the UNFCCC, by managing inventory data in those internationally accepted ways. This will also streamline their own internal interagency reporting. Many of Africa's national economies remain hindered by historical standards due to both economic and political causes, including weak economic and political institutions, insufficient infrastructure in quantity and quality, negative perceptions of international investors, political disturbances, and protracted civil unrest, and equally importantly, the effect of disease (particularly HIV/AIDS). These present daunting challenges to growth in a number of African countries, especially where funds need to be allocated for technological advancement, alternative source of livelihoods and education.

1.3 Existing Gaps and the Contribution of the GHG Project

The Project on land cover mapping and capacity building for the development of greenhouse gas inventories was implemented over a period of 3 years, across nine countries with varying landscapes and characteristics. Its main aim was to produce baseline land cover maps, as well as build the capacity of the participating countries for future mapping tasks. While maps can have many uses, their main purpose was to support the development of the GHG Inventories.

The project was successful on many levels, contributing to filling the gaps in skills, expertise, and working procedures, as well as, strengthening the existing endeavors at the institutional level in the nine participating countries. Training was on various concepts in GIS and remote sensing, with particular emphasis on land cover mapping using remotely sensed datasets, fieldwork exercises using handheld GPS devices, scene identification and downloading of Landsat satellite imagery, development of metadata for archived datasets, the different approaches on the use of a boundary to restrict a study site in a GIS environment among other key concepts.

In many instances, the project contributed to formalizing and fostering collaboration between key government institutions. In Tanzania for example, the team pledged to source for budgetary allocation from their government, in order to continue meetings to advance the agenda on GHG Inventories development. The project offered a meeting platform for representatives from many institutions in each participating country during the workshops. The initial workshop where ancillary data are collected brought together many institutions that are custodians to various datasets. In many countries, it enabled officials from the institutions to be also aware of, and access what their counterparts were keeping. This is because each institution was to declare what datasets they held based on a predefined checklist.

One major strength of the project was its partnership design. Apart from the government agencies upon whose goodwill and relations activities were scheduled and implemented, the project brought together strategic partners in the United States and Africa. At the core of the network was the RCMRD and NASA, under the SERVIR Project, with financial support from the USAID. A Technical Advisory Board (TAB) constituted to provide technical advice on the map development process comprised of representatives from UNFCCC, NASA's Earth Science Office, the University of Colorado, RCMRD and the ICF International. The Colorado State University, the developers of the ALU Tool for GHG Inventories were also a strategic partner.

1.4 Background on the Participating Countries

The nine countries where the project was implemented had varying landscapes, climate and characteristics, but very similar regional setting in Sub-Saharan Africa. From Ethiopia to the north of Eastern Africa to Namibia and Lesotho in the far Southern Africa, all the countries face similar pressures on existing natural resources, a rapidly growing population, inadequate expertise, and facilities in government institutions, as well as overdependence on rainfed agriculture. Most of the countries are landlocked, with relatively high population densities in high resource areas, e.g., Rwanda (Byamukama et al. 2010). According to the African Development Bank (2011), environmental degradation in Ethiopia is mainly driven by high population growth, urbanization, agriculture, and poverty and infrastructure expansion. In most countries agricultural activities contribute to the highest percentage of net GHG emissions. The GHG Inventories (1995–2000) show that Malawi is a net emitter of GHG; and the agriculture, forestry, and land use sector remain key contributors (96 %), based on their second national communication done in 2001. Stiebert (2013) identified the agriculture sector as the main contributor to greenhouse gasses in Rwanda, accounting for about 65 % of non-LULUCF emissions for Rwanda in 2010. This was attributed to emissions from soil cultivation. The countries were also prone to natural disasters and extreme weather events. McSweeney et al. (2010) observed that like many countries in Africa, Zambia was one of the most severely affected countries by the 1997/1998 El Niño phenomenon, suffering flooding and destruction. In Botswana, rainfall is unreliable as some two-thirds of the country is covered by the Kalahari Desert sands, and is not suitable for agricultural production. However, Botswana's natural resources consist of the unique Okavango Delta, some wetlands, arable land, woodlands mineral resources, and wildlife and mineral resources Wingqvist and Dahlberg (2008). Tanzania too is well endowed with natural resources, being home to the Serengeti, a prominent tourist destination in Africa (Agrawala et al. 2003).

Even for a dry country like Namibia, the variety of ecosystems, natural habitats, and wildlife species are remarkable. Internationally known biodiversity "hotspots" include the Sperrgebiet, located in southern Namibia in the Succulent Karoo floral kingdom, and the rugged Namib Escarpment, which is part of Africa's great western escarpment. The main land uses in Namibia include farming for crops and animal husbandry, conservation on state protected areas, communal conservancies, resettlement; government agriculture farms, and mining (MET 2010). Another similarity in the countries was the rapid rate of deforestation. According to the World Bank (2015), the rate of deforestation in Uganda is very high at 2.3 %, well above the Sub-Saharan and world average of 0.6 %. Biomass fuel such as firewood, charcoal, animal dung, and crop residues remains the main energy source for Ethiopia, and as such, environmental degradation is mainly driven by high population growth, urbanization, agriculture, and poverty and infrastructure expansion (César and Ekbom 2013; African Development Bank 2011).

1.5 The Datasets and Rationale

The Landsat imagery were used in this project. Landsat images presented numerous spatiotemporal advantages compared to other satellite datasets available for land cover mapping, primarily because they provided consistency and reliability in national coverage for the selected countries and epochs. They are also freely available and downloadable where internet connection is fairly stable. The Landsat lifespan over the last four decades has been continuous with new releases being launched before the lifespan of the previous launches, making it a reliable data source that is useful in tracking land cover and land use change over time.

In this work, the interpretations derived from image classifications were supported by ancillary data, since the accuracy of the information gathered-based solely on spectral variability is often insufficient (Gercek 2004). This project accessed a wide range of ancillary datasets including the literature, layers of spatial map-based thematic data, topographical data, as well as, nonspatial data. These were especially vital, where land cover classes had low spectral separability. It was observed that initial classification accuracy was especially high, where the country had shared plenty of ancillary data, and where the project technicians had utilized it as required.

2 Working Structure in the Countries

2.1 Regional Scope

The project initially covered six countries: Malawi, Rwanda, Tanzania, Zambia, Namibia, and Botswana. In 2014/2015, three more countries were included: Ethiopia, Uganda, and Lesotho.

2.2 Implementation Structure for a Country

2.2.1 Inception Meetings

The main objective of holding these meetings was to create awareness of the project to all relevant stakeholders at the same time taking advantage of the meetings for definition of national land cover classification schemes. Through this it was possible to define the classification schema to be used for the country that not only conforms to the guidelines of the UNFCCC and IPCC but also meets the national definition and stakeholder requirements as far as land cover maps are concerned.

2.2.2 Ancillary Data Collection Process

The main objective for this process was to ensure that consistent and accurate information was made available for proper definition of national land cover classification schemes. In addition, gathering existing and or historical land cover land use maps and previously collected ground reference data was an important step in this process. Finally, the project ensured that enough relevant ancillary datasets were available for classification of Landsat imagery to the required classes/categories.

2.2.3 Maps and Analysis

Scheme I maps depict the six IPCC land cover categories: forestland, grassland, wetland, cropland, settlements, and otherland. Scheme II maps on the other hand represented a subcategorization of Scheme I maps and were country specific. They were defined by the country and could be rolled back to Scheme I maps. Mapping was done at RCMRD with respect to the two Classification schemas developed, ancillary data, using Landsat Images. Tables 1 and 2 present details of Scheme I and Scheme II classes, as defined in this work.

Table 1 Scheme I Land Cover classes for Malawi, Rwanda, Zambia, Namibia, Botswana, Tanzania, Ethiopia, Uganda, and Lesotho

Index	Land cover	Used in
1.	Cropland	All
2.	Forestland	All
3.	Grassland	All
4.	Wetland	All
5.	Settlement	All
6.	Otherland	All

Table 2 Scheme II Land Cover Classes, which were defined based on country preferences and are unique in each country

No.	Country	Classes
1.	Namibia	Forestland, Woodlands, Planted Forest, Grassland, Savanna Grassland, Shrubland, Annual Cropland, Perennial Cropland, Wetland, Water Body, Settlement, Rock Outcrop, Bare Soil, Desert Dune, Desert Sand
2.	Botswana	Dense Forest, Moderate Forest, Sparse Forest, Woodlands, Planted Forest, Open Shrubland, Closed Shrubland, Open Grassland, Closed Grassland, Cropland, Wetland, Water Body, Settlement, Otherland, Pans
3.	Malawi	Dense Forest, Moderate Forest, Sparse Forest, Planted Forest, Open Shrubland, Closed Shrubland, Open Grassland, Closed Grassland, Annual Cropland, Perennial Cropland, Wetland, Water Body, Settlement, Otherland
4.	Zambia	High Dense Forest, Medium Dense Forest, Open Dense Forest, Planted Forest, Open Shrubland, Closed Shrubland, Open Grassland, Closed Grassland, Annual Cropland, Perennial Cropland, Wetland, Water Body, Settlement, Otherland
5.	Rwanda	Dense Forest, Moderate Forest, Sparse Forest, Planted Forest, Open Shrubland, Closed Shrubland, Open Grassland, Closed Grassland, Annual Cropland, Perennial Cropland, Wetland, Water Body, Settlement, Otherland
6.	Ethiopia	Dense Forest, Moderate Forest, Sparse Forest, Woodlands, Open Shrubland, Closed Shrubland, Open Grassland, Closed Grassland, Annual Cropland, Perennial Cropland, Wetland, Water Body, Settlement, Rock Outcrop, Bare soil, Lava Flow, Salt Pan
7.	Tanzania	Dense Forest, Moderate Forest, Sparse Forest, Woodlands, Planted Forest, Open Bushland, Closed Bushland, Open Grassland, Closed Grassland, Annual Cropland, Perennial Cropland, Wetland, Water Body, Settlement, Rock Outcrop, Bare soil, Salt Crust, Snow/Icecap
8.	Lesotho	Natural Forest, Planted Forest, Shrubland, Grassland, Orchards, Annual Cropland, Wetland, Water Body, Settlement, Otherland
9.	Uganda	Dense Forest, Moderate Forest, Sparse Forest, Dense Woodlands, Moderate Woodlands, Sparse Woodlands, Planted Forest, Open Shrubland, Closed Shrubland, Open Grassland, Closed Grassland, Annual Commercial Cropland, Perennial Commercial Cropland, Wetland, Water Body, Settlement, Otherland

2.2.4 Field Validation

Once the first draft of the maps was finalized, a fieldwork session was organized within the country. Field validation for the later epochs was done in collaboration with the stakeholders in the country to collect reference points for accuracy assessment and for further refinement of the maps.

2.2.5 Peer Review and Technical Advisory Board

The initial map review was done internally at RCMRD through a project team meeting, and then a broader RCMRD technical review meeting was convened to assess issues highlighted at the project team review level. Once the comments from these had been sufficiently addressed, the map products were submitted to the specific country technical team for further reviews. The feedback obtained from this was used to further refine the maps before further validation was done to check the accuracy and have the initial outputs. These were then submitted to the Technical Advisory Board for further comments. Often, the TAB members provided its input via teleconferences after having reviewed the products individually. This was basically the final review process, after which the maps were corrected and packaged for dissemination. However, the window for country review did not close as in many instances the comments kept coming and necessary discussions or corrections were effected as required.

2.2.6 Dissemination Workshops

This workshop was the final output of the project and had the following key objectives:

- To hand over map products developed by RCMRD to the Government
- To introduce online platforms for sharing the map products.
- To discuss other possible uses of the land cover maps, apart from the primary goal of supporting the development of GHG Inventory
- To offer expert training in the areas of land cover mapping, as a follow up to earlier training held at RCMRD.
- To share technical knowledge and interact with the stakeholders at a formal project wrap up forum.

2.2.7 Visualization Platform

All final land cover maps developed were uploaded onto a visualization platform developed to foster efficiency of data access, wider coverage, and sustainability in dissemination efforts. Two tools referred to as *Compare* and *Statistics* were provided on the platform. The *Compare* tool allows the user to visually compare land covers of the same schema across different epochs visually while the *Statistics* tool enables the user to derive statistics based on an administrative jurisdiction provided for in the country. The platform also allows a user to download the files as well as their styles or legends, and detailed metadata should further processing be required.

- http://geoportal.rcmrd.org/layers/
- http://apps.rcmrd.org/landcoverviewer/

2.3 Impact and Way Forward

Provision of the land cover maps in an interactive process which also addressed capacity building needs of the relevant agency within then respective countries has had the following short and long-term impacts:

- Availability of a readily available dataset for use in the agriculture, forestry, and land use sectors.
- Accessibility of regionally consistent land cover datasets for GHG estimation processes to enable countries developed comparable products.
- Capacity development in land cover map development was a significant step toward sustainability of such initiatives.
- Creation or activation of a network of institutions with common objectives (preparation of input maps for GHG processes), within and among countries has fostered synergy and collaborations.

Stakeholders involved in the project in the different Countries: Ministries/Departments of:

(1) Forestry
(2) Land Management
(3) Surveys and mapping
(4) Environment
(5) Land use and/or planning
(6) Land use
(7) Natural resources
(8) Wildlife
(9) Agriculture
(10) Universities
(11) International organizations
(12) Climate and meteorology
(13) Water and wetlands
(14) Private sector.

3 Land Cover Mapping Methodology

3.1 Data Acquisition

The large volumes of Landsat images required for this project were initially supplied by GDI Consulting and Training, a US consultancy firm, having applied initial image corrections. Later on, RCMRD required more images and downloaded them directly from USGS website from time to time. This project also accessed a wide range of ancillary data from each country, which included literature, layers of map-based thematic data, topographical data, as well as, non-spatial data. For each country, these were provided by the government agencies, departments and ministry officials, during the workshop on ancillary data collection.

3.2 Preprocessing

According to Shahrokhi (2004), problems with image pixels including position and perturbation in value can lead to feature misinterpretation. It is therefore important to employ manual or automatic techniques to correct for such defects on the radiometric, atmospheric, and geometric integrity of the satellite images. Radiometric quality is influenced by sensor characteristics and detector responses and includes striping, drop lines, other noise elements, and missing bands or data in image data sets. Atmospheric quality on the other hand is dependent upon image acquisition conditions and includes haze and cloud cover, while geometric quality is determined by sensor characteristics and also satellite situation such as attitude, position, velocity, and perturbations. The Earth's surface relief also affects geometric quality of the images. Compromise in geometric quality is shown by lack of band to band co-registration as well as image to map co-registration.

Encountering such problems on a project involving regional land cover mapping over a long period of time was almost inevitable. However, no major corrections were carried out on the images acquired from USGS in recent times since they were already preprocessed and included a UTM projection with a WGS84 datum. However, selected ancillary data required to be projected to the Landsat imagery's UTM projection and WGS84 datum, in order to enable overlays. The project had the advantage of selecting images over a 3-year window period, and therefore a replacement image was the most immediate remedy whenever a defective image was encountered. However, image enhancement techniques were applied as needed primarily to improve detection and identification of the various land cover classes, and minimize the effect varying spectral characteristics. The images were clipped to the national boundary, usually buffered to include a 10 km distance, in case other versions of the boundary emerged later on.

3.3 Supervised Classification

The project applied the maximum likelihood Classifier, which basically evaluates the variance and covariance of each class, in order to determine where to assign a pixel, and is proven to give the best results (Al-Ahmadi and Hames 2008). The supervised classification method relies heavily on the prior knowledge of the analyst. With sufficient information from visual interpretation and other ancillary data, the analyst therefore would select 'training sites' by digitizing around a homogenous feature. This was done for every visually identified and distinct land cover class. Severally, it was found necessary to collect training sites for further segmentation of subclasses, to be later combined. For instance, where the class "Annual Cropland" would have variations in interpretations, corresponding to various stages of crop cultivation and growth, as well as different crop species. These were initially classified as different classes using different training sites.

While there have been a number of comparative studies which found a few other methods to have better classification results than supervised classification (Kim 1993; Liu 2002), supervised classification has numerous advantages. It is faster once the training sites have been collected, as compared to manual digitization methods. This makes it suitable for large scale projects such as the regional land cover mapping tasks in this project. The accuracy of the results is also increased by maintaining "pixel hygiene" in the training sites. In addition, separability analyses as well as contingency matrix can be carried out to check that the classes do not overlap in statistics. The training sites are also reusable, for example where an image was replaced with that of nearly the same time. Compared to unsupervised classification, the information classes are generated as predetermined by the analyst. Supervised classification is a commonly used method in many countries and capacity building efforts were largely aided by the fact that most participants had used it even though in a geographically smaller project.

In order to ensure uniformity and seamlessness among the Landsat scenes, chain classification was adopted. This involves selecting common training sites on two adjacent scenes in the overlap area. These are then used on both scenes to reduce edge differences between them. Alternatively, training sites from the two images can be harmonized and applied for the classification of the preceding image.

3.4 Post Classification

Image post classification refers to the series of operations carried out after an image has been initially classified. The post classification processes used in this work included combining classes, color coding, and edge matching for neighboring scenes among others. Due to the emergence of subcategories of almost each particular land cover class, class combination was carried out. Further, there was the need to reduce the noise from scattered pixels, these were filtered away based on logical operations, to reduce the "salt and pepper effect", usually associated with pixel based classification methods. After rigorous quality checks and comparisons, a legend was assigned and the classified images mosaicked. The legend information, including specific names and colors were stored in a country-specific decision tree. These were then coded using the decision tree and a national mosaic created. Where only a small region of the image mosaic needed corrections, it was subset and corrected using the classification editor, an extension in the ENVI Software. It is worth noting that the settlements class was generated in a different way. Due to its spectral similarity to bare lands and cropland, it was manually digitized on Google

Earth, set at the corresponding time, rasterized to later mosaic onto the classification using a decision tree rule.

3.5 Quality Checks and Control

Numerous quality checks and control were applied to classification output, in relation to the interpreted satellite imagery, Google Earth imagery, available ancillary data, and ground truth point data. These included both the internal checks within the project team and at the institutional level, as well as by the respective country technical staff and the independent project Technical Advisory Board. Each of these reviews at least for the initial six countries, ended up with very particular recommendations to be carried out by the project team. The corrections were discussed and assigned to individual project staff or a team to address. With corrections and refinements at each country, the errors got fewer and team expertise grew.

3.6 Validation

3.6.1 Sampling

Ground reference data generally cannot be collected for large portions of the entire project area, therefore samples are frequently used (Lillesand and Kiefer 1994). Ground reference data is collected to train the computer to recognize the various land cover categories latent in the imagery and to assess the categorical accuracy of the resulting classification, as it was in this project, because there was only one field data collection activity. Since the ground-referencing activity was limited in time and resources, thus unable to extensive coverage within a country, there was a complementary unbiased approach, to take care of temporal differences between the image dates, and the time of field data were collected. The approach used required that stratified random points were generated within the boundary of the country. Approximately, a thousand points were generated for each country. A buffer of 30 m was generated from the points, and the resulting polygons overlaid on Landsat images, Google Earth, and other high resolution images for the independent interpretation.

3.6.2 Accuracy Assessment

The accuracy assessment was carried out using ground-referencing data, and point data from the interpretation of high resolution images including Google Earth. The latter was especially important for the year 2000 accuracies, for which ground-referencing data were not applicable as the data were obtained in the years

2012–2014. The centroids of the independently interpreted polygons were combined with selected ground-referencing points for accuracy assessment. This was applied to the 'current year' classification; 2010 or 2014. However, ground truth data were not applied in the accuracy assessment of the first epoch; 2000 for most countries and 2003 for Ethiopia. The accuracy assessment of the developed map was done through the construction and analysis of an error matrix in ENVI software, and the acceptable minimum threshold for overall accuracy according to USGS classification was 75 %.

4 Output

The project was tied to various outputs, some of which were modified or further clarified in the course of implementation. Primarily, baseline land cover maps were required for each of the nine countries. These were delivered in three epochs of the years 1990 (for Malawi and Rwanda), 2000 and 2010. However, after the initial two countries, it was agreed that maps for subsequent countries would be developed for two epochs representing years 2000 and 2010. The last three countries to be mapped (Ethiopia, Uganda, and Lesotho), preferred different mapping epochs. These were: Ethiopia, 2003 and 2008; Uganda, 2000 and 2014; and Lesotho, 2000 and 2014. Since the maps were made in two schemas, a total of forty maps were developed for the nine countries. Two epochs per year of mapping, implying four maps per country except Malawi and Rwanda for which six maps were made.

The maps, alongside project reports, training manuals and visualization software were delivered to the countries officially during the dissemination workshops, packaged in an interactive DVD. They were also shared with the country point of contact alongside the raw images interpreted, in portable computer devices. After dissemination, the maps were also made available online, and the link (http://apps.rcmrd.org/landcoverviewer/) to access the data was shared with the technical experts and the wider user community. It was expected that with the official handover, the map products and related reports be available for public consumption as was envisioned in the project goal. Another key output of the project was institutional capacity building on the mapping methods and allied technology, including the use of handheld GPS devices in ground-referencing.

4.1 The Land Cover Maps

This subsection presents the twenty maps developed for the nine countries, alongside the accuracy summaries achieved. The maps were generated by the interpretation of Landsat images with a spatial resolution of 30 m, and the required minimum accuracy was 80 % for the Schema I maps, and 75 % for the Schema II maps. Scheme II maps are not presented.

Table 3 Two sets of maps (Scheme I and II) per year per country were produced as shown

No.	Country	1990	2000	2003	2008	2010	2014
1.	Rwanda	x	x			x	
2.	Botswana		x			x	
3.	Ethiopia			x	x		
4.	Lesotho		x				x
5.	Malawi	x	x			x	
6.	Tanzania		x			x	
7.	Namibia		x			x	
8.	Zambia		x			x	
9.	Uganda		x				x

Table 3 presents a summary of maps generated for each country, and the respective years for which they were done. Table 4 presents overall accuracies and Kappa Coefficients from accuracy assessments.

Figures 1, 2, 3, 4, 5, 6, 7, 8, 9 shows Scheme I maps developed for Botswana, Ethiopia, Lesotho, Malawi, Namibia, Rwanda, Tanzania, Uganda, and Zambia (Figs. 10, 11 and 12).

4.2 Ancillary Data Collection Workshops

For each of the nine countries, there was an initial workshop with the main objective of sharing existing ancillary data within the country, to aid in the mapping processes. Additionally, an agreement was made on the legend of the nationally defined land cover classes, for the Schema II maps. Combined statistics on attendance are presented in Table 5. These workshops were conducted within the specific countries, and event facilitators have not been included in the counts.

4.3 Capacity Building Workshops

Among the project activities, there were two capacity building workshops arranged for each country, the first one was to sponsor two participants for a 2 week 'on the job training' at RCMRD. This happened during the map interpretation period, and participants were able to offer useful insights into the mapping experience. Some countries sponsored additional participants for the training including Botswana and Zambia. Uganda on the other hand preferred that the 'on the job training' be conducted in their country to attract more participants.

The second planned training workshop was the 1 week training in the country. This was done during the dissemination workshop week, and the participants

Table 4 Overall percentage accuracies and Kappa coefficients for maps developed for nine countries

		Epoch											
		1990		2000		2003		2008		2010		2014	
	Map schema	I	II	I	II	I	II	I	II	I	II	I	II
Botswana	Overall accuracy	–	–	91.5	75.5	–	–	–	–	91.05	75.94	–	–
	Kappa coefficient	–	–	0.792	0.67	–	–	–	–	0.845	0.702	–	–
Ethiopia	Overall accuracy	–	–	–	–	88	76.52	86.68	82.28	–	–	–	–
	Kappa coefficient	–	–	–	–	0.79	0.711	0.82	0.798	–	–	–	–
Lesotho	Overall accuracy	–	–	92.5	85	–	–	–	–	–	–	85.18	77.85
	Kappa coefficient	–	–	0.839	0.787	–	–	–	–	–	–	0.777	0.718
Malawi	Overall accuracy	85.81	78.33	82.74	77.65	–	–	–	–	83.09	77.1	–	–
	Kappa coefficient	0.79	0.734	0.766	0.726	–	–	–	–	0.774	0.725	–	–
Namibia	Overall accuracy	–	–	89.29	75.51	–	–	–	–	88.73	76.95	–	–
	Kappa coefficient	–	–	0.825	0.706	–	–	–	–	0.814	0.718	–	–
Rwanda	Overall accuracy	82.2	79.8	82.74	77.65	–	–	–	–	81.3	76.41	–	–
	Kappa coefficient	0.753	0.754	0.766	0.726	–	–	–	–	0.741	0.715	–	–
Tanzania	Overall accuracy	–	–	86.63	79.48	–	–	–	–	80.18	75.38	–	–
	Kappa coefficient	–	–	0.806	0.76	–	–	–	–	0.731	0.718	–	–
Uganda	Overall accuracy	–	–	90.27	86.34	–	–	–	–	–	–	84.22	79.11
	Kappa coefficient	–	–	0.87	0.844	–	–	–	–	–	–	0.784	0.76
Zambia	Overall accuracy	–	–	89.05	75.51	–	–	–	–	85.53	76.46	–	–
	Kappa coefficient	–	–	0.827	0.706	–	–	–	–	0.793	0.717	–	–

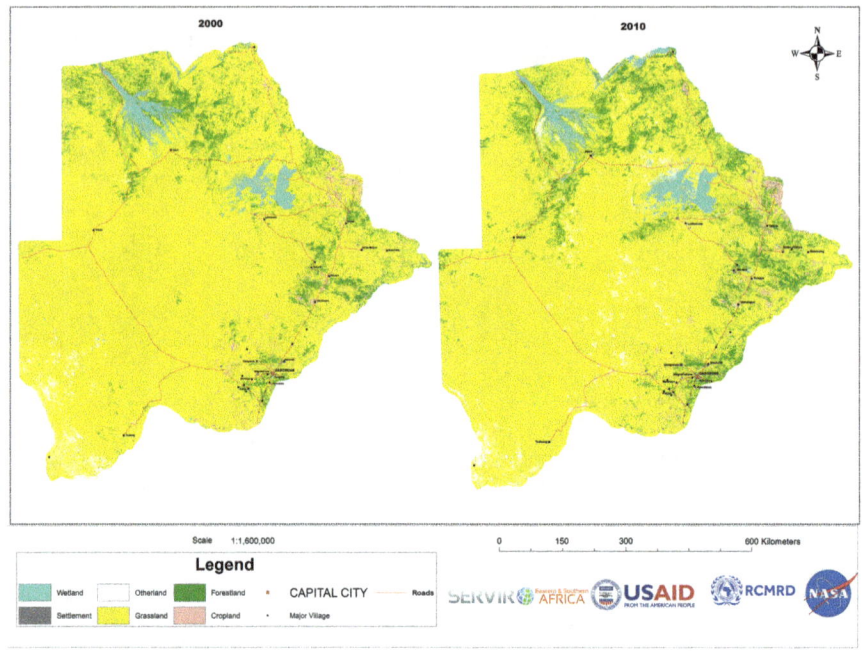

Fig. 1 Botswana Scheme I Land Cover Maps for 2000 and 2010

ranged between 10 and 15 per country. There were other training events whose statistics are not included herein. For example, before the start of each ground-referencing activity, the participants were trained on the sampling methods and the use of handheld GPS devices in fieldwork for 2 days. Another important event was the 2 week workshop at RCMRD on land cover mapping in the East and Southern Africa Region organized by the US Department of Interior in March 2014. This was attended by many participants from the project countries, and they spent the second week in the Project Lab learning and assessing the maps for their respective countries. The numbers for all the capacity building workshops are summarized in Table 6.

4.4 Dissemination Workshops

The final major project activity for each country was the data dissemination workshop, an event during which high ranking officials in the respective governments were invited. The dissemination workshops were held in the same week as the in country training. Statistics on the in country trainings are presented in Table 7.

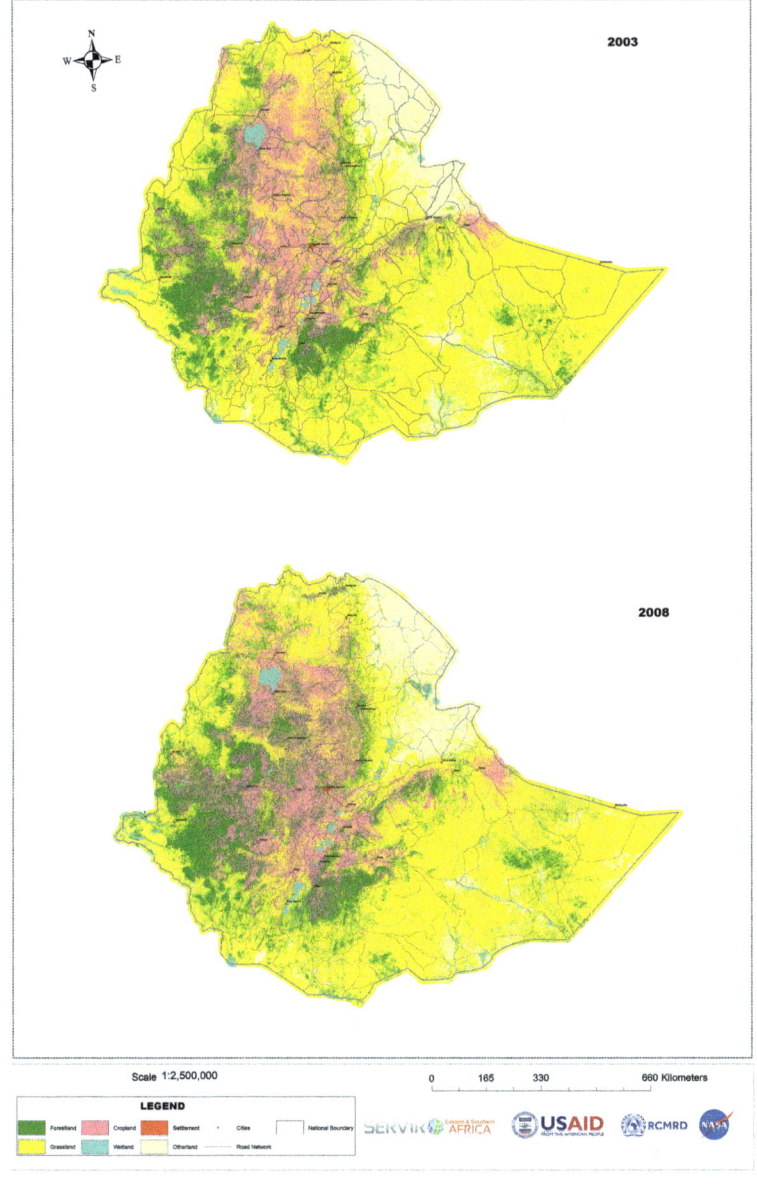

Fig. 2 Ethiopia Scheme I Land Cover Maps for 2003 and 2008

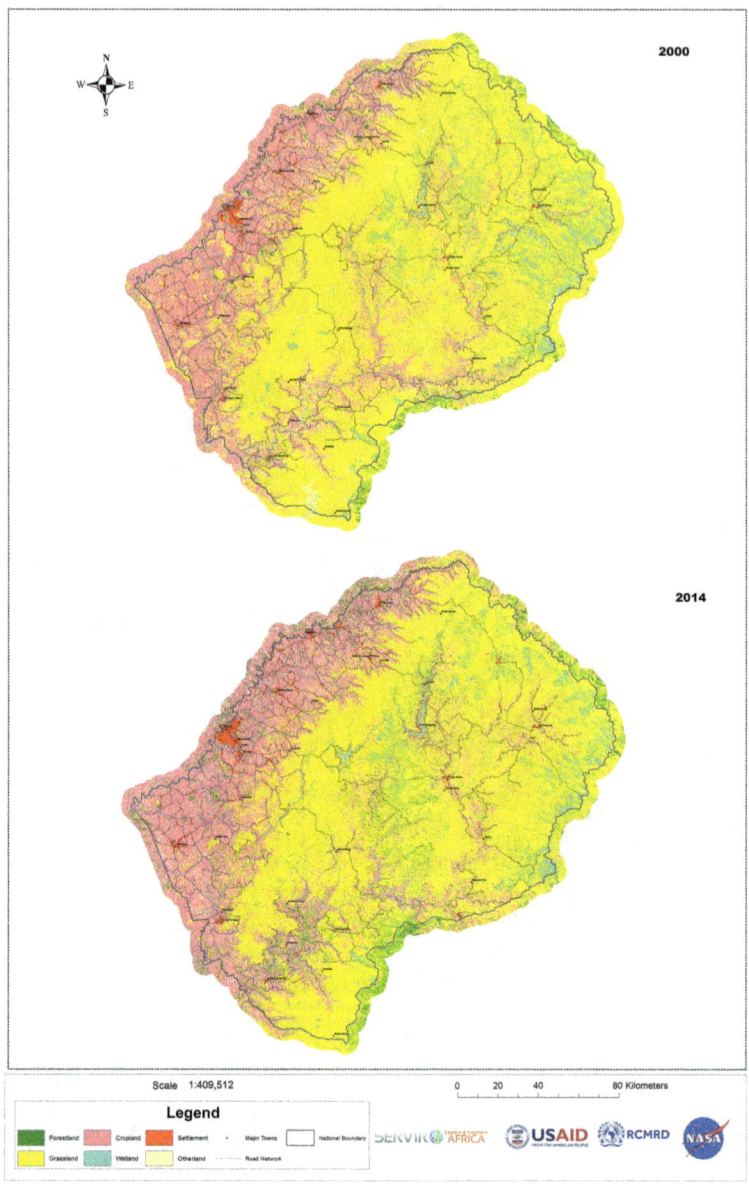

Fig. 3 Lesotho Scheme I Land Cover Maps for 2000 and 2014

Land Cover Mapping for Green House Gas Inventories in Eastern ... 103

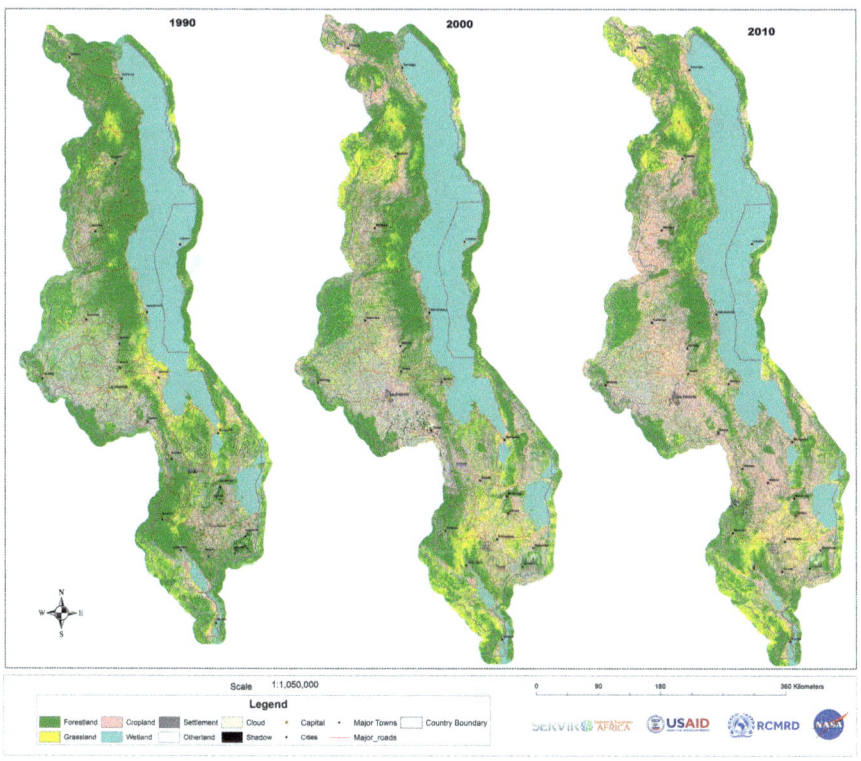

Fig. 4 Malawi Scheme I Land Cover Maps for 1990, 2000 and 2010

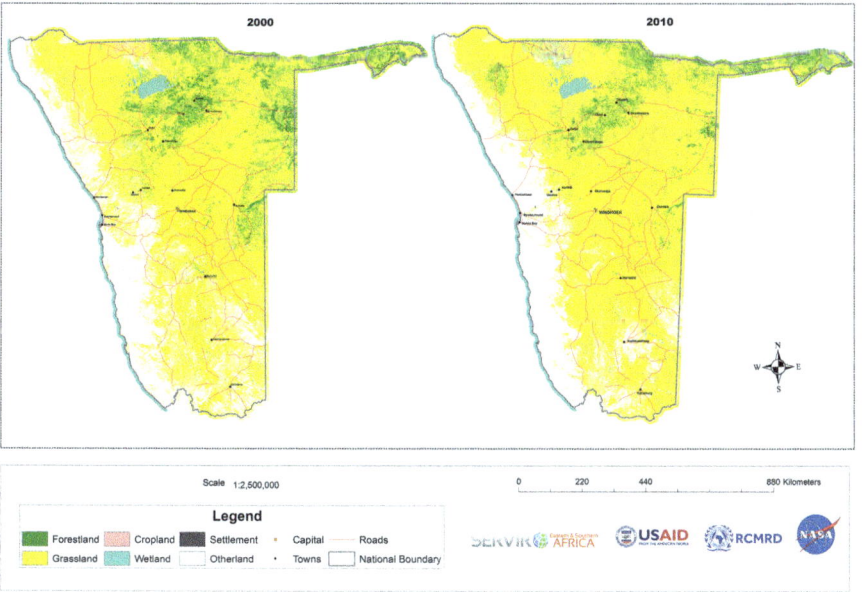

Fig. 5 Namibia Scheme I Land Cover Maps for 2000 and 2010

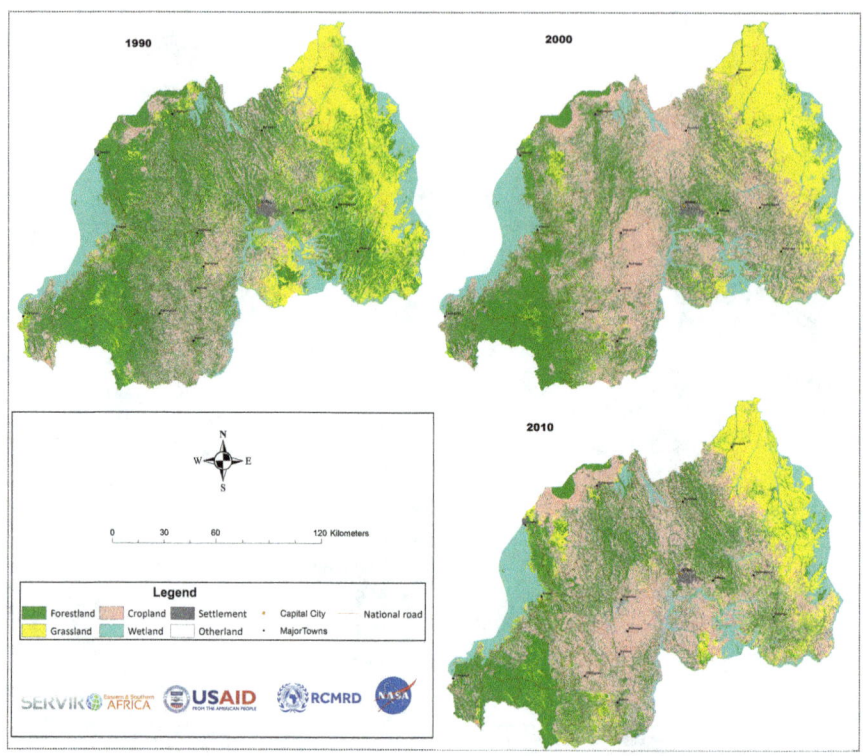

Fig. 6 Rwanda Scheme I Land Cover Maps for 1990, 2000 and 2010

4.4.1 The Interactive DVD

During the dissemination workshop, the interactive DVD was introduced, as well as the links to the online map viewer.

The interactive DVD contained:

- Land Cover Maps
- Technical Reports
  - Training Manual
  - Accuracy Reports

4.4.2 The Web Portals

All Land cover maps in either Scheme I or II are downloadable from the locations listed below, including the respective styles and metadata.

(1) http://geoportal.rcmrd.org/layers/
(2) http://apps.rcmrd.org/landcoverviewer/

Land Cover Mapping for Green House Gas Inventories in Eastern …

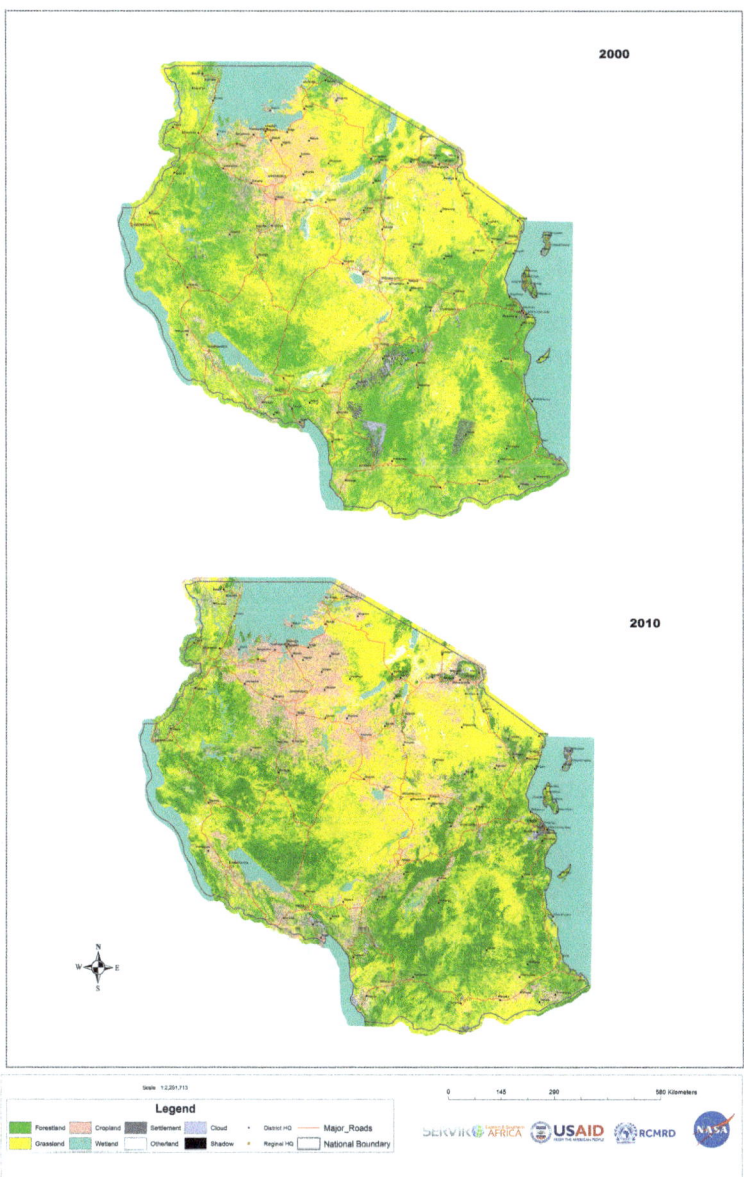

Fig. 7 Tanzania Scheme I Land Cover Maps for 1990, 2000 and 2010

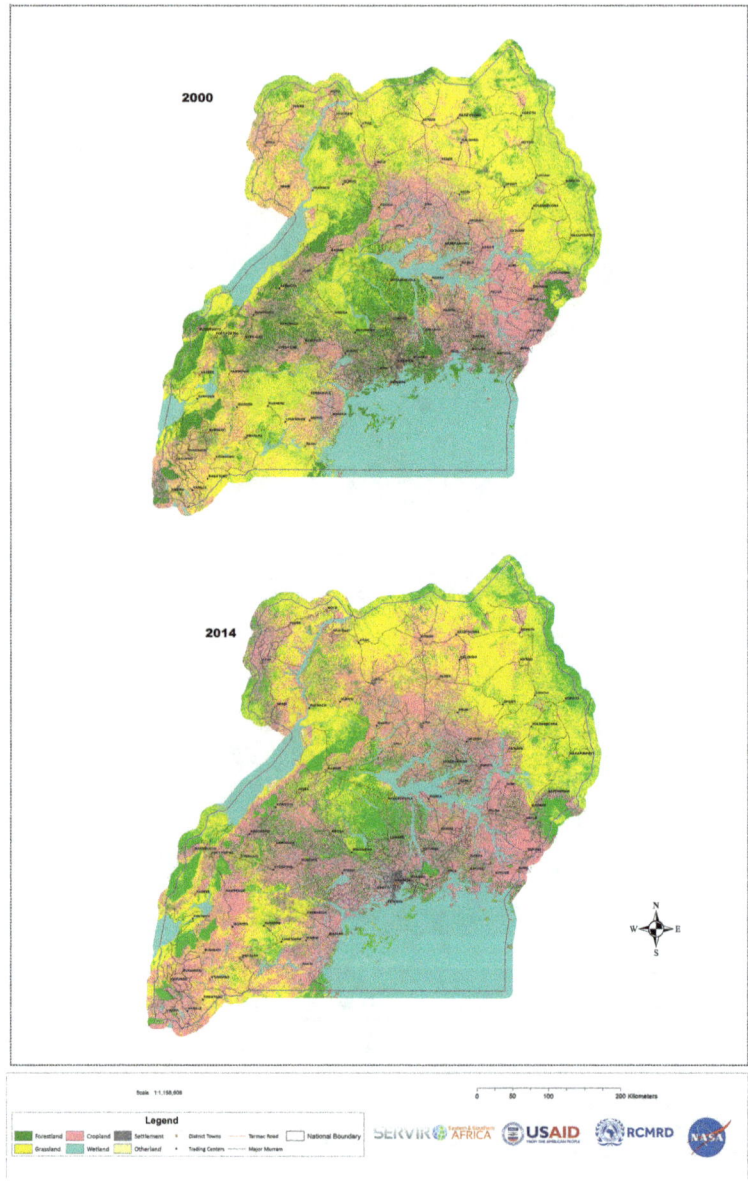

Fig. 8 Uganda Scheme I Land Cover Maps for 2000 and 2014

Land Cover Mapping for Green House Gas Inventories in Eastern … 107

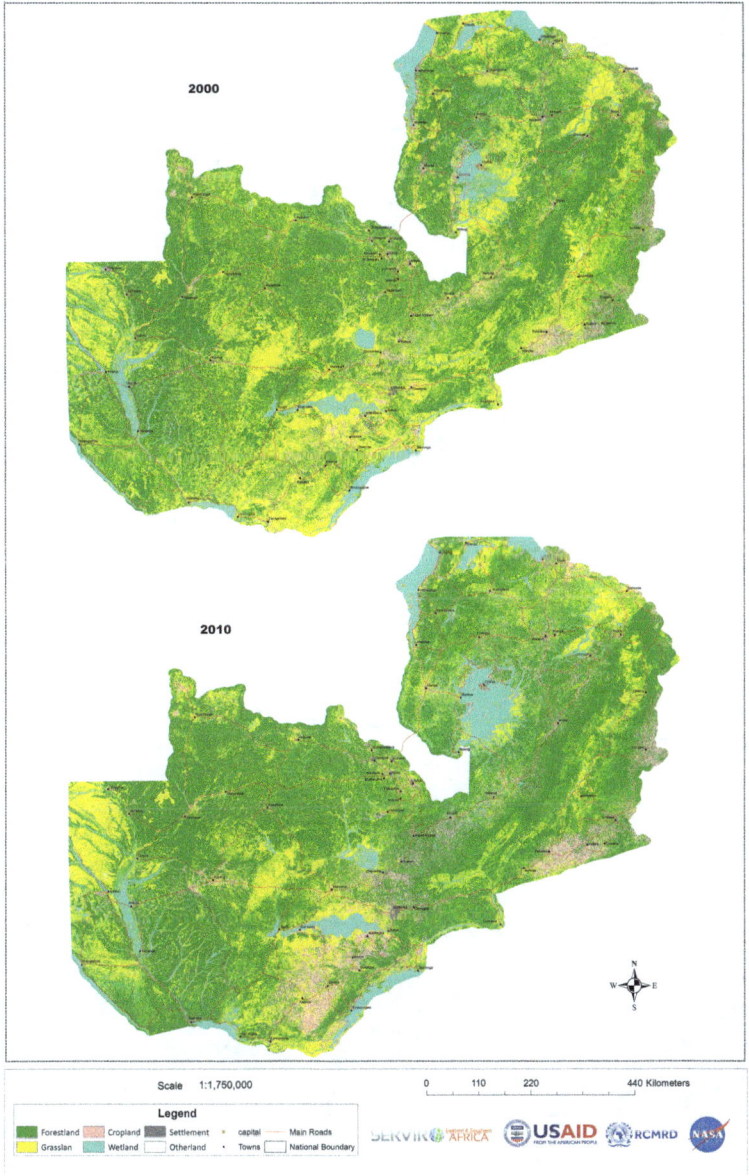

Fig. 9 Zambia Scheme I Land Cover Maps for 2000 and 2010

Fig. 10 Synergy meeting at Surveys and Mapping, Botswana, during the Ancillary Data Collection Workshop

Fig. 11 The Ethiopia Data Dissemination Workshop. The workshop was graced by the State Minister for Forest and Environment in Ethiopia, Ato Kebede Yimam, the Director General of RCMRD, Dr. Hussein Farah, and the Director General of Ethiopia Mapping Agency, Ato Sultan Mohammed

Fig. 12 First Page of the Interactive DVD (Lesotho)

Table 5 Number of persons attending the ancillary data collection workshops, and the respective institutions that they represented

	Country	Number of participants	Institutions represented
1.	Namibia	11	Ministry of Environment and Tourism (MET), Namibia Statistics Agency (NSA), Ministry of Agriculture, Water and Forestry—Department of Forestry and National Centre for Remote Sensing, United Nations Development Programme (UNDP)
2.	Malawi	14	Department of Environmental Affairs, Le Bocage, Mt. OryMoka, Mauritius, Land Resources Conservation Department, Lilongwe City Council, Surveys Department, Department of Animal Health and Livestock Development, Ministry of Natural Resources, Energy and Environment, Forestry Research Institute of Malawi, Crop Production Department
3.	Rwanda	16	Rwanda Environmental Management Authority (REMA), Rwanda Agriculture Board (RAB), University of Rwanda (NUR), World Agro-forestry Centre (ICRAF), National

(continued)

Table 5 (continued)

	Country	Number of participants	Institutions represented
			Institute of Statistics of Rwanda (NISR), Rwanda Natural Resources Authority (RNRA)
4.	Botswana	9	University of Botswana (UB), Surveys and Mapping, Meteorology Department, Statistics
5.	Tanzania	12	Vice President's Office (VPO)-Division of Environment, National Bureau of Statistics (NBS), Tanzania Meteorological Agency (TMA), Ministry of Agriculture-National Food Security, Ministry of Livestock and Fisheries, University or Dares Salaam- Institute of Resource Assessment
6.	Zambia	13	Centre of Energy for Engineering, Ministry of Lands, Energy and Water Department-Department of Survey, Department of Forestry and Department of Natural Resources and Environmental Protection, University of Zambia, Ministry of Agriculture, FAO, Zambia Wildlife Authority (ZAWA)
7.	Uganda	9	National Forest Authority (NFA), National Environment management Authority (NEMA), Ministry of Water and Environment, Uganda Wildlife Authority (UWA)
8.	Ethiopia	31	Ethiopia Mapping Agency (EMA), Ministry of Environment and Forestry (MoEF), Central Statistical Agency of Ethiopia (CSA), National Population Council (NPC), Ethiopian Wildlife Conservation Authority (EWCA), MPC, National Meteorology Agency (NMA)
9.	Lesotho	14	Ministry of Local Government, Department of Land Use and Department of Land Surveys and Physical Planning, Ministry of Forestry, Meteorology, Maseru City Council

Table 6 A summary of number of persons who participated in the 'On job training' and In country training workshops per country, and the respective institutions they represented

	Country	Number of participants	Institutions represented
1.	Namibia	18	Ministry of Environment and Tourism (MET), Namibia Statistics Agency (NSA), Ministry of Agriculture, Water and Forestry—Department of Forestry and National Centre for Remote Sensing, UNDP, National Remote Sensing Center-Forestry
2.	Malawi	17	Department of Environmental Affairs, Le Bocage, Mt. OryMoka, Mauritius, Land Resources Conservation Department, Lilongwe City Council, Surveys Department, Department of Animal Health and Livestock Development, Ministry of Natural Resources, Energy and Environment, Forestry Research Institute of Malawi, Crop Production Department

(continued)

Table 6 (continued)

	Country	Number of participants	Institutions represented
3.	Rwanda	19	Rwanda Environmental Management Authority (REMA), Rwanda Agriculture Board (RAB), University of Rwanda (NUR), World Agro-forestry Centre (ICRAF), National Institute of Statistics of Rwanda (NISR), Rwanda Natural Resources Authority (RNRA)
4.	Botswana	20	University of Botswana (UB), Surveys and Mapping, Meteorology Department, Statistics
5.	Tanzania	14	Vice President's Office (VPO)-Division of Environment, National Bureau of Statistics (NBS), Tanzania Meteorological Agency (TMA), Ministry of Agriculture-National food security, Ministry of Livestock and Fisheries, University or Dares Salaam-Institute of Resource Assessment
6.	Zambia	19	Centre of Energy for Engineering, Ministry of Lands, Energy and Water Department-Department of Survey, Department of Forestry and Department of Natural Resources and Environmental Protection, University of Zambia, Ministry of Agriculture, Food and Agriculture Organization (FAO), Zambia Wildlife Authority (ZAWA)
7.	Uganda	20	National Forest Authority(NFA), National Environment management Authority (NEMA), Ministry of Water and Environment, Uganda Wildlife Authority (UWA)
8.	Ethiopia	12	Ethiopia Mapping Agency (EMA), Ministry of Environment and Forestry (MoEF), Central Statistical Agency of Ethiopia (CSA), National Population Council (NPC), Ethiopian Wildlife Conservation Authority (EWCA), MPC, National Meteorology Agency (NMA)
9.	Lesotho	16	Ministry of Local Government, Department of land Use and Department of Land Surveys and Physical Planning, Ministry of Forestry, Meteorology, Maseru City Council

Table 7 A summary of number of persons participating in the dissemination workshops in each of the countries, and their respective institutions

	Country	Number of participants	Institutions represented
1.	Namibia	15	Ministry of Environment and Tourism (MET), Namibia Statistics Agency (NSA), Ministry of Agriculture, Water, National Remote Sensing Center-Forestry, Ministry of Lands and Resettlement (MLR)
2.	Malawi	15	Department of Environmental Affairs, Land Resources Conservation Department, Lilongwe City Council, Survey's Department, Department of Animal Health and Livestock Development, Ministry of Natural Resources, Energy and Environment, Forestry Department

(continued)

Table 7 (continued)

	Country	Number of participants	Institutions represented
3.	Rwanda	16	Rwanda Environmental Management Authority (REMA), Rwanda Agriculture Board (RAB), RHA, University of Rwanda (NUR), Rwanda Natural Resources Authority (RNRA), Center for GIS-University of Rwanda
4.	Botswana	10	University of Botswana (UB), Surveys and Mapping, Department of Town and Regional Planning (DTRP), Ministry of Agriculture
5.	Tanzania	12	Vice President's Office (VPO)-Division of Environment, National Environment Management Council (NEMC), Ministry of Agriculture, Sokoine University of Agriculture (SUA), Institute of Resource Assessment (IRA), Ministry of Livestock and Fisheries, Ardhi University (ARU), University or Dares Salaam- Institute of Resource Assessment, Tanzania Forest Agency (TFS)
6.	Zambia	11	Centre of Energy, Environment and Engineering, Ministry of Lands, Energy and Water Department-Department of Survey, Department of Forestry and Department of Natural Resources and Environmental Protection, Biocarbon Partners, Ministry of Agriculture, FAO REDD+, National Remote Sensing Center (NRSC)
7.	Uganda	11	National Forest Authority(NFA), National Environment management Authority (NEMA), Ministry of Water and Environment, Uganda Wildlife Authority (UWA), Makerere University
8.	Ethiopia	10	Ethiopia Mapping Agency (EMA), Ministry of Environment and Forestry (MoEF), Central Statistical Agency of Ethiopia (CSA)
9.	Lesotho	14	Ministry of Local Government, Department of land use and Department of Land Surveys and Physical Planning, Ministry of Forestry, Meteorology, Maseru City Council

5 Discussion and Conclusion

The 'GHG' Project was successfully implemented over a period of 3 years, starting in March 2012 and ending in September 2015. Its main objective was to develop baseline land cover maps for nine countries, while also building the capacities of government agencies in such relevant areas. This mandate was achieved in the nine countries: Botswana, Ethiopia, Lesotho, Malawi, Namibia, Rwanda, Tanzania, Uganda, and Zambia. In total 40 maps were produced, four maps for all countries except Malawi and Rwanda who got six maps each. These were mapped for two time slices, basically the year 2000 and 2010, and an additional 1990 for Malawi and Rwanda. Each time slice had two maps; the priority Schema I map which comprised of the six IPCC land cover categories, and a Schema II map which comprised of some 12–15 land cover categories, defined by the specific country. The main purpose of the maps developed under this project was to support national processes in reporting to the UNFCCC on GHG emission estimates. However, many countries have also found additional uses that the maps can offer, especially the nationally defined Schema II map products.

Over 120 government officials representing about 50 institutions in the region were trained in the course of the project. A network of about 50 key institutions was established courtesy of the numerous project activities implemented, including the workshops on capacity building, ground-referencing exercises, on the job trainings, training, and dissemination workshops. The project was also able to leverage support from other agencies to support capacity building efforts at RCMRD. A good example was a USGS led workshop (March 2013) on land cover mapping in the Eastern and Southern Africa, during which participants from project countries spent an additional 1 week in the project lab for capacity building. The project also delivered a process manual to guide the trainees which bears descriptive step by step processes of map-creation activities, with illustrations and screen shots. A country-specific copy was shared with every trained participant. The project also attracted the interest of undergraduate and graduate students, who gladly took up internships from time to time, some of whom were international students. During the dissemination workshops, the maps and related project reports, as well as manuals were shared in a country-specific interactive DVD. Each participant got a copy. Additionally, the maps and related documentation for each country were released to the public via the online platform.

Some of the challenges encountered during the implementation phases include loss of project time due to unforeseeable time lags associated with staff turnover as well as the assumption that all the countries required the same mapping effort, and were ready to adopt the project implementation schedules. The project team also encountered technical challenges from time to time, prompting the team to be innovative and creative, in order to meet the strict project deadlines. These challenges were eventually overcome through the cooperation of leadership at various levels and institutions.

The project also was one of its kind in the region, and that enabled it to identify a number of trends regarding the regional environment. There are significant changes in the land use and land cover in Eastern and Southern Africa, with the most affected categories being the cropland and settlement whose areas are steadily increasing, as well as the forestland and shrub land which are declining. However, the biophysical conditions in the countries covered were very diverse, with equally varying land cover characteristics and change patterns. It was also evident that the cooperation of national government was aided in part by the long standing cordial relations with the RCMRD. This relationship facilitated sharing of ancillary data, requests for invitation letters to their countries, and invitations to participate in project activities. The requirement for voluntary reporting to the UNFCCC and the much needed technology also made the institutions to look forward to the project.

The 'GHG' Project used Landsat data to a large extent, as the primary source of Earth Observation data for mapping work. The Landsat archive is one of the key NASA Earth Observation data sources that have significantly improved data access for research and public utility projects. In this work, access to Landsat data provided a number of cutting edge advantages important for such a project: (1) the Landsat data were freely available and presented one of the most comprehensive land cover EO datasets in history; (2) the project leveraged on a wealth of practical experience in analyzing and interpreting the Landsat data, drawn from extensive and diverse applications in land cover projects; (3) like in a number of other NASA satellites, the Landsat program provides continuity, presenting a remarkable opportunity for both historical and future comparisons of land cover classes in the same areas, with EO data from the same constellation of satellites; (4) the complimentary information derived from high resolution EO imagery such as Worldview (whose acquisition was supported by the SERVIR Project) greatly improved the interpretation of land cover classes in this project; (5) the spinoff tools and technologies emanating from working with the freely available EO data demonstrate the enormous potential for spurring innovation that can generate practical solutions for societal benefit. Finally, it is evident that supporting access and application of NASA EO data (e.g., Landsat, MODIS, ISERV), as well as from other space agencies (e.g., Sentinel) for tackling environmental related challenges in developing countries can generate remarkable results whose benefits will promote change in the future.

Having successfully implemented this project, we recommend that: (1) facilitating the transfer of methods, lessons learnt, and technology developed under this project to other countries in the region would greatly improve regional skills transfer; (2) additional capacity development efforts in the countries would strengthen the institutions' capabilities in land cover mapping; (3) development of process manuals using open source software, including R and Quantum GIS, would be beneficial to agencies within the region who work with lean budgets; (4) the nine countries that participated in the project would further benefit from this initiative by leveraging on the capacity built to access additional funding to advance their reporting of GHG emission estimates to the UNFCCC; (5) exploring alternative mapping methods and software can spur creativity and faster map compilations; (6) offering additional support to the countries to use the data with the different available toolkits (e.g., ALU tool), alongside other sector data to aid the GHG emission estimation processes, would further improve their national capacities. Finally, systematic and concerted efforts that build upon existing capacity from such a project can spur policy interventions which can guide operational use of EO data in addressing societal environmental challenges.

References

AfDB. (2011). Federal Democratic Republic of Ethiopia, Country strategy paper 2011-2015. African Development Bank.
African Institute for Policy Development. (2012). Population Dynamics, Climate Change, and Sustainable Development in Africa. Retrieved from https://www.afidep.org/?wpfb_dl=17.
African Bank. (2007). Chapter 4. Africa's Natural Resources: The Paradox of Plenty.
Agrawala, S., Moehner, A., Hemp, A., Van Aalst, M., Hitz, S., Smith, J., Meena, H., Mwakifwamba, M. S., Hyera, T., & Mwaipopo, O. U. (2003). Development and Climate Change in Tanzania: Focus on Mount Kilimanjaro. Organisation for Economic Co-operation and Development.
Al-Ahmadi, F. S., & Hames, A. S. (2008). Comparison of Four Classification Methods to Extract Land Use and Land Cover from Raw Satellite Images for Some Remote Arid Areas, Kingdom of Saudi Arabia. JKAU; Earth Science, Vol. 20 No. 1, pp. 167–191 (2009 A.D./1430 A.H.).
Bégué, A., Vintrou, E., Ruelland, D., Claden, M., & Dessay, N. (2011). Can a 25-year trend in Soudano-Sahelian vegetation dynamics be interpreted in terms of land use change? A remote sensing approach. *Global Environmental Change, 21*, 413–420.
Boden, T. A., Marland, G., & Andres, R. J. (2011). Global, Regional, and National Fossil-Fuel CO_2 Emissions. Carbon Dioxide Information Analysis Center, Oak Ridge National Laboratory, U.S. Department of Energy, Oak Ridge, Tenn., U.S.A. doi:10.3334/CDIAC/00001_V2011.
Byamukama, B., Carey, C., Cole, M., Dyszynski, J., & Warnest, M. (2010). National Strategy on Climate Change and Low Carbon Development for Rwanda. Baseline Report. Retrieved from: http://cdkn.org/wp-content/uploads/2010/12/FINAL-Baseline-Report-Rwanda-CCLCD-Strategy-super-low-res.pdf.
Cabral, A. I. R., Vasconcelos, M. J., Oom, D., & Sardinha, R. (2011). Spatial dynamics and quantification of deforestation in the central-plateau woodlands of Angola (1990–009). *Applied Geography, 31*, 1185–1193.
César, E., & Ekbom, A. (2013). Ethiopia Environmental and Climate Change policy brief. Sida's Helpdesk for Environment and Climate Change. Available: www.sidaenvironmenthelpdesk.se/.
Energy Information Administration (EIA). (2009). Office of Integrated Analysis and Forecasting. U.S. Department of Energy.
Food and Agriculture Organization of the United Nations (FAO). (2010). Global Forest Resources Assessment 2010. Main report.
Gercek, D. (2004). Improvement of image classification with the integration of topographic data. In *Proceedings of ISPRS Congress*, Vol. XXXV, part B8, 12–23 July 2004, Istanbul, Turkey (International Society of Photogrammetry and Remote Sensing). Available: http://www.isprs.org/proceedings/XXXV/congress/yf/papers/929.pdf.
Helmut, J. G., & Lambin, E. F. (2002). Proximate Causes and Underlying Driving Forces of Tropical Deforestation. Oxford Journals: Science & Mathematics: BioScience 52(2), 143–150.
Kim, C. (1993). Limitations of supervised classification and possible application for forest damages using airborne MSS data. Geoscience and Remote Sensing Symposium, 1993. IGARSS'93.
Lillesand, T. M., & Kiefer, R. W. (1994). *Remote sensing and image interpretation* (3rd ed., 750 pp). New York: Wiley.
Liu, X. (2002). Supervised Classification and Unsupervised Classification ATS Class Project Report: Available: https://www.cfa.harvard.edu/~xliu/presentations/.
McSweeney, C., New, M., & Lozano, G. (2010). UNDP Climate Change Country Profiles: Zambia. Available: http://www.geog.ox.ac.uk/research/climate/projects/undp-cp/index.html?country=Zambia&d1=Reports.
Mertz, O., Ravnborg, H. M., Lövei, G. L., Nielsen, I., & Konijnendijk, C. C. (2007). Ecosystem services and biodiversity in developing countries. *Biodiversity and Conservation, 2007*(16), 2729–2737.

Nkonya, E., Koo, J., Kato, E., & Guo, Z. (2013). Trends and Patterns of Land Use Change and International Aid in Sub-Saharan Africa. WIDER Working Paper 2013/110, United Nations University, World Institute for Development Economics Research (UNU-WIDER).

Republic of Namibia. Ministry of Environment and Tourism, MET. (2010). National Policy on Climate Change for Namibia.

Shahrokhi, S. M. (2004). An automated system for environmental monitoring and disaster warning. Iran Space Agency (ISA). Available: http://www.unoosa.org/pdf/sap/2004/iran/presentations/jalayerian.pdf

Stiebert, S. (2013). Republic of Rwanda: greenhouse gas emissions baseline projection: Published by the International Institute for Sustainable Development.

Wingqvist, G. O., & Dahlberg, E. (2008). Botswana Environmental and Climate Change Analysis. Sida INEC. Available: http://www.sida.se/globalassets/global/countries-and-regions/africa/botswana/environmental-policy-brief-botswana.pdf.

World Bank. (2015). Uganda Country Profile. Available: http://www.worldbank.org/en/country/uganda/overview/.

Part II
Thematic Perspectives of Satellite Earth Observation Data for Societal Benefit

Role of Earth Observation Data in Disaster Response and Recovery: From Science to Capacity Building

Guy Schumann, Dalia Kirschbaum, Eric Anderson and Kashif Rashid

Abstract Risks from natural hazards such as floods, droughts, earthquakes, and landslides are rising due to increasing populations living in more marginal areas and climatic variability, but our ability to provide warnings and mitigation strategies at short, medium, and long timescales is often challenged by the lack of ground observations in the most vulnerable areas. Satellite remote sensing offers unique global observational capabilities that can provide key insight into the multi-faceted topics of disaster hazard and risk assessment, response, and recovery in a way that ground-based systems cannot do alone. From the vantage point of space, satellite platforms can provide estimates of important hazard-related variables, but have varying degrees of accuracies and spatial resolutions. In some cases these data are used to support direct disaster response such as maps showing the spatial extent of the disaster or impact analyses from detecting pre- and post-event changes on the landscape. Examples of such direct support include the disastrous flood events in Malawi in January 2015 and in the southwestern United States in May and June 2015, and the devastating high-magnitude earthquake that hit Nepal in April 2015 (National Planning Commission 2015).

G. Schumann (✉)
Remote Sensing Solutions, Inc., Monrovia, CA, USA
e-mail: gjpschumann@gmail.com

G. Schumann
School of Geographical Sciences, University of Bristol, Bristol, UK

D. Kirschbaum (✉)
NASA Goddard Space Flight Center, Greenbelt, MD 20771, USA
e-mail: dalia.b.kirschbaum@nasa.gov

E. Anderson (✉)
University of Alabama in Huntsville, Huntsville, AL 35805, USA
e-mail: eric.anderson@nasa.gov

E. Anderson
NASA Marshall Space Flight Center, Huntsville, AL 35805, USA

K. Rashid
United Nations World Food Programme, Rome, Italy

© Springer International Publishing Switzerland 2016
F. Hossain (ed.), *Earth Science Satellite Applications*,
Springer Remote Sensing/Photogrammetry, DOI 10.1007/978-3-319-33438-7_5

1 Overview

Timely responses during such disasters are often enabled by data and tools that were developed with data from Earth observing satellites leveraging the rapid dissemination of satellite information. For example, NASA and the Dartmouth Flood Observatory use direct observations of flood water extent from instruments like the moderate resolution imaging spectroradiometer (MODIS) on NASA's Terra and Aqua satellites and these data are available within hours of acquisition. This method complements modeling efforts such as the Global Flood Monitoring System (GFMS, http://flood.umd.edu) that uses near-real-time satellite-based precipitation information to model potential flood inundation, allowing scientists to compare results and ultimately improve their flood estimates. The availability of optical imagery such as from EO-1, Landsat, or commercial imagery from Digital Globe allows for rapid mapping of landslides following a major trigger, such as the Gorkha Earthquake in Nepal in April, 2015. The mapping efforts provided important information to aid in disaster response and recovery efforts.

Looking to the future, the challenge and question is how best to link Earth Observation (EO) imagery and products together to provide more robust hazard assessment and monitoring systems. Ensuring sustainable value and use of the many platforms, data, and tools offered by satellites and the scientific community requires an intimate engagement between the latter and a wide range of stakeholders (Hossain 2015). This issue has become imminent in recent years, as the strong interplay between human activity and nature drives change on almost all continents. There have been success stories that took advantage of the science and observations afforded by satellites to have spectacular societal impacts or provide unprecedented assistance during major disasters.

Despite the gradual development of more mature remote sensing technology and satellite missions for routine environmental monitoring, there is a general lack of capacity building needed in most regions to take fullest advantage of the tremendous influx of satellite environmental data that are and will become available in the near future. In order to unlock the observational capability of satellites to enhance and accelerate societal applications around the world, scientists, stakeholders, and humanitarian and development agencies need to collaborate closer and ensure sustainable synergies. These partnerships and relationships take time to develop and must be nurtured to most effectively transition scientific research to real-world applications. In reality, these collaborations are typically slow and can prove difficult to establish. For disaster assessment and response, there is a wealth of timely data that can significantly contribute to improving rapid hazard response and recovery; however, few communities are able to take full advantage of these remote sensing data and products. Therefore, a goal looking to the future is to collaborate and coordinate information from an ever increasing number of satellite missions and work with the spectrum of users to more effectively convert data and science into actionable products for better decision-making.

This chapter outlines several examples, where the intersection of remote sensing data and science has had a significant impact on applications and decision-making and has been used to assist relief services during disasters. The following sections describe the role Earth Observation (EO) data play in landslide monitoring, earthquake disaster response (using the Nepal 2015 disaster), and in assisting flood relief services in the Lower Zambezi (in collaboration with the United Nations World Food Programme). We recapture these success stories and report current and future prospects as well as challenges.

2 Landslide Hazard Assessment and Monitoring Using Remotely Sensed Data

2.1 Introduction

Mass movements, including debris flows, landslides, mudflows, rockfalls, etc., (herein referred to as landslides) occur in every country on earth and cause thousands of fatalities and significant destruction each year (Petley 2011; Kirschbaum et al. 2015c). Landslides can range in size from a few meters to several kilometers. They occur over a broad range of lithologies, morphologies, hydrologic settings, and climate zones and can be triggered by intense or prolonged rainfall, earthquakes, rapid freezing and thawing of the surface, and anthropogenic activities, among others (Fig. 1). The location, size, and timing of landslide events can be extremely challenging to forecast or evaluate because of the local scales at which they occur and the complex interactions these events have with triggering events.

EO data have played an important role in advancing the mapping and hazard assessment capabilities over local to global scales. However, there remain some limitations to effectively applying these data to landslide mapping, monitoring, and hazard assessment. Building capacity to better utilize EO data for landslide assessment has been developing over the past decade, and significant advancements have been made to better utilize remote sensing data for landslide hazard

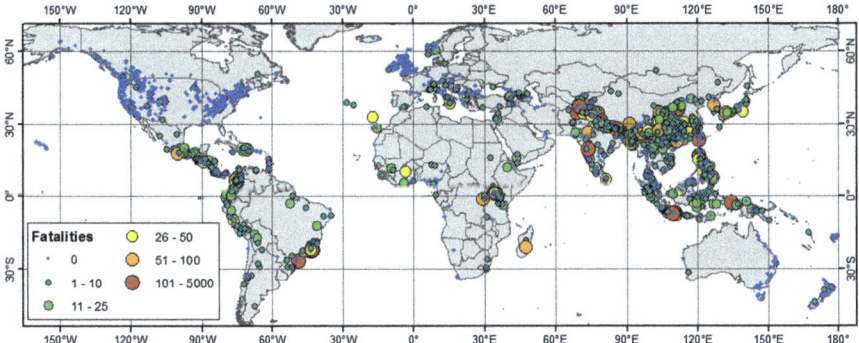

Fig. 1 Global Landslide Catalog (2007–2015) developed at NASA showing number of fatalities for reported events. The Global Landslide Catalog was compiled from media reports, online disaster databases and other sources and contains nearly 7000 events (see Kirschbaum et al. 2010)

assessment. However, there remain some challenges in effectively applying EO data for actionable landslide hazard information. This section provides a few examples of current activities in landslide mapping, monitoring, and hazard assessment and suggests some of the opportunities and challenges of the landslide hazard mapping community in utilizing and building capacity to use EO data.

2.2 Current and Past Approaches

There are many different approaches to landslide hazard assessment and monitoring. One approach to estimating landslide behavior is using a deterministic model to focus on site-specific analysis at a local (hillslope) scale (e.g. Iverson 2000; Baum et al. 2010). In this approach, one needs to account for the way that extreme rainfall interacts with topography, how the water infiltrates the surface and how the subsurface responds to the increased presence of water. These different interactions can vary significantly over one to several meters. The surface and triggering information needed to successfully monitor and assess the hazard is typically obtained from in situ sources (e.g., rain gauges, soil moisture sensors, local high-resolution digital elevation models) and is rarely available from EO data at the resolution needed to conduct the analyses.

Statistical landslide hazard mapping approaches relate potential causal factors (e.g., slope, soil type, distance to drainage, geology) with past local or regional landslide inventories using methods such as logistic regression (e.g. Dai et al. 2004; Ayalew and Yamagishi 2005; Mathew et al. 2009), artificial neural networks (e.g. Ermini et al. 2005; Melchiorre et al. 2008; Pradhan and Lee 2010), or frequency ratio (e.g., Lee and Pradhan 2007; Kirschbaum et al. 2012). These approaches usually produce a static susceptibility map that provides information on the relative or probabilistic potential for landslide activity in a specific area. These studies have ranged from local to global scales and in many cases rely on EO data for homogenous surface inputs.

A third approach that focuses on landslide hazard monitoring considers the timing of landslides triggered by rainfall using an archive of previous events. Techniques consider the intensity of rainfall over short to prolonged periods by relating the intensity and duration (I-D) of a storm with previous landslide occurrence. I-D thresholds have been derived statistically and empirically on the global (Caine 1980; Hong et al. 2006; Guzzetti et al. 2008), regional (Dahal and Hasegawa 2008; Brunetti et al. 2010; Saito et al. 2010), and local scales (Larsen and Simon 1993; Saito et al. 2010). Typically, these I-D thresholds utilize in situ networks of rain gauges but some studies have considered the feasibility of applying satellite-based precipitation to estimate thresholds over broader areas (Rossi et al. 2014; Kirschbaum et al. 2015a).

A common theme in all landslide studies is the need for landslide inventories with sufficient information to validate landslide monitoring and hazard assessment systems. Unfortunately, there is a dearth of this information in general and openly available data in particular. Unlike monitoring networks for hurricanes or earthquakes, global systems have not been created to routinely identify the location, timing, and extent of landslide events. Different types of landslide maps can be prepared depending on the purpose of the inventory and extent of the study area. Traditional landslide inventory methods include obtaining a series of aerial photos and conducting field surveys to compile a database of landslides for a particular area over time. Optical remote sensing data has been applied to assess the landslide density, areal extent, and frequency (Petley et al. 2002; Hervas et al. 2003). Other sensors, such as Synthetic Aperture Radar techniques, are being used more often to locally evaluate displacement of landslides and changes over time (Mazzanti 2011). Freely available EO data provide a significant potential capability to this field for improved and more systematic landslide inventory mapping and analysis.

2.3 Application of EO Data in Landslide Studies

The increased availability, accessibility, and resolution of remote sensing data has provided new opportunities to explore issues of landslide mapping, hazard assessment, and monitoring at a range of spatial scales. Table 1 highlights some of the EO datasets that have been utilized for landslide hazard assessment and monitoring from NASA's fleet. Of the surface variables responsible for slope failures, elevation (and its derived products) remains the most important variable in nearly every model for landslide hazard assessment and monitoring. In terms of NASA data, the Shuttle Radar Topography Mission (SRTM, Farr et al. (2007)) has produced data that have now been released at 3-arc s which can be used to derive slope, curvature, aspect, distance to drainage, etc. State variables that provide both the preconditions and triggering conditions for landslides include soil moisture information from the Soil Moisture Active Passive (SMAP) mission and precipitation from GPM. The presence of past burned areas can increase the susceptibility for landslides due to vegetation loss. This data can be obtained from the moderate resolution imaging spectroradiometer (MODIS) both for past burned areas and active fires.

Finally, an important component of all landslide studies is a robust landslide inventory to validate the landslide model or identify previous instability. Optical imagery such as Landsat, EO-1, or ASTER can provide valuable information over the landscape with the advantages of repeat overpasses (Landsat) or the ability to task specific areas (EO-1 and ASTER). High-resolution commercial imagery provides additional information to classify specific landslides on a slope, but can be limited by the availability and cost. Lastly, SAR capabilities have been used to identify elevation changes from repeat overpasses, but the techniques are not yet widely used for larger area landslide inventory mapping.

2.4 Regional Landslide Hazard Assessment and Monitoring in Central America

One of the challenges with utilizing remotely sensed EO data within a landslide hazard assessment or monitoring system is the scale at which the study is

Table 1 Example of some NASA and other EO data and products useful for landslide hazard assessment (acronyms need to be spelled out—SRTM, SMAP, MODIS, GPW V3)

Data type	EO data set	Resolution	Extent	Source
Elevation	Shuttle Radar Topography Mission (SRTM)	30 arc-s (~90 m) and 3-arc s (~30 m)	65° N-S	http://www2.jpl.nasa.gov/srtm/
Forest cover and loss	Global Forest Change 2000–2013	30 m/99.6 %	Global	http://glad.umd.edu/projects/gfm/
Rainfall	Global Precipitation Measurement (GPM) mission Integrated Multi-satellitE Retrievals for GPM (IMERG)	0.1°, 30-min	65° N-S	http://pmm.nasa.gov
Soil moisture	Soil Moisture Active Passive (SMAP)			
Population	LandScan and Gridded Population of the world, version 3 (GPW V3)	1 km	Global	http://web.ornl.gov/sci/landscan/; http://sedac.ciesin.columbia.edu/data/collection/gpw-v3
Active fire and burned areas	Moderate Resolution Imaging Spectroradiometer (MODIS)	1 km	Global	http://modis-fire.umd.edu/

undertaken. Studies are frequently limited to an area where there is sufficient in situ information to fully parameterize the model. We present a case study here for a regional landslide "nowcasting" system that has been developed over Central America utilizing primarily NASA EO data.

2.4.1 Model Description

The Landslide Hazard Assessment for Situational Awareness (LHASA) model was implemented in Central America and the Caribbean by integrating a regional susceptibility map and satellite-based rainfall estimates into a binary decision tree, considering both daily and antecedent rainfall (Kirschbaum et al. 2015a). LHASA produces a pixel-by-pixel nowcast in near-real time at a resolution of 30 arc s to identify areas of moderate and high landslide hazard. The main goal of this system is to provide a set of tools at the regional level to characterize areas of potential landslide hazard that emergency response agencies, other in country groups or international aid organizations can use to improve their situational awareness and focus attention in areas that may need support.

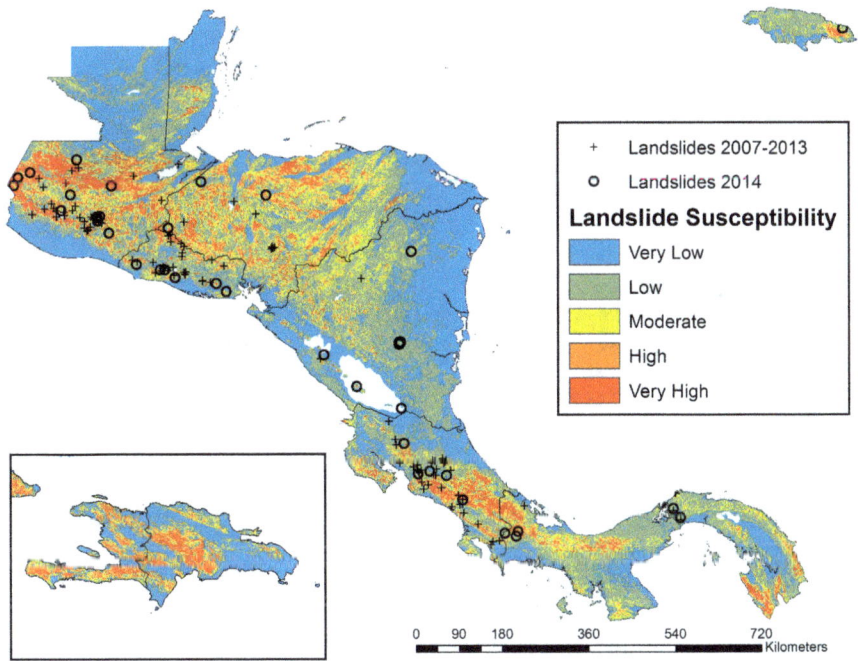

Fig. 2 Landslide susceptibility map for Central America and the Caribbean showing very low to very high susceptibility. *Source* Kirschbaum et al. (2015a)

The landslide susceptibility map going into the LHASA model was developed for Central America and the Caribbean islands by combining three globally available datasets (slope, soil type and road networks) and one regional dataset (fault zones) using a fuzzy overlay methodology (Kirschbaum et al. 2015b). This primarily heuristic model allows for flexibility both in testing a range of different contributing variables as well as incorporating information from landslide inventories that greatly vary in their size, spatiotemporal scope, and collection methods. The resulting susceptibility map provides a relative indication of landslide susceptibility across the over 700,000 km^2 considered at a resolution of 30 arc s (Fig. 2). We tested a range of satellite-derived products including SRTM DEM, forest cover derived from MODIS and AVHRR. After a modified sensitivity analysis, slope, topsoil clay content, presence of roads, and distance to fault zones were included as variables in the susceptibility map.

To provide an indication of potential timing of the landslide activity, TRMM Multi-satellite Precipitation Analysis (TMPA) data was evaluated over its entire archive at the time (2000–2014). An antecedent rainfall index (ARI) was calculated to account for pre-event soil moisture and both daily rainfall and antecedent rainfall percentiles were computed for each 0.25° × 0.25° pixel. Using landslide information obtained from the Global Landslide Catalog (Kirschbaum et al. 2010, 2015c), the study then computed precipitation thresholds based on the occurrence

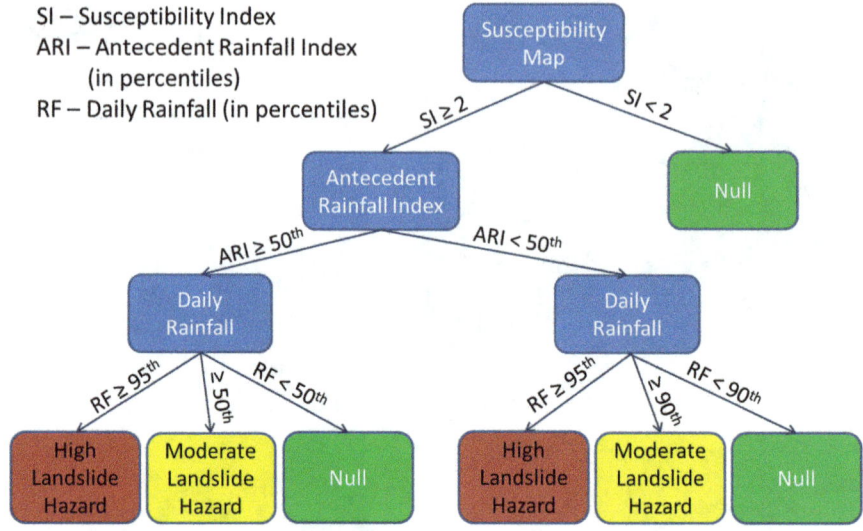

Fig. 3 LHASA Decision tree model. *Source* Kirschbaum et al. (2015a)

of rainfall at the reported landslide locations. A decision tree framework was established to generate moderate hazard, high hazard, or Null nowcasts based on the thresholds for susceptibility, ARI, and daily rainfall (Fig. 3).

The model is updated every day and is currently running in a prototype mode at: http://ojo-streamer.herokuapp.com/meso. This system provides information on all of the input variables (precipitation, susceptibility, and resulting nowcasts), so that individual users can diagnose the conditions at their location of interest as well as see what LHASA generates for potential nowcasts.

2.4.2 Applications and Challenges

The landslide system developed for Central America has a flexible framework that is customizable for other regions with a small level of effort. Currently, this system has been applied in a prototype form in Nepal and Peru. This workflow is also currently being expanded to a global model utilizing a similar approach.

This system provides an overview of potential landslide activity at both regional and soon global perspectives that currently does not exist. By coupling an indication of susceptibility with the intensity of recent rainfall at that pixel, it provides a regional perspective that can provide valuable situational awareness to a range of different end users including the international aid organizations such as the

International Federation of the Red Cross,[1] global disaster centers such as the Pacific Disaster Center,[2] the intelligence and military communities, and others. The framework in which the LHASA system is housed can also allow users to easily extract the underlying data from this model including remotely sensed precipitation, susceptibility, and landslide nowcasts in a variety of different formats.

The challenges in this system lie in the broad regional and global frameworks on which this system was designed. Due to its coarse spatial resolution ($\sim 0.01°$ or approximately 1 km), it has limited applicability for local emergency response groups who may need more detailed, slope-specific information. There is always the potential for increasing the resolution of this product in some areas, however, a major challenge stems from the current resolution of triggering data, namely precipitation, as well as the relative dearth of landslide inventories for model calibration and validation.

2.5 Conclusions

Landslide hazard assessment and monitoring is rarely undertaken at regional or global scales due to the prevailing methodologies that predominately utilize in situ data as well as the techniques that require detailed landslide inventories and triggering data. Of all the limitations of conducting landslide hazard assessment over larger areas and making full use of remotely sensed data, the most challenging element to overcome is the paucity of landslide inventory data beyond site-specific local studies. Optical imagery from EO-1, Landsat imagery can fill a critical need to better identify large landslides. Commercial imagery, such as from DigitalGlobe, provides higher resolution data to map an area. While optical methods are at the mercy of cloudless skies, SAR capabilities can penetrate clouds and allow for detection of changes on a landscape on the order of centimeters. All of these remote sensing capabilities have been used before for landslide detection; however, there is still a need for established, globally acknowledged mapping standards and guidelines to better utilize remote sensing imagery for landslide inventory assessment. Systems such as LHASA demonstrate the feasibility of applying remotely sensed data in a regional or global context to increase awareness, better understand the triggering conditions and ultimately improve response to landslide activity.

---

[1] www.ifrc.org/.
[2] http://www.pdc.org/.

3 Nepal 2015 Earthquake: Addressing International Development Challenges Through Scientific Networks

3.1 Introduction

Reaping the societal benefits that Earth Observation Systems (EOS) takes concerted efforts on the part of scientists and managers to understand the specific needs of beneficiaries. It also requires a clear understanding of entry points for EOS products and tools in decision-making processes. This not only requires a well-informed user group (potential beneficiaries) but also a group of applied scientists and capacity builders who can translate societal needs into possible data- or application-informed solutions. At the nexus of international development and applied sciences is SERVIR, a joint effort led by the U.S. National Aeronautics and Space Administration (NASA) and the U.S. Agency for International Development (USAID), along with international organizations around the world who have great capacity in connecting challenges posed by environmental, climate change, and disaster management with scientific data and products. A capacity building project, SERVIR strengthens developing countries' abilities to access and apply Earth observation data to manage challenges in the areas of food security, water resources, land use change, and natural disasters.[3]

To do this, SERVIR has created a network of scientists, technologists, trainers, and decision-makers, who collaborate to use Earth observations and geospatial technologies to codevelop solutions to environmental challenges. Foci of SERVIR efforts are in regional hubs, which are international organizations or consortia that have been selected for their history and experience of using GIS and remote sensing for environmental challenges. SERVIR hubs function to apply appropriate science and cutting edge technologies to plug into or become part of existing decision support systems. Active SERVIR regions include Eastern & Southern Africa, led by the Regional Centre for Mapping of Resources for Development (RCMRD; http://www.rcmrd.org/), the Hindu Kush Himalaya region, led by the International Centre for Integrated Mountain Development (ICIMOD; http://www.icimod.org/), and the Lower Mekong region served by a consortium of the Asian Disaster Preparedness Center (ADPC; http://www.adpc.net/), Spatial Informatics Group, Stockholm Environment Institute, and Deltares. USAID regional and bilateral missions play an active role in funding the hubs and articulating the development challenges, while a Science Coordination Office (SCO, previously called a the SERVIR Coordination Office (CO)) at NASA provides scientific and technical backstopping to hubs and manages an Applied Sciences Team (AST). This AST works to address end-user needs in SERVIR regions in direct collaboration with hubs. SERVIR addresses a wide variety of thematic areas, focusing on food security, land cover/land use and ecosystems, water and water-related disasters, and weather and climate.

---

[3]https://www.servirglobal.net/.

Given its position to accelerate applications of EO data in decision-making contexts in many regions of the world, SERVIR has been leveraged in many cases to provide technical input into emergency mapping and disaster response support, often in the form of rapid satellite image acquisition and interpretation (Boccardo 2013). Some of the earliest and strongest demands for information have come from fire mapping needs (Lewis 2009). Applications and customizations of satellite-based fire information are often taken up very quickly by environmental management and disaster management communities, likely for two reasons: first, the high applicability and accessibility of the Fire Information for Resource Management System (FIRMS) (Davies et al. 2009) and MODIS active fire products (Giglio 2010; Justice et al. 2011), and second, a close understanding of the challenges that forest and fire managers face. To further facilitate design and operationalization of such systems, applications have been cofunded and codeveloped between SERVIR hubs and with government ministries and departments themselves. Summaries and examples of disaster response support provided in Mesoamerica, Eastern and Southern Africa, and the Hindu Kush Himalaya can be found in Bajracharya et al. (2014), Flores Cordova et al. (2012), Gurung et al. (2014), Graves et al. (2005), Hardin et al. (2005), Macharia et al. (2010) and Wang et al. (2011). Throughout the communication and collaboration that occurs during needs assessments, product development, and iterations thereof, a vibrant network of scientists, technologists, managers, trainers, development professionals, and decision-makers, have been formed. The remainder of this section focuses on the description on one of the many subnetworks that we argue allowed for a successful integration of Earth observations into disaster response support.

3.2 Role of Earth Observations in the 2015 Nepal Earthquake Facilitated by SERVIR Network

The 25 April 2015 magnitude 7.8 earthquake and aftershocks (including the 12 May 2015 magnitude 7.3 aftershock) in Nepal led to over 9000 deaths, widespread damage to buildings and infrastructure, over 4000 landslides, avalanches, and other economic ramifications (National Planning Commission 2015; Kargel et al. 2016). Immediately following the earthquake, ICIMOD and the participants in the SERVIR network came into play to provide EOS-derived products to inform decision-makers the extent and magnitude of the impact, particularly around the state of geohazards (ICIMOD 2015).

Before the earthquake, though, a preexisting network of science data product providers and end users existed (Fig. 4). During the earthquake, ICIMOD and the NASA/SERVIR Coordination Office leveraged the preexisting network to connect ICIMOD and government agencies in Nepal with an even broader network of science data product providers (Fig. 5). Some 40 volunteer landslide mappers in the U.S., organized by the University of Arizona, NASA, and USGS; and another 40 or so volunteer mappers in Nepal, organized by ICIMOD, were able to put EOS

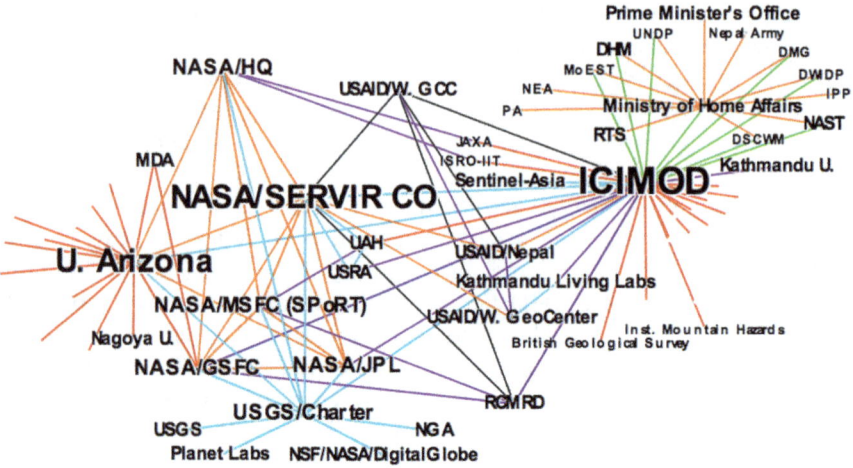

Fig. 4 Network connections before earthquake response (prior to April 2015). *Orange* Disaster response coordination, *Purple* Science/technical collaboration, *Gray* Project management connection, *Blue* Imaging needs communicated and images provided, *Red* Landslide mapping, *Green* Value-added products demanded and provided

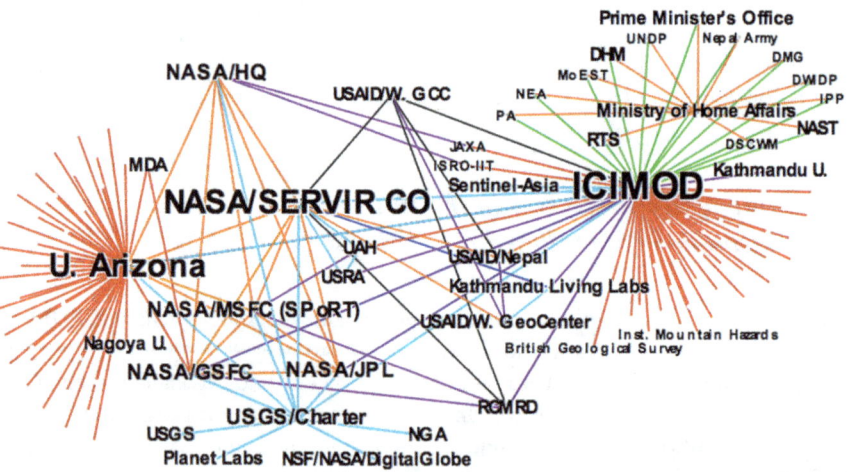

Fig. 5 Network connections during earthquake response (April–June 2015). The demand for situational awareness that satellite images and their analysis provide mushroomed during the earthquake response. The surge in landslide mapping efforts (*red*) produced information and products that were readily transferred through a previously existing network, allowing them to reach government and nongovernment end users in Nepal

products to immediate use by plugging into this strong network that ICIMOD and SERVIR had developed over the previous 5 years. The main audiences for value-added products were the Ministry of Home Affairs, Prime Minister's Office, and numerous other government agencies in Nepal. USAID offices in Washington, D.C., and in Nepal provided crucial funding and project management support, while the NASA/SERVIR CO connected Earth science data users in Nepal to a vast network of Earth scientists within NASA and outward through its affiliates.

Figures 4 and 5 depict lenses of NASA/SERVIR and ICIMOD for the event response. It is important to note that there were even more collaborators with ICIMOD from other countries, more connections between Nepal and the US, more connections between ICIMOD and other Nepal agencies, and more connections between International Charter and international space agencies, than depicted here.

What is the lesson learned here? When it comes to reaping benefits of EO data around the world, networks are not formed overnight. Meaningful responses to disasters rely on existing networks of scientific and technical collaboration, as presented in this brief case study. As of late October 2015, 5760 ad hoc products and 5840 images have been contributed to the USGS Hazard Data Distribution System, the US's platform for organizing and disseminating its contributions to the Charter (USGS 2015). The absorption and use of such a volume of products and satellite images require significant capacity to translate these intermediate outputs into actionable information.

3.3 Conclusions

Much is left to be done toward standardization for international disaster response (McCann and Cordi 2011), but certainly a strong history of collaboration and vibrant network of collaborators eases the connections between how Earth observations can play a role in meeting the demands during disaster response. Even though the network analysis presented is from specific project and organizational lenses, it is telling to see how many "connections" there were before the crisis. Also, many more connections were made during the crisis, but it is difficult to know whether the same quantity and quality of connections could have been made during the crisis response had there not been the previous network connectivity. Without such prior connections and capacities, we hypothesize that there would not have been as effective a response from the EO side.

4 Mapping and Predicting Flood Hazard in the Lower Zambezi for the Humanitarian Aid Sector

4.1 The Situation

The Zambezi River is the fourth largest river in Africa and is an important source for biodiversity, agriculture and hydroelectric power. The river flows into the Indian Ocean and through eight countries (Zambia, Angola, Namibia, Botswana,

Table 2 Flood data for all countries in the ZRB over the last 20 years

Year	Occurrences	Total deaths	Affected	Injured	Homeless	Total affected	Total damage in °000
1995	5	24	5350	0	23300	28650	0
1997	4	118	808028	0	2104	810132	0
1998	5	101	1,319,600	0	0	1,319,600	20,789
1999	2	23	72,000	0	0	72,000	12,400
2000	11	955	4,913,776	28	108,800	5,022,604	508,100
2001	9	218	1,945,904	5	200	1,946,109	46,300
2002	5	26	397,540	0	500	398,040	0
2003	12	73	565,500	3	12,825	578,328	200,000
2004	7	30	558,345	13	1,700	560,058	0
2005	6	18	53,373	12	63,500	116,885	0
2006	7	13	2,300	28	37,725	40,053	8,490
2007	16	209	2,223,362	0	12,159	2,235,521	171,000
2008	7	151	188,780	15	1,942	190,737	0
2009	10	223	1,246,395	0	5,065	1,251,460	0
2010	8	38	261,185	31	78,875	340,091	0
2011	16	290	798,491	201	6,876	805,568	12,000
2012	4	14	91,385	0	0	91,385	0
2013	8	285	380,120	76	0	380,196	30,000
2014	9	50	145,555	2	20,000	165,557	20,000
2015	5	432	768,881	645	595	770,121	0
Total	156	3,291	16,745,870	1,059	376,166	17,123,095	1,029,079

Note that a value of 0 may denote data not available. *Source* EMDAT http://www.emdat.be/advanced_search/index.html

Zimbabwe, Mozambique, Malawi, and Tanzania) in the southeast of the continent. Some of the most important wetlands in Africa are linked to the Zambezi River, while agricultural production is governed by the variability in river storage and flows. The basin encompasses humid, arid, and semiarid regions, with the flow regime being controlled by seasonal rainfall that causes the area to be seasonally flooded. Estimates of population within the Zambezi River Basin (ZRB) range from 30 to 40 million people, the majority of whom live in rural areas.[4] Very frequently, almost every year, moderate to high-magnitude floods put millions of people and their livelihoods at risk (Table 2). Over the past two decades, large floods have affected an estimated 17 million people (see Table 2) and led to significant crop damages with persisting consequences.

---

[4]World Bank: http://siteresources.worldbank.org/INTAFRICA/Resources/Zambezi_MSIOA_-_Vol_1_-_Summary_Report.pdf; WWF: http://wwf.panda.org/about_our_earth/about_freshwater/rivers/zambezi/.

The countries within the ZRB and in particular the downstream country of Mozambique with its vast delta area, suffer from weak infrastructure and resources and thus lack the foundational data required to establish effective flood management, mitigation, and relief services plans, let alone an operative flood forecasting system. Despite some local flood forecasting efforts in the ZRB, according to the World Meteorological Organization (WMO) there is currently no integrated flood warning system in the basin, primarily due to poor communication facilities and limited exchange of information and data in real time (WMO 2009). Furthermore, flow and water level measurement stations are exiguous in most countries in the region.

This general lack of information regarding the most serious and regular natural hazard in the region exacerbates the challenges faced by humanitarian emergency response to flood events. The United Nations World Food Programme (WFP) is the food aid arm of the UN and one of the largest providers of disaster assistance to the region. In addition to rapid emergency food aid, WFP also provides additional logistics and emergency telecommunications assistance to organizations that are part of the overall flood response. Reliable maps showing the extent of rivers and floodplain inundation as well as associated effects on critical infrastructure provide maximum situational awareness in support of preparedness, response, and recovery activities.

Ideally, these types of situational or logistics operational planning maps (Fig. 6) should be delivered in near-real time to WFP field officers on the ground. At present, the only consistently and freely available flood maps delivered in near-real time are provided by a unique global flood monitoring system funded by NASA using daily images from the Moderate Resolution Imaging Spectroradiometer (MODIS) onboard NASA's Terra spacecraft (as illustrated in Fig. 7). With composite MODIS images, NASA produces cloud-free maps showing the precise locations of flooded areas where there may be concomitant (and significant) population displacement. NASA also makes the raw GIS data available for download enabling overlay of the MODIS analysis on operational planning maps. These maps are used by WFP staff to pinpoint the worst-hit areas. With the new composite images, there is a dramatic increase in mapping accuracy as well as a daily timeline of the progression or recession of the flood waters. Observations of floodplain inundation over time allow disaster relief agencies like WFP to identify serious flood events and allocate resources and direct operations accordingly.

The challenge faced by both the humanitarian aid sector and the scientific community is twofold: (1) the data received by the flood response teams need to be actionable information (timely, clear, simple to understand and compatible with standard mapping software such as ESRI) and (2) should not provide a deterministic, but a probable range of floodplain inundation variables (volume, area, depths), preferably with long lead times, in the form of a simple early warning system. This would allow humanitarian aid agencies such as WFP to more effectively allocate and preposition resources that would ensure the most efficient aid distribution and emergency response. Also, it would allow local and national authorities in the region to ensure sustainable management in the areas of food security and water

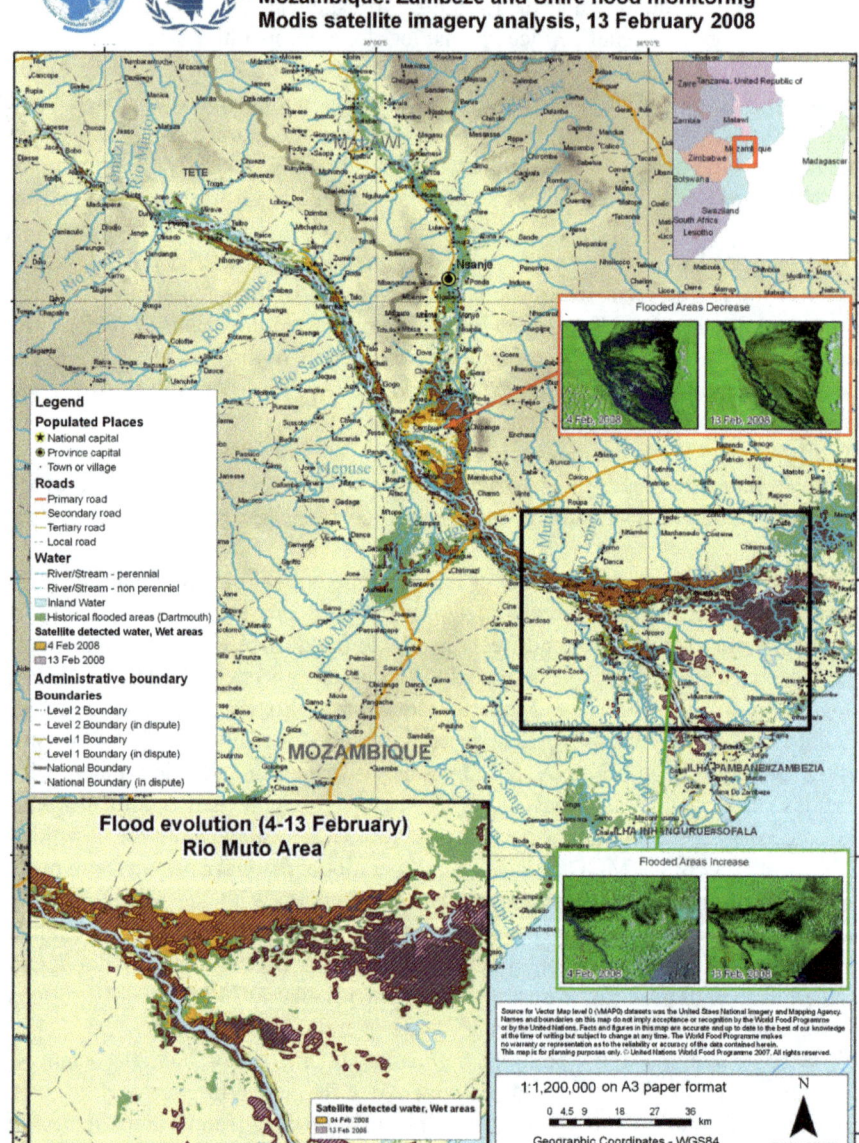

Fig. 6 Example of a map sent to field response teams in the humanitarian aid sector, in this case by WFP. This map shows flood evolution in the Lower Zambezi based on MODIS satellite imagery for the event in February 2008. The map was produced by the Information Technology for Humanitarian Assistance, Cooperation and Action (ITHACA) Institute in Turin, Italy. ITHACA is one of the top-level academic institutions who are enabling WFP to take advantage of recent advances in remote sensing technology. With this type of help, WFP has rapid access to satellite imagery analysis of disaster areas. The detailed maps help guide and inform its humanitarian operations

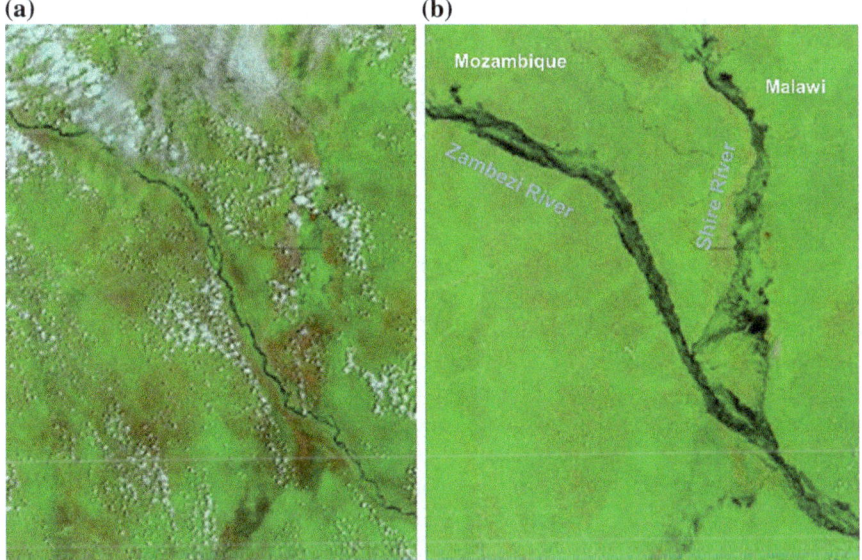

Fig. 7 Inundation surrounds the Zambezi River in the image in **b**, captured by the MODIS on February 10, 2007. Mozambique was experiencing its worst floods in 6 years when the Zambezi overtopped its banks in January and February 2007, reported the United Nations Office for the Coordination of Humanitarian Affairs (OCHA). As of February 12, 2007, an estimated 29 people had died and 60,000 had been evacuated from the riverbanks. These images show the lower Zambezi where it meets the Shire River flowing south from Malawi, one of the most severely affected regions in Mozambique. The image in **b** provides a remarkably cloud-free view of the floods, while the image in **a**, taken on December 31, 2006, shows the region before the rain started in January. Images such as these are provided by the MODIS Rapid Response Team and the Dartmouth Flood Observatory (http://floodobservatory.colorado.edu) on a daily basis (© NASA Earth Observatory)

supply. Most importantly, it would provide the potentially affected communities time to prepare accordingly.

4.2 The Challenge

4.2.1 The Scientific World: Thriving to Do It Right

With flood frequency likely to increase as a result of altered precipitation patterns triggered by climate change, there is a growing demand for more data and, at the same time, improved flood inundation modeling. This is essential for the development of more reliable flood forecasting systems over large scales that account for errors and inconsistencies in both observations and modeling. It is clear that there is an ever increasing abundance of forecast models and data that predict the magnitude, frequency, and impacts of extreme weather events, such as floods. Recent and ongoing advances in Numerical Weather System (NWS) model development (ECMWF models, NOAA NCEP models, and WRF models) can provide forecasts of rainfall fields and even streamflow at a high spatial resolution that can deliver accurate information at the appropriate regional and basin scale for hydrology.

However, predicting flood inundation in 2-D in near-real time and particularly with long lead times (i.e. 10 + days) as required by flood relief services is still a major challenge and only a couple of flood inundation models are actually capable of delivering data at (local) scales appropriate for flood disaster management and emergency response, let alone in near-real time. Over the last few decades, there have been major advances in the fields of remote sensing, numerical weather prediction, and flood inundation modeling (Pappenberger et al. 2005). At the same time, there are currently attempts to roll out models on a continental to global scale (Thielen et al. 2009; Alfieri et al. 2013; Pappenberger et al. 2012; Paiva et al. 2011, 2013; Winsemius et al. 2013). In this context, see in particular Sampson et al. (2015) for an illustration of the type of 2-D flood hazard model used in Schumann et al. (2013) for flood inundation forecasting in the Lower Zambezi.

These models, with a few exceptions, predict at a point discharge level with relatively little attention to accuracy at the appropriate inundation model grid scale. Typical grid resolutions of continental or global scale models dealing with flood inundation processes are in the order of a few tens of square kilometers (Yamazaki et al. 2011; Pappenberger et al. 2012), which is often too coarse to resolve inundation pattern details necessary to understand associated local risks. The type of hydrodynamic model presented by Paiva et al. (2013) in an Amazon River case study moves certainly in the right direction in terms of large-scale hydrology and in-channel hydrodynamics but it employs a simple fill operation for the floodplain with prediction of storage volume only. Since the model lacks floodplain hydraulics, it cannot reproduce inundation area dynamically. In addition to these shortcomings, most models employed in flood forecasting and mapping use simple flow routing schemes that only account for the kinematic force term in hydrodynamics but ignore diffusion and backwater effects.

Consequently, during many major flood events particularly in coastal areas, such as Cyclone Eline associated with the devastating floods in Mozambique in 2000 or the Hurricane Sandy and Katrina events in the U.S., commonly used flood models are inadequate. This is worsened by the fact that oftentimes downstream coastal-ocean boundaries in inland inundation models are not represented correctly or missing altogether. The main reason for this is not because such estuary models do not exist but rather that accurate representation of boundary conditions is needed at very high spatial resolution. Advances in computing only made this possible very recently. In addition, there is a gap between the coastal and riverine scientific communities as these two fields employ computer models that, despite solving very similar hydrodynamic equations, require different forcing and input data.

In an attempt to overcome some of the challenges noted, Schumann et al. (2013) built a first large-scale flood inundation forecasting model for data-poor regions and

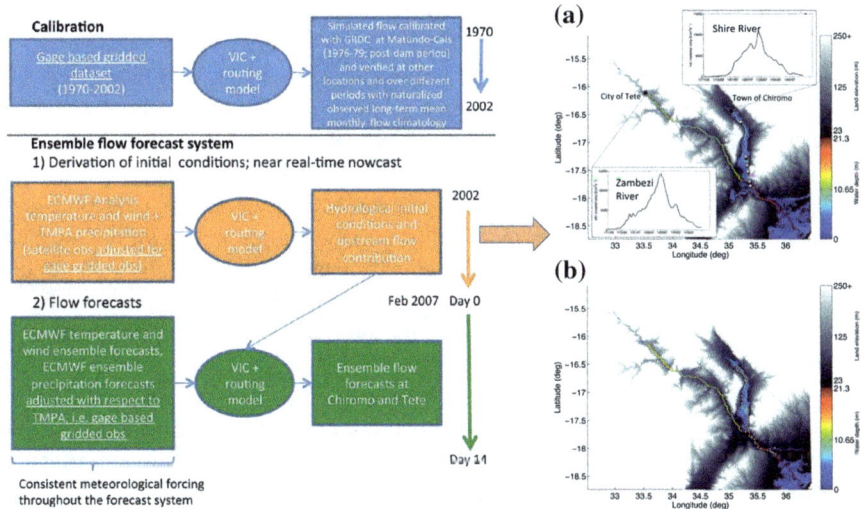

Fig. 8 Schematic of the calibration of the VIC hydrology model and the setup of the forecast flow generation (*left panel*). Illustration of the LISFLOOD-FP flood simulation (1 km resolution) at the time of the (**a**) ICESat-1 overpass and (**b**) Landsat overpass in February 2007, using the VIC baseline hydrographs shown. ICESat-1 water level locations and Landsat flood edge points are also shown. The *color shading* represents flooded area and water depths in m (*right panel*). Modified from Schumann et al. (2013)

tested it on the Lower Zambezi basin. In their application, the Variable Infiltration Capacity (VIC) hydrology model (Liang et al. 1994) is forced with medium-range weather forecast reanalysis data to provide flow at entry points to a flood model. The flood model applied is the computationally efficient 2-D hydrodynamic model, LISFLOOD-FP (Bates et al. 2010), complemented with a subgrid channel formulation (Neal et al. 2012) to generate flood inundation variables for the Lower Zambezi basin. Their forecast system (Fig. 8) showed good performance levels in both in-channel water levels during the calibration phase (18 cm in error) and flooded area predictions during the validation phase (86 % fit compared to a Landsat image of the 2007 event).

4.2.2 The Decision-Making World: Delivering Actionable Information

It is clear that there is an ever increasing abundance of extreme event data, from all kinds of models and observational systems, as well as other types of geospatial data sets available to describe and quantify the processes, magnitude, frequency, and impacts from extreme weather and climate change, such as floods. The scientific community is working to provide an enormous volume of valuable geospatial datasets to the public that can deliver information at various temporal and spatial resolutions spanning the entire natural process of an extreme event. This information can be in the form of event reanalysis, probabilistic forecast or scenario projections.

However, this wealth of information is completely under-utilized by emergency response teams, due to a number of reasons, most of which relate to its relative novelty: (1) limited time and personnel capacity to understand, extract, process, and handle new types of geospatial datasets; (2) limited near-real-time data accessibility, bandwidth, and sharing capacity; (3) incompatibility between user platforms and geospatial data formats; (4) data availability may be simply unknown and/or data latency may be inadequate; and (5) limited understanding by scientists and engineers about end user information product and timing needs; and (6) limited feedback from end users as to the usefulness and accuracy of the information when the data is actually made available.

In order to address this frequently encountered mismatch between data availability and end-user needs, geospatial data layers of extreme event prediction relevant to stakeholders should be delivered in an easily accessible format, through a user friendly web-based interface such as one that could be provided by Google, Inc., Amazon.com, Inc. or Environmental Systems Research Institute, Inc. (Esri) for instance.

4.3 Meeting Midway

4.3.1 Innovation

Flood prediction is a major component of any integrated flood management plan, which in turn constitutes an essential part of efficient water resources prediction and management. The Global Flood Working Group (GFWG), a consortium of top scientists and decision-makers concerned with flooding, has identified the need for better flood forecasting up to 30-days lead time as one of their primary objectives and scientific pillars, especially in data-poor regions. A possible way forward may be to develop a simple and robust satellite-assisted operational early warning system that can be used to predict with a long lead time (>30 days) a possible range of anomalous flood water conditions that can manifest themselves in the form of inundation area, volume, and depth above the mean condition. In order to achieve optimal use of such a system for emergency flood relief services, the modeling framework proposed for the Lower Zambezi area in collaboration with the humanitarian aid sector (WFP) and through space agency funding support (NASA, 13-THP13-0042) needs to address the following aspects, which makes it unique and quite different from existing (flood) early warning systems:

- Establishing a very long term record of forecast flood inundation that can be queried using observed antecedent and present satellite soil moisture data (e.g. from satellite missions such as ESA's SMOS or NASA's SMAP) as well as forecast rainfall data or forecast flows, or indeed any relevant near-real time

observed or gauged flood variables for that matter. In fact, one may decide to use data assimilation techniques, such as presented by Neal et al. (2009) for example, to augment the model (forecast) accuracies. Figure 4 (Pillar 1) schematizes one possible modeling framework.
- This flood inundation early warning system or 'dynamic library' can be applied without the need to run models and provides not a deterministic but a probable range of flood inundation with long lead times.
- The proposed system computes actual floodplain flow processes by solving the shallow water equation, using LISFLOOD-FP after Bates et al. (2010) and Neal et al. (2012), and outputs actual flood inundation at high spatial resolution (1 km). Moreover, it is tuned towards a specific high-risk region and the decision-making needs of one of the largest humanitarian organizations (WFP), which is to look at possible flood inundation projections and not only in-channel flow predictions. This is important since floodwaters exiting the channel will spread across the terrain and inundate the floodplain where the risk and hazard then inevitably become localized. Note that most existing regional to global so-called flood (forecasting) models are often hydrologic models and simulate in-channel flow only, and oftentimes with only relatively basic water flow algorithms.
- Easily transitional to end user and flexible for a wide range of water-related issues. The proposed system is tailored specifically towards easy implementation in a variety of decision-making operations.

This is the first time such a system is presented for the ZRB and potentially operated. Currently, neither any operational flood early warning systems nor flood forecasting model dynamically computes inundation patterns at acceptable resolutions across floodplains in data-poor regions. Furthermore, given that many flood model runs are performed a priori using a wide range of possible flows, many inundation scenarios can easily be uploaded to a server made accessible to humanitarian agencies as well as the scientific community. This allows data queries to be performed easily and files to be downloaded as needed. Also, such a system would be straightforward to scale up to a global level (following for example a model setup similar to that described in Sampson et al. 2015) to address other flood prone hotspots which require almost yearly attention by emergency responders (Fig. 9).

4.3.2 The Way Forward

The Current Situation

In order to respond most efficiently to flood events in the region, WFP needs a reliable long lead-time forecast of high-resolution floodplain inundation for its decision-making process. At WFP, most of the advanced remote sensing information is processed/managed at HQ level in Rome, Italy. Data and maps are made

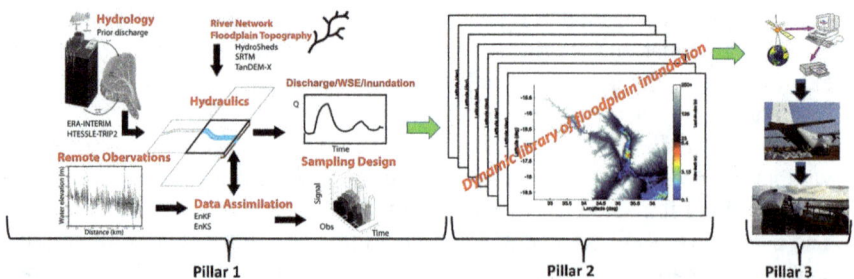

Fig. 9 Schematic of the flood inundation forecasting system proposed for the flood prone regions in the Lower Zambezi basin. Pillar 1 (personal communication, Neal (2014)) presents one possible modeling framework that can be adopted. Pillar 2 illustrates the 'dynamic' library that will be the output of the modeling in Pillar 1. This library of possible flood inundation scenarios can then be queried by decision-makers at WFP in our case (Pillar 3)

available to field offices in order to help with strategic planning for the humanitarian response. Since computer resources are fairly limited at both HQ and field level, the WFP relies at the moment heavily on support from both the remote sensing community and the weather forecast community, primarily in the form of NRT flood mapping from MODIS, real-time satellite precipitation as well as rainfall forecasts and the GDACS Global Flood Detection System (Version 2) (GDACS GFDS) run by the EC-JRC in collaboration with the UN and Dartmouth Flood Observatory (DFO). However, none of these systems provide forecast on floodplain inundation despite it being such valuable information. The main reason for this is that the types of 2-D hydrodynamic models (i.e. flood models) needed to simulate floodplain inundation at high enough spatial resolutions (at least 1 km grids) cannot currently be run within an operational forecast mode since these models typically have a high computational burden and require very accurate topographic boundary conditions.

A New Approach: Combining Global Flood Model Layers with Satellite Data for Decision-Makers

Facilitating Satellite Data Use for Decision-Makers: The 'Early Bird Catches the Worm'

The NASA Early Adopter Programs (see the chapter on Early Adopter for more details on page 231) are designed to provide specific support to so-called 'Early Adopters' in prelaunch applied research to facilitate feedback on NASA mission products (e.g. SMAP, http://smap.jpl.nasa.gov/science/early-adopters) *pre-launch*, and accelerate the use of products *post-launch*. Working with WFP as a SMAP early adopter, Schumann et al. (2014) started to look into simplifying model output delivery, extending forecast lead time and augmenting performance. As outlined earlier, in this particular setup, they used ECMWF archived forecast rainfall to compute flows that generate daily inundation patterns over a period of about 10 years using the coupled VIC-LISFLOOD-FP forecast model. These simulations allow generating a library of flood model outputs (inundated area, floodplain water volume), essentially from a long time series of rainfall data. They then used this library to correlate and log-regress ECMWF rainfall, satellite soil moisture, and model output floodplain inundation volume. Subsequently, the regression model is applied as a predictor for flood inundation variables and conditions. Using both forecast rainfall and soil moisture conditions to predict flood inundation volume has much higher skill than using rainfall as a sole predictor (correlation of 0.88 vs. 0.49). In the Mozambique test case, their regression model had a relative bias of 17 %, with a relative error in predicting the 2007 flood event of 33 %. In other words, based on rainfall forecasts and satellite soil moisture observations that WFP would have access to, a simple regression can be queried to predict a range of plausible floodplain inundation volume and area contained in the library, with a long lead time. This 'dynamic library' could be extended and hosted by the JRC-UN GDACS GFDS platform, which WFP can easily access and is already familiar with.

Enabling Free and Seamless Data Access with High-End Web-Based Data Analysis Platforms

Current efforts can be augmented by including satellite and other observed or modeled data layers from national agency data centers into web-based data analysis platforms (e.g. Google Earth Engine) thereby enhancing deliverables that plan to make satellite-based flood maps and global flood model simulations easily accessible through such online platforms. Multiple data layers, such as for instance rainfall products from the Global Precipitation Measurement (GPM) mission or soil moisture fields, can be handled to feed in seamlessly with the already planned data layers, i.e., flood hazard data from DFO and maps from a global high-resolution flood hazard model (Fig. 10). The result would be a multilayer flood event hazard chain ranging from a flood driver layer (i.e., precipitation from the GPM mission) through flood onset layers (soil moisture products) to flood event hazard layers (from NRT MODIS combined with global flood model maps). In addition to making all this available on the platforms such as provided by Google, Inc., one could envisage employing new ICT technologies to deliver these layers seamlessly and tailored to targeted, more local, stakeholders. New mature geospatial technologies can leverage current data system capabilities such as provided by NASA's ECHO, NSIDC and EOSDIS (Worldview) for example, and allow interoperability between multiple interactive map viewers on both mobile and traditional computing platforms. These solutions are now sustainable and extensible thereby increasing the efficiency for decision-makers and enabling new users to benefit from Earth science data.

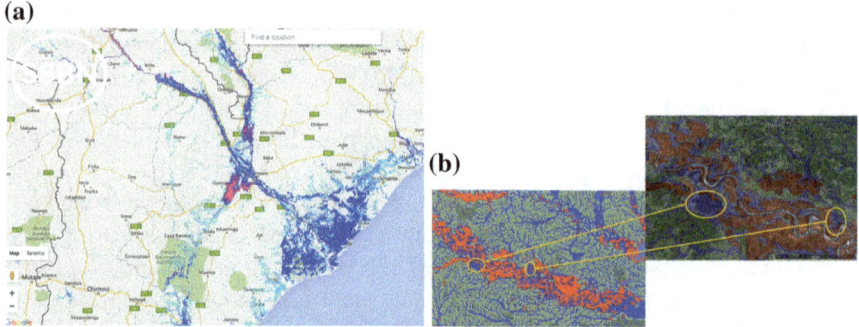

Fig. 10 Illustration of one example of new and innovative solutions/approaches. **a** High-resolution map layers of global flood hazard from 1:100 year return period flow based on the LISFLOOD-FP model (© University of Bristol/SSBN Ltd) made freely available on Google's visualization platform for noncommercial use. Here the area of the Lower Zambezi is shown with flood depth ranging from 1 m (*light blue*) to 5 m (*pink*). **b** MODIS near-real-time flood map (© DFO) of a river near the northern border of Texas during the devastating floods in May/June 2015 (in *red*) overlain on the 1:100 year flood hazard layer from the global flood model shown in **a** (in *blue*). Note that combining satellite observations and models allows to overcome commonly encountered limitations in real-time flood mapping, such as obstruction by cloud and/or forest cover as depicted in the image in the *top panel* in **b**

5 Outlook

The case studies reported in this chapter illustrate great potential and possible ways forward on how the scientific community can engage with the stakeholder community to address most urgent issues, locally, regionally, and globally using EO data. Such connections could allow meaningful discussion that has up to now been largely uncoordinated, oftentimes only taking place on a voluntary effort basis. Such connections are however necessary to globalize and accelerate societal applications that utilize satellite data. This need is further reinforced by the fact that although societal benefits from satellite observations share a number of common features, what works for one region may not necessarily work for another region, even when the problems are similar.

It is clear that the satellite technology and science communities must engage with the stakeholder community to discuss what is possible and most urgently needed. These communities must determine priorities in order to scale up the data and efforts most efficiently in order to benefit societal applications in the best possible way (Hossain 2015). In anticipation of even more data and applications from Earth observing satellites, the Earth science community should engage into identifying key applications alongside key scientific issues. Using these key applications as a guide, the Earth science community needs to establish strong connections with regional stakeholder communities from all around the world, allowing for more rapid dissemination and discovery of EO data at the local level.

Currently, the scientific community has varying perspectives on how these technologies and activities should be pursued in the coming years for the benefit of decision-making and societies in general. Since it will undoubtedly take time and effort to reach consensus, communities should begin to build stronger collaborations and engage to address the challenges that lie ahead.

References

Alfieri, L., Burek, P., Dutra, E., Krzeminski, B., Muraro, D., Thielen, J., & Pappenberger, F. (2013). GloFAS—global ensemble streamflow forecasting and flood early warning. *Hydrology and Earth System Sciences, 17*, 1161–1175. doi:10.5194/hess-17-1161-2013.
Ayalew, L., & Yamagishi, H. (2005). The application of GIS-based logistic regression for landslide susceptibility mapping in the Kakuda-Yahiko Mountains. *Central Japan, Geomorphology, 65*, 15–31. doi:10.1016/j.geomorph.2004.06.010.
Bajracharya, B., Murthy, M. S. R., & Shrestha, B. (2014). SERVIR HIMALYA: Enabling improved environmental management and livelihoods in the HKH. *ISPRS-International Archives of the Photogrammetry, Remote Sensing and Spatial Information Sciences, 1*, 1277–1281.
Bates, P. D., Horritt, M. S., & Fewtrell, T. J. (2010). A simple inertial formulation of the shallow water equations for efficient two-dimensional flood inundation modelling. *Journal of Hydrology, 387*(1–2), 33–45.
Baum, R. L., Godt, J. W., & Savage, W. Z. (2010). Estimating the timing and location of shallow rainfall induced landslides using a model for transient, unsaturated infiltration. *Journal of Geophysical Research, 115*(F03013). doi:10.1029/2009JF001321.
Boccardo, P. (2013). New perspectives in emergency mapping. *European Journal of Remote Sensing, 46*, 571–582.
Brunetti, M. T., Peruccacci, S., Rossi, M., Luciani, S., Valigi, D., & Guzzetti, F. (2010). Rainfall thresholds for the possible occurrence of landslides in Italy. *Natural Hazards and Earth Systems Sciences, 10*, 447–458.
Caine, N. (1980). The rainfall intensity: Duration control of shallow landslides and debris flows. *Geografiska Annaler Series A, Physical Geography, 62*(1/2), 23–27.
Dahal, R. K., & Hasegawa, S. (2008). Representative rainfall thresholds for landslides in the Nepal Himalaya. *Geomorphology, 100*, 429–443. doi:10.1016/j.geomorph.2008.01.014.
Dai, F. C., Lee, C. F., Tham, L. G., Ng, K. C., & Shum, W. L. (2004). Logistic regression modelling of storm-induced shallow landsliding in time and space on natural terrain of Lantau Island, Hong Kong. *Bulletin of Engineering Geology and the Environment, 63*, 315–327. doi:10.1007/s10064-004-0245-6.
Davies, D. K., Ilavajhala, S., Wong, M. M., & Justice, C. O. (2009). Fire information for resource management system: Archiving and distributing MODIS active fire data. *IEEE Transactions on Geoscience and Remote Sensing, 47*(1), 72–79.
Ermini, L., Catani, F., & Casagli, N. (2005). Artificial neural networks applied to landslide susceptibility assessment. *Geomorphology, 66*(1–4), 327–343. doi:10.1016/j.geomorph.2004.09.025.
Farr, T. G., Rosen, P. A., Caro, E., Crippen, R., Duren, R., et al. (2007). The shuttle radar topography mission. *Reviews of Geophysics, 45*(RG2004). doi:10.1029/2005RG000183.
Flores Cordova, A. I., Anderson, E. R., Irwin, D., & Cherrington, E. (2012). Contributions of SERVIR in promoting the use of space data in climate change and disaster management. In *63rd International Astronautical Congress 2012. Conference Proceedings*. Naples, Italy.
Giglio, L. (2010). *MODIS collection 5 active fire product user's guide version 2.4*. Science Systems and Applications, Inc.

Graves, S., Hardin, D., Sever, T., & Irwin, D. (2005). Data access and visualization for SERVIR: An environmental monitoring and decision support system for Mesoamerica. In *Earth Science Technology Conference*.

Gurung, D. R., Shrestha, M., Shrestha, N., Debnath, B., Jishi, G., Bajracharya, R., et al. (2014). Multi scale disaster risk reduction systems, space and community based experiences over HKH region. *ISPRS—International Archives of the Photogrammetry, Remote Sensing and Spatial Information Sciences, 1*, 1301–1307.

Guzzetti, F., Peruccacci, S., Rossi, M., & Stark, C. P. (2008). The rainfall intensity–duration control of shallow landslides and debris flows: An update. *Landslides 5*, 3–17. doi:10.1007/s10346-007-0112-1.

Hardin, D., Graves, S., Sever, T., & Irwin, D. (2005). Visualizing Earth science data for environmental monitoring and decision support in Mesoamerica: The SERVIR project. In *AGU Spring Meeting Abstracts,1*.

Hervas, J., Barredo, J. I., Rosin, P. L., Pasuto, A., Mantovani, F., & Silvano, S. (2003). Monitoring landslides from optical remotely sensed imagery: The case history of Tessina landslide, Italy. *Geomorphology, 54*, 63–75. doi:10.1016/S0169-555X(03)00056-4.

Hong, Y., Adler, R., & Huffman, G. (2006). Evaluation of the potential of NASA multi-satellite precipitation analysis in global landslide hazard assessment. *Geophysical Reseach Letters, 33* (L22402), 1–5. doi:10.1029/2006GL028010.

Hossain, F. (2015). Data for all: Using satellite observations for social good, *Eos, 96*. doi:10.1029/2015EO037319.

ICIMOD. (2015). *Nepal earthquake 2015: Disaster relief and recovery information platform (NDRRIP)*. Retrieved 2015, October 25, from http://apps.geoportal.icimod.org/ndrrip/.

Iverson, R. M. (2000). Landslide triggering by rain infiltration. *Water Resources Research, 36*(7), 1897–1910.

Justice, C. O., Giglio, L., Roy, D., Boschetti, L., Csiszar, I., Davies, D., et al. (2011). MODIS-derived global fire products. In *Land remote sensing and global environmental change* (pp. 661–679). New York: Springer.

Kargel, J. S., Leonard, G. J., Shugar, D. H., Haritashya, U. K., Bevington, A., Fielding, E. J., et al. (2016). Geomorphic and geologic controls of geohazards induced by Nepal's 2015 Gorkha earthquake. *Science, 351*(6269).

Kirschbaum, D. B., Adler, R., Hong, Y., Hill, S., & Lerner-Lam, A. (2010). A global landslide catalog for hazard applications: Method, results, and limitations. *Natural Hazards, 52*(3), 561–575. doi:10.1007/s11069-009-9401-4.

Kirschbaum, D. B., Adler, R., Hong, Y., Kumar, S., Peters-Lidard, C., & Lerner-Lam, A. (2012). Advances in landslide nowcasting: Evaluation of a global and regional modeling approach. *Environmental Earth Sciences, 66*(6), 1683–1696. doi:10.1007/s12665-011-0990-3.

Kirschbaum, D. B., Stanley, T., & Simmons, J. (2015a). A dynamic landslide hazard assessment system for Central America and Hispaniola. *Natural Hazards and Earth System Sciences*, 1–32.

Kirschbaum, D. B., Stanley, T., & Yatheendradas, S. (2015b). Modeling landslide susceptibility over large regions with fuzzy overlay. *Landslides*. doi:10.1007/s10346-015-0577-2.

Kirschbaum, D. B., Stanley, T., & Zhou, Y. (2015c). Spatial and temporal analysis of a global landslide catalog. *Geomorphology*. doi:10.1016/j.geomorph.2015.03.016.

Larsen, M. C., & Simon, A. (1993). A rainfall intensity-duration threshold for landslides in a humid-tropical environment, Puerto Rico. *Geografiska Annaler Series A-physical Geography, 75*(1/2), 13–23.

Lee, S., & Pradhan, B. (2007). Landslide hazard mapping at Selangor, Malaysia using frequency ratio and logistic regression models. *Landslides, 4*, 33–41. doi:10.1007/s10346-006-0047-y.

Lewis, S. (2009). Remote sensing for natural disasters: Facts and figures. *Science and Development Network*. Retrieved 2015, October 16, from http://www.scidev.net/global/earth-science/feature/remote-sensing-for-natural-disasters-facts-and-figures.html.

Liang, X., Lettenmaier, D. P., Wood, E. F., & Burges, S. J. (1994). A simple hydrologically based model of land surface water and energy fluxes for general circulation models. *Journal of Geophysical Research: Atmospheres (1984–2012), 99*(D7), 14415–14428.

Macharia, D., Korme, T., Policelli, F., Irwin, D., Adler, B., & Hong, Y. (2010). SERVIR-Africa: Developing an integrated platform for floods disaster management in Africa. In *8th International Conference African Association of Remote Sensing of the Environment (AARSE)*, Addis Ababa, Ethiopia, 25–29 October, 2010.

Mathew, J., Jha, V. K., & Rawat, G. S. (2009). Landslide susceptibility zonation mapping and its validation in part of Garhwal Lesser Himalaya, India, using binary logistic regression analysis and receiver operating characteristic curve method. *Landslides, 6*, 17–26. doi:10.1007/s10346-008-0138-z.

Mazzanti, P. (2011). Displacement monitoring by terrestrial SAR interferometry for geotechnical purposes. *Geotechnical Instrumentation News*, (June), 25–28.

McCann, D. G., & Cordi, H. P. (2011). Developing international standards for disaster preparedness and response: How do we get there? *World Medical & Health Policy, 3*, 1–4. doi:10.2202/1948-4682.1154.

Melchiorre, C., Matteucci, M., Azzoni, A., & Zanchi, A. (2008). Artificial neural networks and cluster analysis in landslide susceptibility zonation. *Geomorphology, 94*(3–4), 379–400. doi:10.1016/j.geomorph.2006.10.035.

National Planning Commission. (2015). *Nepal earthquake 2015: Post disaster needs assessment*. Government of Nepal. Retrieved 2016, January 14, from http://www.npc.gov.np/images/download/PDNA_Volume_A.pdf.

Neal, J. (2014). *Personal communication*. Bristol, UK: School of Geographical Sciences, University of Bristol.

Neal, J., Schumann, G., Bates, P., Buytaert, W., Matgen, P., & Pappenberger, F. (2009). A data assimilation approach to discharge estimation from space. *Hydrological Processes, 23*, 3641–3649. doi:10.1002/hyp.7518.

Neal, J. C., Schumann, G., & Bates, P. D. (2012). A subgrid channel model for simulating river hydraulics and floodplain inundation over large and data sparse areas. *Water Resources Research, 48*. doi: 10.1029/2012WR012514.

Paiva, R. C. D., Collischonn, W., & Tucci, C. E. M. (2011). Large scale hydrologic and hydrodynamic modelling using limited data and a GIS based approach. *Journal of Hydrology, 406*, 170–181.

Paiva, R. C. D., Collischonn, W., & Buarque, D. C. (2013). Validation of a full hydrodynamic model for large-scale hydrologic modelling in the Amazon. *Hydrological Processes, 27*, 333–346.

Pappenberger, F., Beven, K. J., Hunter, N., Gouweleeuw, B., Bates, P., De Roo, A., & Thielen, J. (2005). Cascading model uncertainty from medium range weather forecasts (10 days) through a rainfall-runoff model to flood inundation predictions within the European Flood Forecasting System (EFFS). *Hydrology and Earth System Sciences, 9*(4), 381–393.

Pappenberger, F., Dutra, E., Wetterhall, F., & Cloke, H. (2012). Deriving global flood hazard maps of fluvial floods through a physical model cascade. *Hydrology and Earth System Sciences, 16*, 4143–4156.

Petley, D. N. (2011). Global deaths from landslides in 2010 (updated to include a comparison with previous years). *Landslide Blog*. Retrieved from http://blogs.agu.org/landslideblog/.

Petley, D. N., Crick, W. D. O., & Hart, A. B. (2002). The use of satellite imagery in landslide studies in high mountain areas. In *The Proceedings of the 23rd Asian Conference on Remote Sensing (ACRS 2002), Kathmandu*.

Pradhan, B., & Lee, S. (2010). Landslide susceptiblity assessment and factor effect analysis: Backpropagation artifical neural networks and their comparison with frequency ratio and bivariate logistic regression modelling. *Environmental Modelling and Software, 25*, 747–759. doi:10.1016/j.envsoft.2009.10.016.

Rossi, M., Kirschbaum, D., Luciani, S., & Guzzetti, A. C. M. F. (2014). Comparison of satellite rainfall estimates and rain gauge measurements in Italy, and impact on landslide modeling. *Natural Hazards and Earth System Sciences* (in preparation).

Saito, H., Nakayama, D., & Matsuyama, H. (2010). Relationship between the initiation of a shallow landslide and rainfall intensity—Duration thresholds in Japan. *Geomorphology, 118*(1–2), 167–175. doi:10.1016/j.geomorph.2009.12.016.

Sampson, C., Christopher, C., Smith, A. M., Bates, P. D., Neal, J. C., Alfieri, L., & Freer, J. E. (2015). A high-resolution global flood hazard model. *Water Resources Research*. doi:10.1002/2015WR016954.

Schumann, G. J.-P., Neal, J. C., Voisin, N., Andreadis, K. M., Pappenberger, F., Phanthuwongpakdee, N., et al. (2013). A first large scale flood inundation forecasting model. *Water Resources Research, 49*(10), 6248–6257.

Schumann, G. J.-P., Andreadis, K., Niebuhr, E., Rashid, K., & Njoku, E. (2014). A simple satellite and model based index for forecasting large-scale flood inundation in data-poor regions. In *Proceedings of the EGU General Assembly*, 1157, EGU, 28 April–2 May 2014.

Thielen, J., Bartholmes, J., Ramos, M.-H, & De Roo, A. (2009). The European flood alert system part 1: Concept and development. *Hydrology and Earth System Sciences, 13*, 125–140.

USGS. (2015). Hazards Data Distribution System (HDDS) Explorer—USGS. Retrieved 2015, October 28, from http://hddsexplorer.usgs.gov.

Wang, J., Hong, Y., Li, L., Gourley, J. J., Khan, S. I., Yilmaz, K. K., et al. (2011). The coupled routing and excess storage (CREST) distributed hydrological model. *Hydrological Sciences Journal, 56*(1), 84–98.

Winsemius, H. C., Van Beek, L. P. H., Jongman, B., Ward, P. J., & Bouwman, A. (2013). A framework for global river flood risk assessments. *Hydrology and Earth System Sciences, 17*, 1871–1892.

WMO. (2009). *Regional Consultation Meeting on Zambezi River Basin Flood Forecasting and Early Warning Strategy and WMO Information System (WIS) and WIGOS Pilot Project, Final Report, WMO and USAID, Maputo, Mozambique, Africa*, 1–5 Dec. 2009. Geneva, Switzerland: World Meteorological Organization (WMO).

Yamazaki, D., Kanae, S., Kim, H., & Oki, T. (2011). A physically-based description of floodplain inundation dynamics in a global river routing model. *Water Resources Research, 47*, W04501. doi:10.1029/2010WR009726.

Applying Earth Observations to Water Resources Challenges

Christine M. Lee, Aleix Serrat-Capdevila, Naveed Iqbal,
Muhammad Ashraf, Benjamin Zaitchik, John Bolten, Forrest Melton
and Bradley Doorn

Abstract Since 2007, significant strides have been made to build the applied research and Earth observations (EO) capacity building community and develop pathways for NASA and Earth observations to help address challenges in water resources. Water is both a critical research topic (e.g. understanding the global water cycle) as well as a critical resource for civilization. As a result, there is a consensus that information about water availability could be valuable for improved management and for water security. The biggest challenge in developing useful applications is finding a way to translate research products, intended to address research questions, to applications that can yield a societal benefit. This chapter addresses the current challenges and future prospects of earth observing systems in the field of water resources.

C.M. Lee (✉)
NASA Jet Propulsion Laboratory, California Institute of Technology, Pasadena, USA
e-mail: christine.m.lee@jpl.nasa.gov

A. Serrat-Capdevila
University of Arizona, Tuscon, AZ, USA

N. Iqbal · M. Ashraf
Pakistan Council of Research in Water Resources, Islamabad, Pakistan

B. Zaitchik
Johns Hopkins University, Baltimore, USA

J. Bolten
NASA Goddard Space Flight Center, Greenbelt, USA

F. Melton
Cooperative for Research in Earth Science and Technology,
NASA Ames Research Center, Moffett Field, USA

B. Doorn
NASA Headquarters, Washington, DC, USA

1 Overview and Current Status

Scientists from remote sensing and capacity building communities have made significant progress in applying satellite data and hydrologic models to address a range of water resource management challenges. Much of this progress occurred following the release of the Earth Science and Applications from Space ("the Decadal Survey") in 2007, which identified priority areas for Earth Science, among which were Earth observations for water-related missions (NRC 2007). The Decadal Survey also underscored the importance of societal benefit and applications (NRC 2007). The convergence of this movement towards building capabilities to understand the global hydrologic cycle and the increasing awareness and understanding of water security issues (United Nations 2015) uniquely position the Earth observations and remote sensing communities to, through applications, provide direct support to addressing water resources challenges. However, while there is consensus that water availability information could be valuable for improved management and for water security (GEOSS 2014), the process for doing so is fraught with numerous challenges. For example, finding an appropriate way to translate data, originally intended to address research questions, to applications that can yield direct societal benefits, is among the most difficult challenges faced by the research-to-operations community. Barriers to this translation are often related not just to technical capacity, but also institutional, organizational, and resource (such as access to infrastructure) capacity and the need to build a common language among team members around a partnership or project.

Currently, there are 18 NASA Earth observing missions that enable scientific insights into the water cycle and also provide useful information to those facing water-related challenges. One prime example of this is in the United States: the U.S. Drought Monitor (USDM) has been an important activity that integrates various datasets to reflect drought conditions. USDM also utilizes and references satellite data to communicate and illustrate the onset and extent of drought events, including the flash drought in the central U.S. in 2012, and the ongoing multi-year drought in California and Western U.S. (National Drought Mitigation Center et al.).

The following case studies represent science applications development that look to address critical water security issues in East Africa (Mara Basin, Nile River Basin), South Asia, and with a global lens. These case studies prioritize partnership development and capacity building, citing such efforts as essential to enabling improved access to and use of Earth observations-based information for their respective decision contexts.

The two key challenges in working in the water resources domain include the following:

1. Water basins cross political boundaries—that is, where water is contained and available is different from how water is managed and treated. Hydrologic boundaries, which can also vary based on whether one is considering water supply on land surfaces or subsurface, are not the same as management boundaries. This issue exists in the U.S., where water is transported across state boundaries to support cities/populations that are hundreds of miles away from its

watershed of origin. This is also an issue in transboundary water basins in, for example, the Nile River in East Africa or the Mekong in Southeast Asia, which span multiple countries. Critically, management decisions, such as the construction of dams, can impact neighboring countries sharing that basin.
2. Water security is dependent on how much water is available, and water availability is dependent on hydrology (what is physically present and accessible) and demands—i.e., who is using water and for what purpose (drinking, sanitation, agriculture, municipal, and industrial), which is also closely tied to infrastructure and development. Security as it relates to freshwater availability is also extremely vulnerable to weather and climate conditions, including changes in precipitation seasonality and inter-annual variability (Feng et al. 2013), precipitation type/phase (rain vs snow) (Barnett et al. 2005), and extremes such as floods and droughts.

Other critical challenges for this community include

3. Understanding how and what water-related information is useful;
4. Bridging the research to operations gap; and
5. Resources to fund activities on the partner or user end can be difficult to obtain

The following case studies examine how applied scientists and researchers have facilitated the use of Earth observations to develop improved data products and support water management practices.

2 Case Study 1. East Africa Mara Basin (2011-present)

Aleix Serrat-Capdevila

The *SERVIR Water Africa—Arizona Team* Project in the Mara Basin

Summary of Project. As part of the NASA-U.S. Agency for International Development (USAID) SERVIR (Sistema Regional de Visualización y Monitoreo) Program and its Applied Sciences Team (AST), the University of Arizona has been working with government agencies and stakeholders from three basins in Africa to better inform water and environmental management decisions. The project supports basin-level water management by developing monitoring and forecasting tools using Earth observations and meteorological forecasts with rainfall-runoff hydrologic models. Since different satellite rainfall products and models have different strengths and weaknesses, a multi-product and multi-model approach was adopted, with the aim of identifying the most suitable approach to supporting decisions.

The three basins that this project focused on are in Africa: the Tekeze River in Ethiopia and Eritrea, the Upper Zambezi in Zambia and Angola, and the Mara River in Kenya and Tanzania. These basins vary by size, topography, latitudes, and management challenges. In this case study, we focus on the Mara River Basin. The project has a three-pronged approach to provide operational capabilities for hydrologic monitoring and forecasting (Fig. 1), which are as follows:

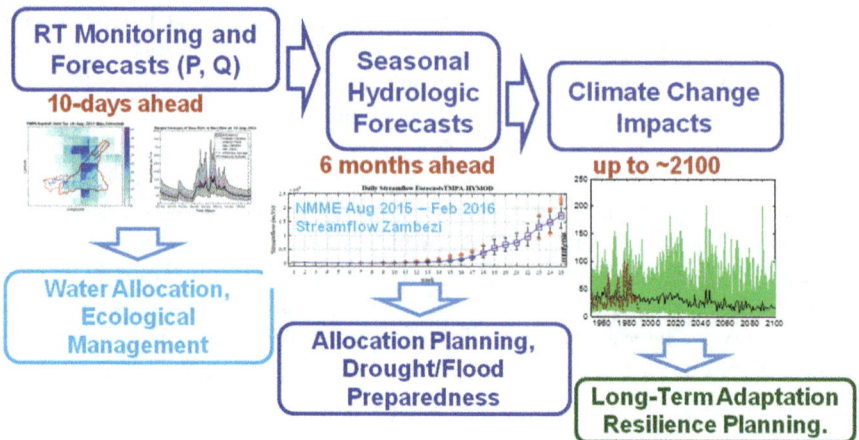

Fig. 1 The three-pronged approach to develop operational monitoring and forecasting capabilities to support water and environmental management activities at different time horizons, ranging from short-term water allocation and ecological management, to seasonal allocation and drought/flood preparedness, and long-term adaptation and resilience planning

(1) Real-time monitoring and short-term forecasting of rainfall and streamflow: use of real-time satellite estimates of precipitation, 7-day meteorological forecasts, and rainfall-runoff models to provide 7–10 day streamflow forecasts.
(2) Medium-term seasonal forecasting: use of downscaled and bias-corrected 6-month ahead seasonal forecasts from sister AST project to run our hydrologic models.
(3) Long-term projections of future climate change impacts on regional water resources, using bias-corrected multimodel projections of rainfall and temperature to drive our hydrologic models.

Description of Target Region. The Mara Basin is located in Kenya and Tanzania between the equator and 2° S, draining into Lake Victoria. Most of its water comes from the highlands of the Mau plateau, where the Mau Forest and tea plantations are, and flows down the Mau escarpment via two main tributaries, the Nyangores and the Amala, which unite to form the Mara River upstream of the grasslands and savannas of the Massai Mara (Kenya) and Serengeti (Tanzania) national parks (McClain et al. 2014). As many other basins in Kenya, the sustainability challenges in the Mara Basin are related to population growth, land-use change, and increased water abstractions, leading to the decrease of ecosystem services and functions. The Mau forest is a natural regulating system that buffers floods, enhances infiltration, and maintains base flows downstream in the game reserves during the dry seasons.

Management Challenges in Agriculture. Human encroachment for wood extraction and expansion of agricultural land (Fig. 2) affect these functions as well as increases the volume of abstracted water upstream in the basin. In addition,

Fig. 2 Land use change in the Mara Basin: the expansion of small-scale agricultural fields (*left*) is a significant pressure on the Mau Forest Reserve (*right*). Tea plantations can be seen, used here as buffer to prevent human encroachment (Image source: Google)

current development plans in the basin include the construction of dams and the expansion of agriculture, a vision that is shared across Africa (UN Water/Africa 2015). As permitted, abstractions have increased in Kenyan basins, and the number of days with very low downstream flows during the dry seasons has progressively increased as well. Poor monitoring and forecasting capabilities are considerable obstacles to the sustainable management of water resources, the long-term viability of agricultural development, and the conservation of natural ecosystems that generate important tourism revenues.

2.1 Understanding User Requirements for Decision Makings

The process for developing useful monitoring and forecasting applications for decision-making starts by understanding the decision-making challenges and context in which these will be used; an important element for success for an applications project is continued engagement with and participation from end users whose decisions could potentially benefit from the application being developed.

To develop an understanding and working partnership with our users (and ultimately understand their management context, challenges, and opportunities to benefit existing decision practices), we approached partnership development through several avenues including

(1) we worked with an in-region partner, who was a hub and bridge to country ministries who were our end users. That partner is the Regional Center for Mapping of Resources for Development (RCMRD—Nairobi SERVIR Hub);
(2) we established regular communications with our end users through meetings and informal discussions;

(3) we supported and participated in a workshop organized by RCMRD, with regional end users, as well as other workshops.

2.2 Workshops with RCMRD and Other Stakeholders

The RCMRD workshop was critical to understanding decision challenges and cultivating relationships/partnerships with regional end users. Workshop participants were officers from water-related government agencies such as the Water Resources Management Authority (WRMA), Water Resources Department (WRD) of the Ministry of Environment, Water and Natural Resources, Kenya Meteorological Department (KMD), and other institutions and collaborators (UNESCO Regional Hydrologist, USAID officers, International Center for Tropical Agriculture, Jomo Kenyatta University, and others). During the workshop, participants provided written responses for the following four questions:

1. What management decisions are you confronted within your workplace?
2. What accuracy/precision is required in the information for each one of these decisions?
3. What information are you using currently for those management and planning decisions?
4. What improvements in the information you are using would make the greatest difference?

We also participated in a multi-day workshop in Bomet, within the Mara basin, with numerous local agency and stakeholders representatives, where we presented and sought feedback for monitoring and forecasting applications that were under development.

This workshop enabled our understanding of critical contextual information related to water abstractions in this region:

- There is a need to better understand hydrologic processes, water resources availability, and the functioning of the system in the basin.
- The basin currently does not have reservoirs and is not regulated
- The decision that can be informed by tools and applications is the issuing of permits for water abstractions
- Livelihood, land-use changes, water quality, and ecosystem services were also raised as important issues
- Data products that were graphically or visually represented were essential to communicating understanding of water resources

Participants were faced with the challenge: if the river is over-abstracted, how can new permits and existing abstractions be coordinated? This is a difficult situation, particularly in times of water scarcity that is also subject to changing seasonal and inter-annual variability. With a better understanding and ability to monitor water resources, abstractions could potentially be better regulated, or "optimized," with a balance be found between much-needed environmental flows during the dry season, which support the Masai Mara and Serengeti ecosystems, and other uses.

Bringing together stakeholders and sectors to build awareness and familiarity with remote sensing information and applications was very valuable for this project. Furthermore, we were able to begin providing a basis for a common language and an expanded understanding of the basin system (from the scientists' perspectives and from the partners' perspectives). These interactions also facilitated the identification of future opportunities to collaborate and evolve the partnership and project. Other decision and planning contexts that this project could provide value to includes the planning and constructions of future reservoirs, preparing for drought and floods, and ecosystems protections (particularly during dry season).

In response to learning these additional priorities and feedback, we were able to adapt our model development efforts and translated our calibration approach to give equal importance to simulating low flows as well as high flows. We also took into account stakeholder feedback about data latency challenges and automated our operational online broadcasting platform to update rainfall estimates in as products become available.

In addition to learning about areas of needs, our partnership also allowed us to communicate regularly about how to effectively use the application outputs e.g. under what conditions and in what scenarios. For example, our rainfall-runoff applications did not always appear to be well-suited for flood alerts, as the relative errors in satellite estimates can accumulate during flood events, impacting magnitude and timing-of-event characteristics. We also emphasized the importance of using knowledge of uncertainty as a way to understand how precise simulated products are over a given timeframe. As generate practice, we shared guidance for how to use knowledge of uncertainty in conjunction with deterministic models.

Currently, the project provides bias-corrected and downscaled rainfall monitoring displays over its pilot basins, including in the Mara (Fig. 3); multimodel/product simulations and bias-corrected streamflows, with a good uncertainty characterization in model runs; and a merged forecast, probabilistically assimilating the individual model–product combination forecasts and their uncertainty (Roy et al. 2016). The streamflow monitoring and 7–10 day forecasts are operational in the Mara basin and the Zambezi, and they are being implemented in the Tekeze basin. The seasonal forecasting system is operational in the Upper Zambezi. The monitoring displays and forecasts can be found for the three pilot basins at the project site www.swaat.arizona.edu with interactive visualizations for the streamflow forecasts (Fig. 4), for improved access and communication with agencies and stakeholders.

Conclusions. The project is entering its fourth year as of the end of 2015, during which these tools will be fully transferred to RCMRD (the SERVIR Hub in Nairobi) and interested agencies, while we also continue to run and broadcast the forecasts online.

The face-to-face interactions were essential to build trust, strengthen connections with key individuals, and build a foundation for future work, new data acquisition, results sharing, and implementation. The continued dialog with users and stakeholders enables the project to incorporate feedback on an on-going basis and understand how useful a technical will be and how it will be used after its delivery. As such, tracking the trajectory of communications can, in itself, be a

Fig. 3 Real-time bias-corrected satellite estimates over the Mara Basin. In the case of TMPA (*left*) and CMORPH (not shown), estimates at 0.25° were bias-corrected and downscaled using the CHIRPS reference dataset at 0.05°. In the case of PERSIANN-CCS, estimates were bias-corrected and re-gridded from their original 0.04°–0.05° to obtain a consistent input to rainfall-runoff models. *Figure rendered by T. Roy*

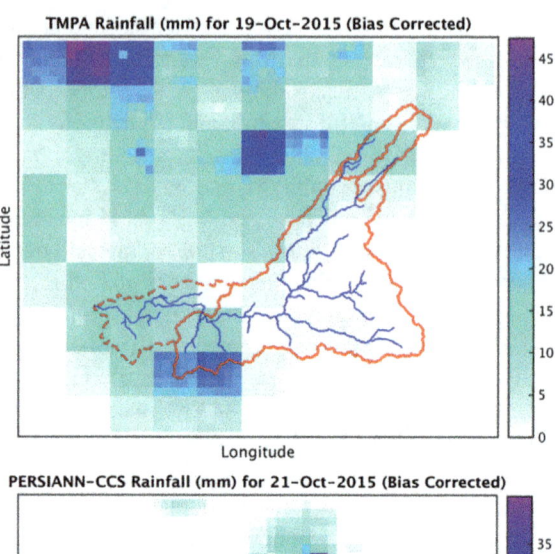

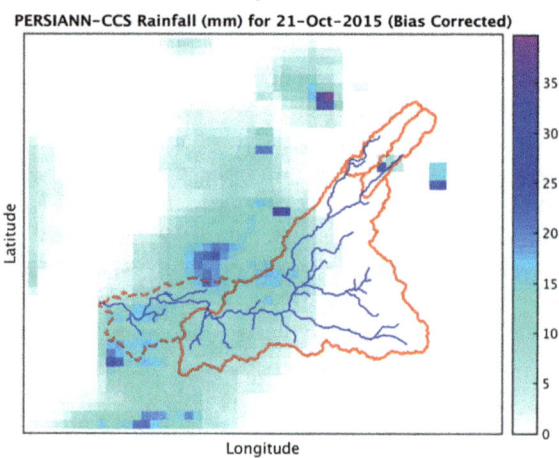

reflection of how valuable the tool is to the stakeholders. Ideally, such tools can be evolving applications that are fine-tuned as new remote sensing products become available and management needs evolve.

3   Case Study 2. Africa Nile River Basin (2009–2014)

Benjamin Zaitchik

NASA's Project Nile—Distributed hydrological information for water management in the Nile Basin

Summary of Project. This work aimed to apply NASA tools in support of scientifically informed water management in nations that share the Nile basin (Fig. 5). Over the course of the project (2009–2014), U.S. scientists at Johns Hopkins

Applying Earth Observations to Water Resources Challenges 155

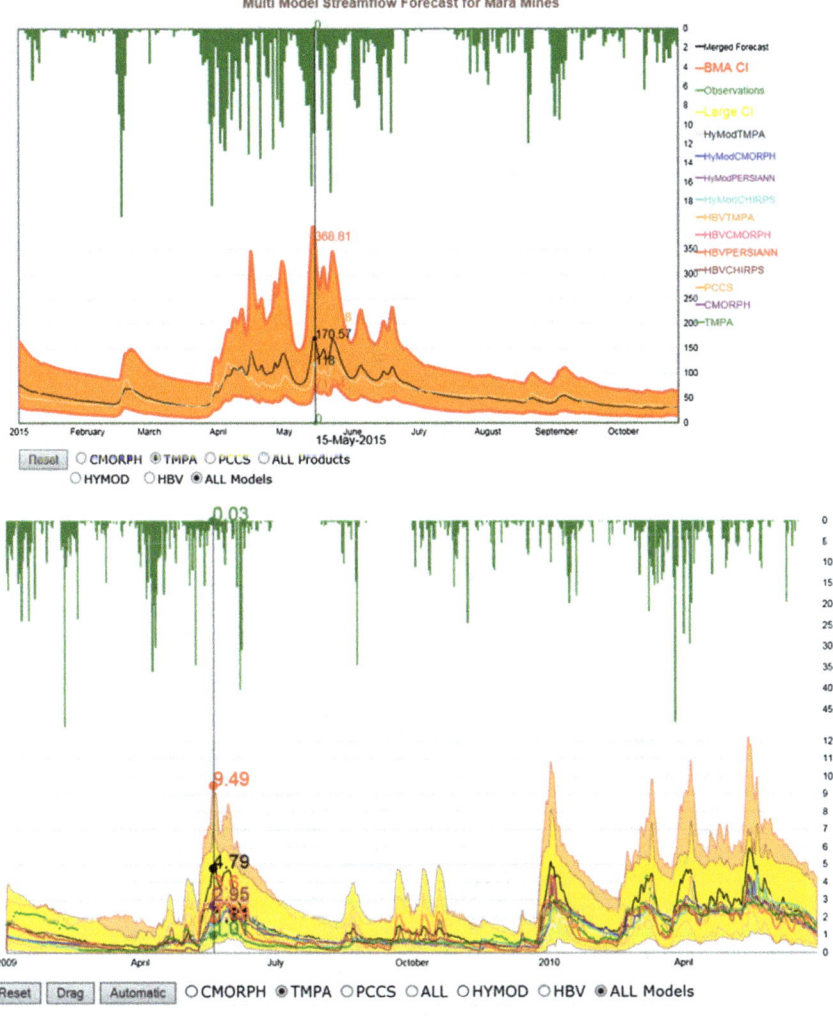

Fig. 4 Experimental interactive displays of streamflow simulations, with 95 % confidence intervals and a merged forecast from Bayesian model averaging. Aggregated basin precipitation is shown in the upper part of the visualization. Note that the user can select the model and product combinations to be shown. *Visualizations by M. Durcik and simulation work by T. Roy, A. Serrat-Capdevila, J. Valdes and H. Gupta*

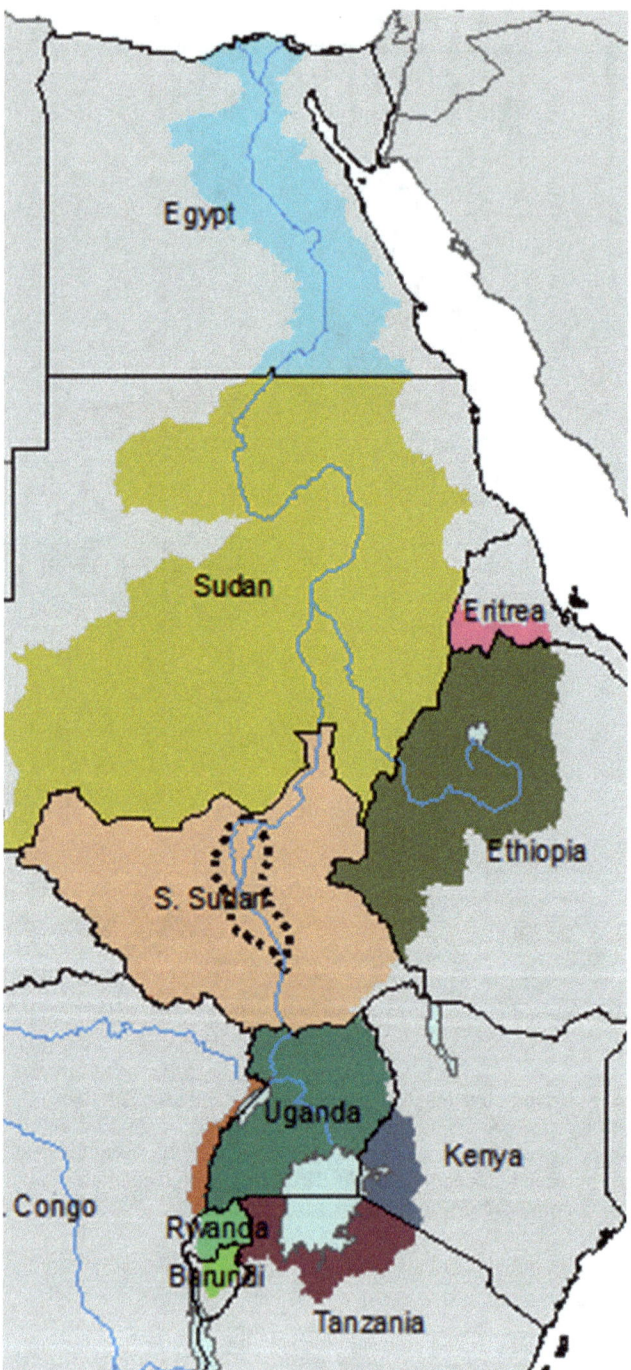

Fig. 5 The Nile basin, which spans eleven countries. The *Project Nile* domain included the entire basin, and at partner request the project's LDAS and ALEXI products were extended to include the full territory of all of these countries except for the Democratic Republic of Congo

University, NASA, U.S. Department of Agriculture (USDA), and the University of Wisconsin worked with partners at USAID, the Nile Basin Office Eastern Nile Regional Technical Office (NBI-ENTRO), the Ethiopian National Meteorological Agency and Environmental Protection Agency, and Addis Ababa University. The team developed satellite-based land cover maps, produced satellite-derived evapotranspiration estimates using a new Meteosat implementation of the USDA Atmosphere Land Exchange Inverse model (ALEXI), and implemented a Land Data Assimilation System (LDAS) customized to match identified information needs.

The work led to improved water balance estimates across the Nile Basin, enhancements to NASA modeling tools, and ongoing collaborations of flood and drought response and climate change adaptation in the Nile and surrounding regions. The research accomplishments of the project are represented in a number of peer-reviewed publications (Anderson et al. 2012a, b; Berhane et al. 2014; Foltz et al. 2013; Satti et al. 2015; Shortridge, et al. 2015; Simane et al. 2013; Wilusz et al. in review; Yilmaz et al. 2014; Zaitchik et al. 2012). Water resource applications addressed in these publications include drought monitoring (Anderson et al. 2012a, b), wetland mapping (Wilusz et al. in review), estimates of irrigated water use (Yilmaz et al. 2014), climate vulnerability analysis (Foltz et al. 2013), agricultural land classification (Simane et al. 2013), and hydroeconomic optimization (Satti et al. 2015).

NASA Data and Transboundary Water Challenges. *Project Nile* exemplified the potential that NASA products have to inform decision making in a contentious transboundary basin. In 2011, LDAS and ALEXI outputs were featured in the Nile Basin Initiative Donors Meeting in Copenhagen and the Ethiopian National Adaptation Plan of Action (NAPA) was submitted to the United Nations Framework Convention on Climate Change to the Nile Basin Initiative's Atlas of the Nile Basin. In each case, the objectivity, trans-national consistency, and non-political nature of NASA-based analysis facilitated open discussion of hydrology of the basin. Demonstrations of decision support were also performed with the Ethiopian Ministry of Water, Irrigation and Energy, in which *Project Nile* LDAS was used to drive a Water Evaluation and Planning (WEAP) system model for the Lake Tana basin. Data were also provided to decision makers at Egyptian Ministry of Water Resources and Irrigation, with a similar analysis in mind. There were discussions of making the *Project Nile* LDAS available as a near real-time data product distributed through NASA data servers, but given geographic overlap with other NASA projects it was decided that LDAS innovations accomplished in *Project Nile* could be fed into other LDAS systems that cover East Africa, including FEWSNET LDAS (FLDAS), which is now distributing a similar analysis through NASA data portals.

Additional Benefits of *Project Nile*. *Project Nile* has been leveraged for numerous follow-on collaborations across East Africa and, in the case of ALEXI, all of Africa. From a capacity building perspective it has yielded lasting relationships with project partners in Ethiopia, Sudan, Egypt, Kenya, and Uganda, who continue to engage to learn about new NASA products and to improve their modeling skills. We have held workshops for government scientists and academics from Ethiopia, Sudan, Kenya, Rwanda, and Uganda on satellite-based land cover mapping, the theory and use of ALEXI, and the application of *Project Nile* LDAS to water resource analysis.

Challenges: Contention Around Water in the Nile Basin. This project also encountered significant challenges that were not rooted in technical or scientific elements of the work but instead were the product of changing geopolitical landscapes. At the time the proposal was written, the Nile Basin Initiative (NBI) was working toward a comprehensive water-sharing framework that would include all member States. During the life of the project, however, the Ethiopian government began construction on the largest hydropower dam in Africa over intense Egyptian objections, resulting in Egypt and Sudan abandoning NBI negotiations for several years, which was then compounded by several transitions in government in Egypt, the founding of South Sudan and its subsequent political turmoil. These events, among others, complicated relationships within NBI. NBI-ENTRO experienced nearly 100 % staff turnover and several changes in strategic vision, while NBI political leadership embraced, then rejected, and subsequently re-embraced the principle of collaborating with NASA for water resource analysis. We worked around these complications by broadening our base of partners at national government agencies, international organizations, and universities in the region, allowing the project to continue providing value to the region project's accomplishments.

Partnerships Lessons Learned. There is no question that the stated capacity building goals of the project suffered due to these institutional difficulties: while *Project Nile* has contributed to the technical understanding of the Nile water balance and its end products have been utilized by decision makers, political barriers and staff turnover within the NBI made it not feasible to achieve operational hand-off to NBI-ENTRO in the timeframe that we were working.

Given this experience, *Project Nile* has shares a few important considerations for other applied sciences projects that may have a focus in transboundary basins where political instability can be a huge facet of the project landscape:

(1) Considerations for partnering with regional organizations (with members states) or working directly with country ministries. In the case of working in transboundary water basins, this consideration is well illustrated by comparing options for partnering organizations: for example, international river commissions in the developing world, with multiple member states, are generally thought to hold the most promise for advancing collaborative water management, as opposed to unilateral development by each member country. However, these organizations are not always fully empowered (both technically and politically) to make substantive progress in these areas relative to member countries. We encountered this tension between regional river commissions and member countries when considering whether to work with the relatively high capacity water ministries in one or more Nile basin countries or to attempt to partner with NBI despite their relative political (and hence technical) fragility. Insomuch as the success of an Applied Sciences project is defined in terms of successful transition of an advanced NASA product, modeling or observation system, in our case, there was an advantage to working with single countries. If the goal is to apply NASA analysis in

partnership to advance scientifically informed water sharing; however, then working with basin commissions may be the more appropriate partner. Indeed, providing a NASA analytical backbone for these organizations has the potential to empower them in the eyes of their member States.

(2) Considerations for partnering with an operational agency or a research entity. Partnering with an academic or research institute may have allowed *Project Nile* to sidestep some of the challenges related to political circumstances that then impeded development of capacity for project hand-off. It is possible that *Project Nile* would have been able to transit an improved model or product to another research group or university with greater institutional and staff stability. This may be because many countries' operational water resource agencies not have the political and financial flexibility to be fully engaged partners: data are proprietary, decision support systems can be resistant to change, and staff and IT time may not be available.

While development and provision of a mature analysis tool has the potential to motivate cost-sharing investments from the end user, our experience with *Project Nile* would suggest an alternate strategy for projects that are higher risk or require preliminary demonstration to win buy-in. For that subset of projects, it can be more valuable and productive to partner with an academic organization or an independent research center, where there is the flexibility and mandate to work on transformative analysis. Such projects are unquestionably "applied" in their objectives, but they fall on the low end of the Applications Readiness Level (ARL) scale and might need partnership from a research group even more than they need immediate access to decision makers. Furthermore, many lower capacity countries turn to universities and research centers for their technical and scientific expertise and in that way these entities can serve as a conduit to an operational partner. Capacity building, thus, that is focused on non-political research institutions would likely have additional benefits for member states. We have found that these secondary partnerships with non-politicized, in-region academic and research centers provided the best data sharing relationships and strongest capacity building legacy for *Project Nile*.

4 Case Study 3 Pakistan: Role of Earth Observation for Operational Groundwater Resource Management in Pakistan: A Capacity Building Perspective from Pakistan (2014-)

Naveed Iqbal

Challenge Summary. Sustainable groundwater resource management is essential to ensure food security and sustainable socioeconomic development of an agrarian country like Pakistan. Surface water availability is scarce and subject to limited storage, increasing population, variable climate conditions, and the occurrence of extreme events (floods and droughts). Together, these factors pose a challenge to managing groundwater sustainably in Pakistan. Other emerging and increasingly complex stressors on Pakistan's water resources are water table depletion, groundwater mining, saline water intrusion, and groundwater quality deterioration.

This situation is further aggravated by limitations in in situ data availability, both spatially and temporally. Furthermore, data collected by various water management agencies in Pakistan are not easily accessible or shared with the research community and policy makers. Currently, different geophysical, isotopic, and groundwater modeling approaches are used for the detailed assessment, analysis, as well as understanding of long-term changes in groundwater dynamics. These tools are used to formulate appropriate strategies under varying climate conditions. These techniques are very detailed but laborious, time-consuming, and costly, as they also involve field surveys. The accuracy of numerical groundwater models is also hampered at large-scale coverage such as in Indus Basin due to paucity of input data.

Remote sensing data can be used as part of a cost-effective approach for conducting hydrologic analyses at various spatial and temporal scales. Its large-scale coverage (regional-to-global), high temporal (days-to-months), spatial resolutions (meters-to-kilometers), and availability of make remote sensing data particularly amenable and valuable for use in data-constrained areas. However, in situ or cal/val activities are required to help ensure remote sensing products are well calibrated and can be used with confidence. Currently, many hydrologists are able to effectively use Earth observations technology for improved water resource management.

Role of Earth Observations. Earth observations have wide applicability in hydrology. Many sensors on-board various satellite platforms are providing useful information pertaining to precipitation, soil moisture, topography, water level measurement in rivers and lakes, terrestrial water storage information, among other variables. These datasets are often freely available at high spatial and temporal scales for regions all over the world. Remote sensing has enabled researchers and managers to conduct more holistic, basin-scale analysis which was not yet possible using traditional approaches based on field surveys and hydrological modeling.

The Gravity Recovery and Climate Experiment (GRACE) satellite observes monthly gravity anomalies at resolutions of 10 micron changes over 200 km, which can be useful for extracting variations in terrestrial water storage (TWS). GRACE is a unique twin gravity satellite that is a joint German Aerospace Center (DLR)-NASA mission that was launched in 2002 (Rodell et al. 2009). GRACE can be used to better understand Total Water Storage (TWS) and calculate groundwater storage (GWS), which is calculated using soil moisture information and GRACE-TWS (Rodell et al. 2009). Monthly groundwater storage changes are useful for the analysis of long-term groundwater system behavior and understanding of groundwater dynamics under climatic implications.

Project Summary. Under a capacity building effort of National Aeronautics and Space Administration (NASA), through University of Washington (UW-Seattle) for Pakistan Council of Research in water Resources (PCRWR), a 6-month professional training on satellite gravimetry (GRACE) application for groundwater

Applying Earth Observations to Water Resources Challenges 161

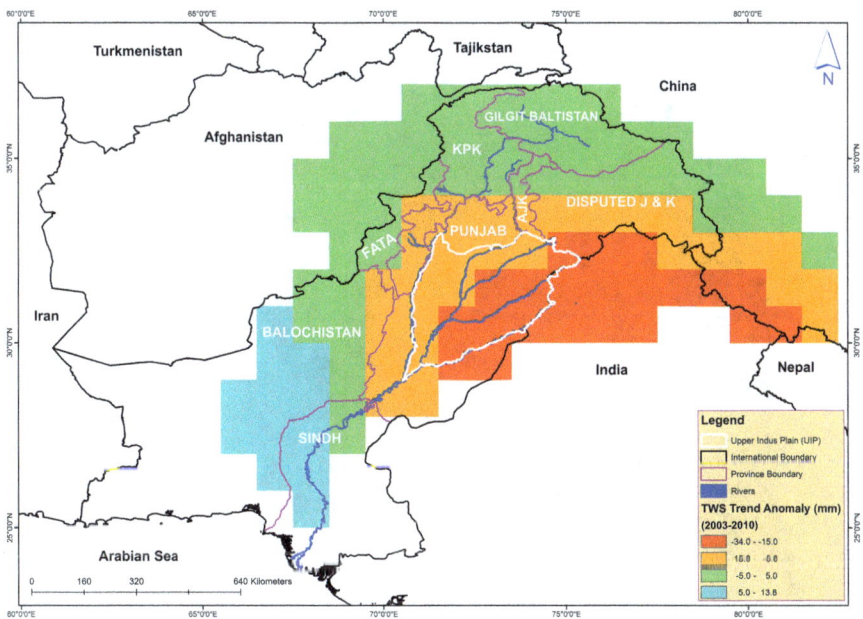

Fig. 6 Variations in terrestrial water storage anomalies over Indus Basin from 2003 to 2010

resource management is conducted at University of Washington (UW). During this specialized training, the effectiveness of GRACE satellite for the groundwater resource management at operational scales is evaluated. The Variable Infiltration Capacity (VIC) hydrological model is applied on Indus Basin for the simulation of monthly soil moisture and surface runoff information at a spatial scale of $0.1° \times 0.1°$ from 2003 to 2010. The GWS changes are inferred from GRACE-TWS using VIC model-generated soil moisture and runoff data.

Variations in TWS range from −34 to −5 mm over Punjab Province from 2003 to 2010 (Fig. 6), potentially due to groundwater use for anthropogenic applications in Punjab Province. Agriculture is the largest user of groundwater and the area is famous as the main food basket of the country due to fertile agriculture land. Punjab Province consists of four riverine flood plains locally known as Thal, Chaj, Rechna, and Bari doabs (area between the two rivers). These doabs are bound by River Indus and its four major tributaries (Jhelum, Chenab, Ravi, and Sutlej). Our analysis utilizes GRACE data and indicates that groundwater storage is being depleted at an average rate of about 12.96 mm/year and 6.78 mm/year in Bari and Rechna doabs and, 7.34 mm/year and 3.54 mm/year km^3/year in Thal and Chaj doabs over the period 2003–2010 (Fig. 7). In Bari and Rechna doabs, the depletion trend is found to be more persistent than in Thal and Chaj doabs due to their intermixed trends of recharge and depletion. Massive flooding in late July 2010 has played a role in significantly decreasing the groundwater depletion rate. It has been observed that GRACE is more sensitive to the significant changes in the groundwater storage variations caused by

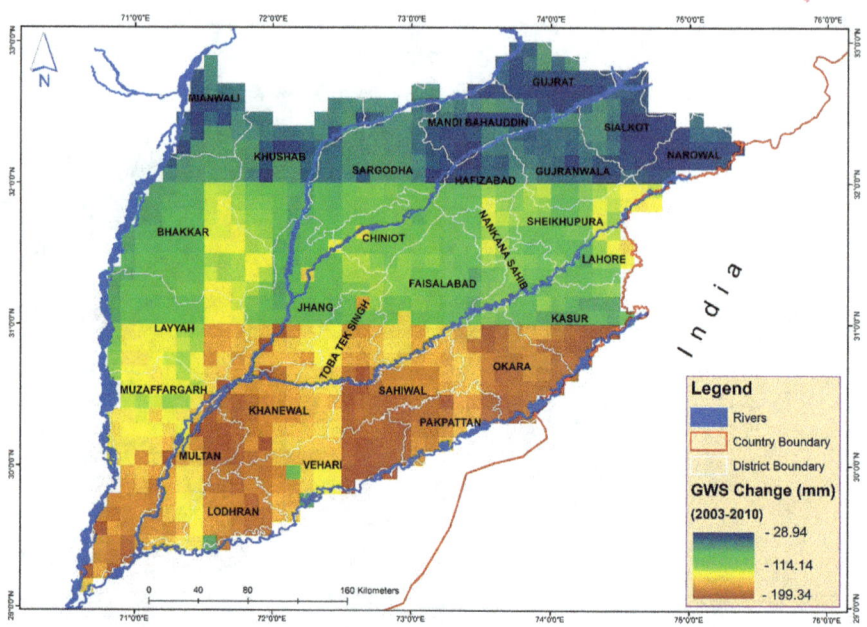

Fig. 7 Changes in groundwater storage anomalies over Upper Indus Basin from 2003 to 2010

depletion or recharge at appropriate scales. Furthermore, its effectiveness increases with large-scale and persistent groundwater storage trends over long periods.

The groundwater storage information is a key parameter for improving groundwater resource management (Jin and Feng 2013). The month-to-yearly spatial and temporal changes in groundwater storage are useful for formulating appropriate groundwater management strategies. It is envisaged that a GRACE-based application is a cost-effective tool for the understanding basin-scale hydrology and monitoring groundwater behavior at operational level. This application will help provide scientific information in a cost-effective way for decision-making and for policy recommendations.

Earth observations are widely available and has been shown to be applicable and useful for natural resources management. An opportunity is presented here, in that the community can now work to maximize the benefit of freely available Earth observations information through applications. Building capacity within the end user community (researchers and managers, for example) is a very important component for translating knowledge to the benefit of mankind. It is an integral link between the data providers and end-user community.

Challenges in Capacity Building for Earth Observations Use. Two major challenges in building capacity to use Earth observations in Pakistan are infrastructure development (hardware and software) and trained personnel. Finding support to establish a properly equipped geospatial lab can also be a challenge, but there have been recent investments at the government and private sector level to help. Many universities have started various satellite remote sensing (SRS) and geographic information science (GIS) programs and are producing skilled technical personnel. The high computation power that is required to run the latest versions of related software and specialized skillsets in the field of Earth observations remain a barrier to being able to fully realize the potential of Earth observations for operational use.

Examples of Capacity Building Activities. Conferences and workshops (with trainings) are a useful approach for generating and expanding awareness of satellite data but a more integrated program, such as a technical exchange or permanent appointment within a trained research team, fully equipped with necessary tools, would enable continued progress in capacity building. Another challenge is retention of skilled personnel, many of whom transition to other positions, creating a gap for a much-needed skillset within the institution.

Capacity building is a continuous process that may require extensive effort in an iterative manner. Furthermore, capacity building is only effective if it is two-way street, where all parties are contributing and gaining equally. A personnel exchange is a type of collaboration that can be extremely effective, more so than a short training course or participation in a conference is. Through these types of concrete, collaborative projects, experts can share their knowledge of applications and help devise solutions for ongoing challenges at the institutional level.

We have observed that the efficacy of capacity building efforts led by researchers at the University of Washington, and supported by NASA, with PCRWR, is very much dependent on ongoing and continued collaborations. The Memorandum of Understanding (MoU) between UW and PCRWR (signed on Nov 2, 2015) will further strengthen the relationships between these two organizations. The future joint activities will also help ensure the sustainability of the current effort. This approach will help PCRWR become more independent by improving its research infrastructure and building a team of specialized professionals.

The operational adoption of GRACE as a tool for groundwater monitoring and resource management and the continuity of GRACE satellite mission in the form of GRACE Follow-on (GRACE-FO) are linked and very important. The expertise on VIC hydrological model for the simulation of soil moisture and surface runoff datasets, assistance for high computation, and relevant softwares are also key requirements for PCRWR. The collaboration between NASA and PCRWR will also help address real-world issues in the field of sustainable water resource management in Pakistan. It will also enable the application of Earth observations as a scientific tool for the social benefit.

Being a federal research and development organization, PCRWR is helping both the groundwater managers and farmers address water resources concerns as well as informing policy recommendations for decision-makers. Different capacity building training programs are being offered by PCRWR on the various water conservation and management techniques for effective water resource management. In this way, capacity building of PCRWR will ultimately help to benefit the operational managers and farmers to achieve the goal of sustainable water resource management.

PCRWR would like to see future joint activities include the use of Earth observations and physical modeling-based tools development for the early warning of floods and droughts, addressing transboundary water issues, real-time soil moisture, and evapotranspiration estimations. Applications that utilize Earth observations for the water cycle budgeting in general could also be of great value in the context of disasters mitigation and sustainable water resource management.

5 Case Study 4: Enhancing Global Crop Assessment of the USDA Foreign Agriculture Service with NASA Soil Moisture Products (2014-)

John Bolten

Project Summary. One example of the operational application of global satellite-based observations for improved decision-making is the NASA-USDA Global Soil Moisture Product employed by the Foreign Agricultural Service (FAS). The development of this global product was envisioned to improve the USDA FAS global crop assessment decision support system via the integration of NASA soil moisture data products. USDA FAS crop yield forecasts affect decisions made by farmers, businesses, and governments by predicting fundamental conditions in global commodity markets. Regional and national crop yield forecasts are made by crop analysts based on the Crop Condition Data Retrieval and Evaluation (CADRE) Data Base Management System (DBMS). Soil moisture availability is a major factor impacting these forecasts and the CADRE DBMS system predominantly estimates soil moisture from a simple water balance model (the Palmer model) based on precipitation and temperature datasets operationally obtained from the World Meteorological Organization and U.S. Air Force Weather Agency.

To improve understanding and prediction of agricultural growth in varying climate regimes, land cover, crop type, and management strategies, it is necessary to recognize the important role that soil moisture plays in the carbon, water, and energy cycles, as well as in regulating crop growth and health. Near-surface soil moisture (i.e., the amount of water in the first few inches of soil) is a critical component of all of these cycles. The volume of water in the first few inches of soil controls infiltration and has a direct influence on the amount of water that reaches lower layers of the soil column. It also constrains land surface temperature by affecting the fraction of latent and sensible heat of the land surface. As a result, observations of near-surface soil moisture are used to help monitor and predict regional flood and drought events (Dai et al. 2004; Komma et al. 2008; Norbiato et al. 2008). In addition, regional variations in soil moisture availability can provide a leading signal for subsequent anomalies in vegetative health and productivity (Adegoke and Carleton 2002; Ji and Peters 2003). Thus, soil moisture has increasingly become a key input to crop forecasting and basin management strategic strategies, as well as many large-scale flood, drought, and agricultural monitoring systems (Mo et al. 2011).

However, very few networks exist due to their high cost of installation and maintenance. In addition, the heterogeneous nature of soil moisture causes in situ methods to be locally restrictive and they do not accurately capture the basin-scale soil-moisture changes that are often needed for water resources management applications (Engman and Gurney 1991). To this end, ground-based observational soil moisture data alone are not sufficient to provide accurate water budget information, and can be greatly improved through the addition of satellite-based regional observations of near-surface soil moisture.

Targeting Decision Making of the USDA Foreign Agriculture Service. The USDA FAS attempts to anticipate the impact of drought on regional agricultural productivity by monitoring soil moisture conditions using a quasi-global soil water balance model (Bolten et al. 2002). However, the accuracy of such models is dependent on the quality of their required meteorological inputs and is thus questionable over data-poor regions of the globe and can be improved through the integration of satellite-based soil moisture observations. In this case, the Palmer soil moisture model implemented by the FAS is driven by daily precipitation and near-surface air temperature observations (Palmer 1965). However, over much of the globe, particularly food insecure areas that are of interest to the FAS, these observations are sparse or do not exist. For example, agricultural areas in South Africa and Brazil are significantly lacking in reliable networks of gage-based precipitation and temperature observations. As a result, the modeled soil moisture over these areas has a high uncertainty, which leads to inaccurate estimates of agricultural productivity. A key challenge for the USDA FAS is to provide monthly global estimates of agricultural productivity over areas like these lacking reliable observations of precipitation, temperature, or soil moisture.

The main objective of the project is to enhance the U.S. Department of Agriculture (USDA) Foreign Agricultural Service (FAS) global crop assessment decision support system via the integration of NASA soil moisture products through the implementation of data assimilation tools. The baseline soil moisture estimates used by USDA FAS are developed using the two-layer Palmer model, which is a physically based hydrologic model forced by daily precipitation and minimum and maximum temperature measurements acquired from the U.S. Air Force Weather Agency. Integration of the satellite observations is done using a 1-D Ensemble Kalman Filter (EnKF) technique set to run with 30-ensemble member (Bolten et al. 2010).

The Palmer model estimates of surface and root-zone soil moisture are derived from the two-layer Palmer water balance model currently used operationally by USDA FAS. The two-layer Palmer model is based on a bucket-type modeling approach as described in *Palmer* (Palmer 1965). It estimates the available water capacity (AWC) of the top model layer, assumed to be 2.54 cm at field capacity, and the AWC of the second layer (i.e., root-zone layer) is calculated using soil texture, depth to bedrock, and soil type derived from the Food and Agriculture Organization (FAO) Digital Soil Map of the World available from the FAO at http://www.fao.org/ag/agl/lwdms.stm#cd1. In this fashion, water-holding capacity for both layers (incorporating near-surface soil moisture and groundwater) ranges between 2.54 and 30 cm according to soil texture and soil depth. Evapotranspiration is calculated from the modified FAO Penman-Monteith method and observations of daily min/max temperature. Further modeling details are available in (Bolten et al. 2010). Required daily rainfall accumulation and air temperature datasets are obtained from the U.S. Air Force Weather Agency (AFWA) Agriculture Meteorological (AGRMET) system which derives a daily rainfall accumulation product based on microwave sensors on various polar-orbiting satellites, infrared sensors on geostationary satellites, a model-based cloud analysis, and World Meteorological Organization (WMO) surface gage observations (Bolten et al. 2010).

The NASA-USDA Global Soil Moisture Product was initially developed using soil moisture observations from the EOS Advanced Microwave Scanning Radiometer (AMSR-E), which provided quasi-global soil moisture observations from 2002 to 2011. The product has transitioned to other satellites as they have become available, including the onset of data availability from the ESA Soil Moisture and Ocean Salinity (SMOS) in 2009 and NASA Soil Moisture Active and Passive (SMAP) in 2015 (Entekhabi et al. 2010; Kerr and Levine 2008). As a result, the system is envisaged to provide over 15-year heritage of global soil moisture which should result in a significant expansion in our ability to study the impact of retrieving surface soil moisture using satellite remote sensing for many agricultural and hydrological applications.

AMSR-E stopped producing reliable observations in October of 2011. The current operational system ingests soil moisture observations acquired from the Soil Moisture Ocean Salinity (SMOS) mission and it is in preparation for switching to data provided by the NASA's Soil Moisture Active Passive (SMAP) mission. Due to the failure of the SMAP active instrument soon after launch, the system is being modified to incorporate soil moisture observations from the Advanced Scatterometer ASCAT instrument onboard the EUMETSAT METOP satellite.

The soil moisture product is based on the implementation of an Ensemble Kalman Filter (EnKF). The EnKF is a nonlinear extension of the standard Kalman filter and has been successfully applied to a number of land surface data assimilation problems (e.g., Reichle 2005). The current system has been found to function adequately using a 30-member Monte Carlo ensemble size (Bolten and Crow 2012) and is being modified to dually assimilate both SMAP and ASCAT soil moisture observations.

The soil moisture products generated by this system have been operationally delivered to USDA FAS since 2013 and have been fully incorporated into the USDA FAS Crop Explorer decision support system since April of 2014 (Fig. 8). Products include surface and subsurface soil moisture, total soil profile soil moisture, and soil moisture anomalies for the surface and subsurface layer. The products have become an invaluable part of the USDA FAS CADRE (Crop Assessment Data Retrieval and Evaluation) and are routinely used by USDA FAS crop analysts and experts to assess and improve their global yield predictions that are included in the

Applying Earth Observations to Water Resources Challenges 167

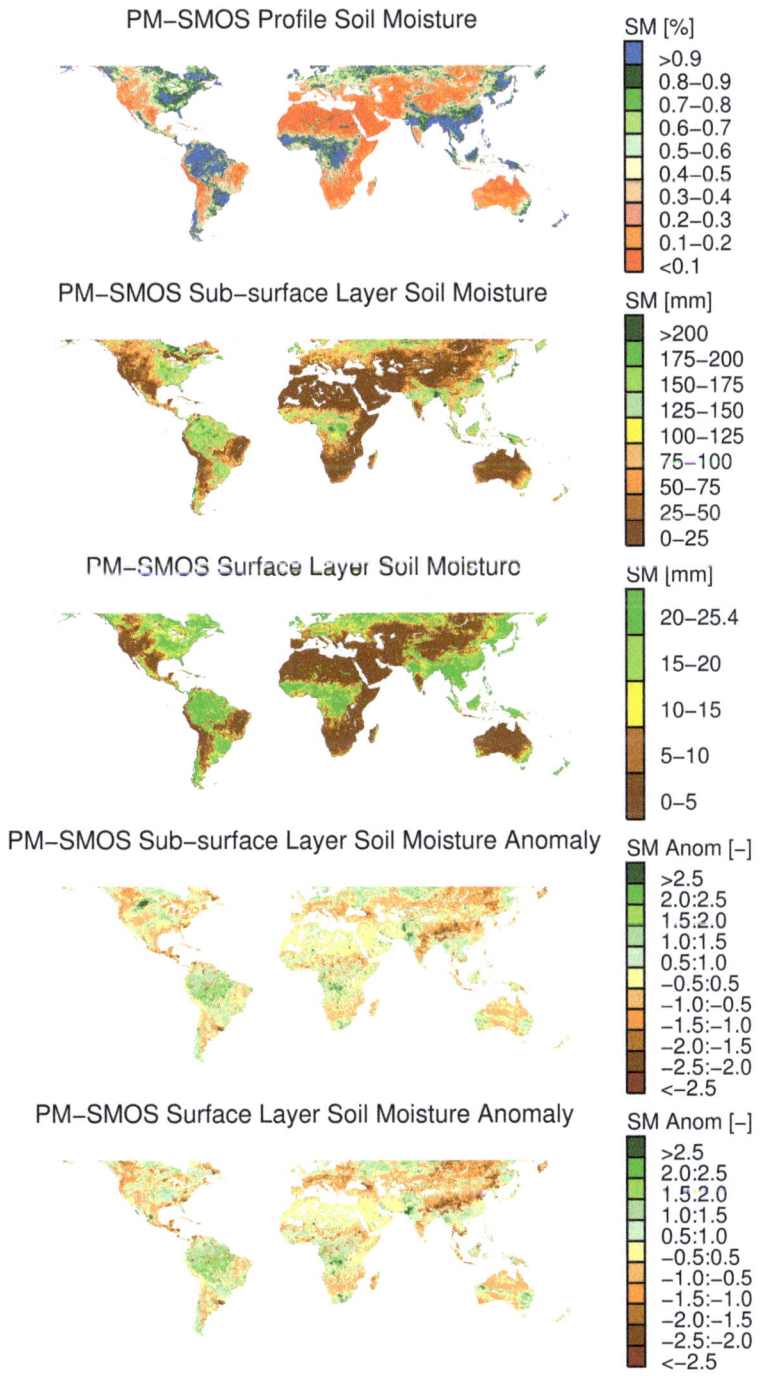

Fig. 8 Examples of the global soil moisture products being operationally delivered to

monthly USDA World Agricultural Supply and Demand Estimate (WASDE) report. Furthermore, since its public release, several additional agencies have expressed interest in utilizing the data including USDA National Agricultural Statistics Service (NASS), which is responsible for assessing the nation's crop conditions and developing yield estimates for the WASDE report as well.

Lessons Learned. One of the keys to success of this project is that it is leveraging off of an existing partnership between NASA and USDA within a user decision context. By addressing specific elements of a partner's existing process for assessing crop productivity, this project is able to provide a quantifiable benefit with its primary focus on enabling improved yield forecasting products, capacity building activities, and a more extensive process to define which decision to target and understand user requirements and constraints.

6 Conclusion

The case studies in this chapter represent a diversity of application activities working to bridge the gap between Earth observations for research and water resource applications needs on an international landscape. They each present different perspectives and approaches with respect to geography, barriers-faced, partnerships, capacity building, and the application targets themselves. Summarized below are the common threads and lessons learned from the case studies.

- **Partnerships are the key to addressing water resource issues, particularly in transboundary water basins**. Most of the projects found success when a common understanding was achieved between the NASA funded investigators and end users. However, as Zaitchik indicated, one key element of success is understanding what types of partners are the best match for the maturity of the application (e.g., research institutions versus a regional coordinating body). Zaitchik also notes that identifying a partner with high potential for being able to sustain the use of a transitioned product is also important, and that this must be weighed against what may appear to be a broader impact (e.g., targeting multiple states through a multilateral organization versus working with specific country ministries.)
- **Capacity building is essential and needs to be a continuous process.** In the case of Serrat-Capdevila's work, and Iqbal's case study, we observe that continuous engagements with partners are critical, and that single, one-off activities such as workshops can have limited impact. In some cases, as pointed out by Iqbal, the next level in capacity building is to develop joint activities that allow members of respective partner organizations to be rooted within each other's teams. Hosting members of the partners' teams within home institutions can help team members develop a good understanding of constraints (both technically and otherwise) and allow continuity in training and access to technical materials and expertise.

- **Potential is huge for Earth observations to benefit end users and to address global or regional water challenges.** In Bolten's work, NASA soil moisture products can greatly benefit USDA FAS, which is employing a global scale model for crop assessment. By targeting an existing element (soil moisture) of an existing process, partner buy-in becomes less of a hurdle since the partner's decision making framework is not being overhauled by a new tool. The potential for Earth observations to provide value to water resources stakeholders, such as research entities or organizations or operational agencies, has also been demonstrated in Serrat-Capdevila, Zaitchik, and Iqbal's work, and targeting various hydrologic variables, such as precipitation, streamflow, evapotranspiration, and groundwater.

Finally, it is also very important to effectively assess and communicate the benefits of this work, and connect those benefits back to the use of Earth observations. Continuing to evaluate and articulate benefits of Earth observations is a vital challenge that is being addressed by the case studies outlined in this chapter and an ongoing focus of the Earth science applications community.

References

Adegoke, J. O., & Carleton, A. M. (2002). Relations between soil moisture and satellite vegetation indices in the U.S. Corn Belt. *Journal of Hydrometeorology*, *3*(4), 395–405. http://doi.org/10.1175/1525-7541(2002)003<0395:RBSMAS>2.0.CO;2.

Anderson, M. C., Allen, R. G., Morse, A., & Kustas, W. P. (2012). Use of Landsat thermal imagery in monitoring evapotranspiration and managing water resources. *Remote Sensing of Environment*, *122*, 50–65. http://doi.org/10.1016/j.rse.2011.08.025.

Anderson, W. B., Zaitchik, B. F., Hain, C. R., Anderson, M. C., Yilmaz, M. T., Mecikalski, J., & Schultz, L. (2012). Towards an integrated soil moisture drought monitor for East Africa. *Hydrology and Earth System Sciences*, *16*(8), 2893–2913. http://doi.org/10.5194/hess-16-2893-2012.

Barnett, T. P., Adam, J. C., & Lettenmaier, D. P. (2005). Potential impacts of a warming climate on water availability in snow-dominated regions. *Nature*, *438*(7066), 303–9. http://doi.org/10.1038/nature04141.

Berhane, F., Zaitchik, B., & Dezfuli, A. (2014). Subseasonal analysis of precipitation variability in the Blue Nile River Basin. *Journal of Climate*, *27*(1), 325–344. http://doi.org/10.1175/JCLI-D-13-00094.1.

Bolten, J. D., & Crow, W. T. (2012). Improved prediction of quasi-global vegetation conditions using remotely-sensed surface soil moisture. *Geophysical Research Letters*, *39*(19), n/a–n/a. http://doi.org/10.1029/2012GL053470.

Bolten, J. D., Crow, W. T., Zhan, X., Jackson, T. J., & Reynolds, C. A. (2010). Evaluating the utility of remotely sensed soil moisture retrievals for operational agricultural drought monitoring. *IEEE Journal of Selected Topics in Applied Earth Observations and Remote Sensing*, *3*(1), 57–66. http://doi.org/10.1109/JSTARS.2009.2037163.

Bolten, J. D., Crow, W. T., Zhan, X., Reynolds, C. A., & Jackson, T. J. (2002). Assimilation of a satellite-based soilmoisture product into a two-layer water balance model for a global crop production decision support system. In S. K. Park & L. Xu (Eds.), *Data assimilation for atmospheric, oceanic and hydrologic applications* (pp. 449–463). Berlin, Heidelberg: Springer Berlin Heidelberg. http://doi.org/10.1007/978-3-540-71056-1.

Chen, J., Famigliett, J. S., Scanlon, B. R., & Rodell, M. (2015). Groundwater storage changes: Present status from GRACE observations. *Surveys in Geophysics.* http://doi.org/10.1007/s10712-015-9332-4.

Dai, A., Trenberth, K. E., & Qian, T. (2004). A global dataset of palmer drought severity index for 1870–2002: Relationship with soil moisture and effects of surface warming. *Journal of Hydrometeorology, 5*(6), 1117–1130. http://doi.org/10.1175/JHM-386.1.

Engman, E. T., & Gurney, R. J. (1991). *Remote sensing in hydrology.* Chapman and Hall. Retrieved from http://www.springer.com/us/book/9789401066709.

Entekhabi, D., Njoku, E. G., O'Neill, P. E., Kellogg, K. H., Crow, W. T., Edelstein, W. N., Van Zyl, J. (2010). The Soil Moisture Active Passive (SMAP) Mission. *Proceedings of the IEEE, 98*(5), 704–716. http://doi.org/10.1109/JPROC.2010.2043918.

Feng, X., Porporato, A., & Rodriguez-Iturbe, I. (2013). Changes in rainfall seasonality in the tropics. *Nature Climate Change, 3*(9), 811–815. http://doi.org/10.1038/nclimate1907.

Foltz, J., Gars, J., Özdoğan, M., Simane, B., & Zaitchik, B. (2013). Weather and welfare in Ethiopia. *Agricultural and Applied Economics Association.*

GEOSS. (2014). GEOSS water strategy: From observations to decisions. Retrieved April 7, 2015, from http://ceos.org/document_management/Ad_Hoc_Teams/WSIST/WSIST_GEOSS-Water-Strategy-Full-Report_Jan2014.pdf.

Ji, L., & Peters, A. J. (2003). Assessing vegetation response to drought in the Northern great plains using vegetation and drought indices. *Remote Sensing of Environment, 87*(1), 85–98. http://doi.org/10.1016/S0034-4257(03)00174-3.

Jin, S., & Feng, G. (2013). Large-scale variations of global groundwater from satellite gravimetry and hydrological models, 2002–2012. *Global and Planetary Change, 106*, 20–30. http://doi.org/10.1016/j.gloplacha.2013.02.008.

Kerr, Y. H., & Levine, D. (2008). Foreword to the Special Issue on the Soil Moisture and Ocean Salinity (SMOS) Mission. *IEEE Transactions on Geoscience and Remote Sensing, 46*(3), 583–585. http://doi.org/10.1109/TGRS.2008.917807.

Komma, J., Blöschl, G., & Reszler, C. (2008). Soil moisture updating by Ensemble Kalman Filtering in real-time flood forecasting. *Journal of Hydrology, 357*(3–4), 228–242. http://doi.org/10.1016/j.jhydrol.2008.05.020.

McClain, M. E., Subalusky, A. L., Anderson, E. P., Dessu, S. B., Melesse, A. M., Ndomba, P. M., Mligo, C. (2014). Comparing flow regime, channel hydraulics, and biological communities to infer flow–ecology relationships in the Mara River of Kenya and Tanzania. *Hydrological Sciences Journal, 59*(3–4), 801–819. http://doi.org/10.1080/02626667.2013.853121.

Mo, K. C., Long, L. N., Xia, Y., Yang, S. K., Schemm, J. E., & Ek, M. (2011). Drought indices based on the climate forecast system reanalysis and ensemble NLDAS. *Journal of Hydrometeorology, 12*(2), 181–205. http://doi.org/10.1175/2010JHM1310.1.

National Research Council. (2007). *Earth science and applications from space: National imperatives for the next decade and beyond.* doi:10.17226/11820

National Drought Mitigation Center, US Department of Agriculture, US Department of Commerce, & National Oceanic and Atmospheric Administration. (n.d.). *United States Drought Monitor.* Retrieved from http://droughtmonitor.unl.edu/.

Norbiato, D., Borga, M., Degli Esposti, S., Gaume, E., & Anquetin, S. (2008). Flash flood warning based on rainfall thresholds and soil moisture conditions: An assessment for gauged and ungauged basins. *Journal of Hydrology, 362*(3–4), 274–290. http://doi.org/10.1016/j.jhydrol.2008.08.023.

Palmer, W. C. (1965). Meteorological Drought. *U.S. Weather Bureau Research Paper 45.*

Parker, D. D., & Zilberman, D. (1996). The use of information services: The case of CIMIS. *Agribusiness, 12*(3), 209–218. http://doi.org/10.1002/(SICI)1520-6297(199605/06)12:3<209::AID-AGR2>3.0.CO;2-4.

Reichle, R. H. (2005). Global assimilation of satellite surface soil moisture retrievals into the NASA Catchment land surface model. *Geophysical Research Letters, 32*(2), L02404. http://doi.org/10.1029/2004GL021700.

Rodell, M., Velicogna, I., & Famiglietti, J. S. (2009). Satellite-based estimates of groundwater depletion in India. *Nature*, *460*(7258), 999–1002. http://doi.org/10.1038/nature08238.

Roy, T., Serrat-Capdevila, A., Gupta, H., & Valdes, J. (2016). A platform for probabilistic multi-model and multi-product streamflow forecasting. Submitted to Water Resources Research.

Satti, S., Zaitchik, B., & Siddiqui, S. (2015). The question of Sudan: A hydro-economic optimization model for the Sudanese Blue Nile. *Hydrology and Earth System Sciences*, *19*(5), 2275–2293. http://doi.org/10.5194/hess-19-2275-2015.

Shortridge, J. E., Guikema, S. D., & Zaitchik, B. F. (2015). Empirical streamflow simulation for water resource management in data-scarce seasonal watersheds. *Hydrology and Earth System Sciences Discussions*, *12*(10), 11083–11127. http://doi.org/10.5194/hessd-12-11083-2015.

Simane, B., Zaitchik, B., & Ozdogan, M. (2013). Agroecosystem analysis of the choke mountain watersheds, Ethiopia. *Sustainability*, *5*(2), 592–616. http://doi.org/10.3390/su5020592.

UN Water/Africa. (2015). The Africa Water Vision for 2025: Equitable and Sustainable Use of Water for Socioeconomic Development.

United Nations. (2015). *Transforming Our World: The 2030 Agenda for Sustainable Development (SDG#6: Clean Water and Sanitation)*.

Wilusz, D., Zaitchik, B., Anderson, M., Hain, C., Yilmaz, M., & Mladenova, I. (n.d.). Monthly monitoring of flooded area in the sudd wetland using low resolution SAR imagery from 2007–2011. *In Review*.

Yilmaz, M. T., Anderson, M. C., Zaitchik, B., Hain, C. R., Crow, W. T., Ozdogan, M., Evans, J. (2014). Comparison of prognostic and diagnostic surface flux modeling approaches over the Nile River Basin. *Water Resources Research*, *50*(1), 386–408. http://doi.org/10.1002/2013WR014194.

Zaitchik, B. F., Simane, B., Habib, S., Anderson, M. C., Ozdogan, M., & Foltz, J. D. (2012). Building climate resilience in the Blue Nile/Abay Highlands: A role for Earth system sciences. *International Journal of Environmental Research and Public Health*, *9*(2), 435–61. http://doi.org/10.3390/ijerph9020435.

Use of Remotely Sensed Climate and Environmental Information for Air Quality and Public Health Applications

William Crosson, Ali Akanda, Pietro Ceccato, Sue M. Estes,
John A. Haynes, David Saah, Thomas Buchholz, Yu-Shuo Chang,
Stephen Connor, Tufa Dinku, Travis Freed, John Gunn,
Andrew Kruczkiewicz, Jerrod Lessel, Jason Moghaddas,
Tadashi Moody, Gary Roller, David Schmidt, Bruce Springsteen,
Alexandra Sweeney and Madeleine C. Thomson

Abstract Earth's environment has direct and dramatic effects on its inhabitants in the realms of health and air quality. The climate, even in an unaltered state, poses great challenges but also presents great opportunity for the mankind to survive and flourish. Anthropogenic factors lead to even greater stress on the global ecosystem and to mankind, particularly with respect to air quality and the concomitant health

---

W. Crosson (✉)
Universities Space Research Association, Marshall Space Flight Center, Huntsville, AL, USA
e-mail: Bill.crosson@nasa.gov

A. Akanda
Civil and Environmental Engineering Department, University of Rhode Island, Kingston, RI, USA

P. Ceccato · T. Dinku · A. Kruczkiewicz · J. Lessel · A. Sweeney · M.C. Thomson
International Research Institute for Climate and Society, The Earth Institute, Columbia University, New York, USA

S.M. Estes
University of Alabama in Huntsville, Huntsville, AL, USA

J.A. Haynes
National Aeronautics and Space Administration, Washington, DC, USA

D. Saah · T. Buchholz · T. Freed · J. Moghaddas · T. Moody · G. Roller · D. Schmidt
Spatial Informatics Group, LLC, 3248 Northampton Ct., Pleasanton, CA, USA

D. Saah
University of San Francisco, Geospatial Analysis Lab, 2130 Fulton Street, San Francisco, CA, USA

T. Buchholz
Gund Institute for Ecological Economics, University of Vermont, 617 Main Street, Burlington, VT, USA

Y.-S. Chang · B. Springsteen
Placer County Air Pollution Control District, 110 Maple Street, Auburn, CA, USA

issues. While the use of remote sensing technology to address issues is in its infancy, there is tremendous potential for using remote sensing as part of systems that monitor and forecast conditions that directly or indirectly affect health and air quality. This chapter discusses current status and future prospects in this field and presents three case studies showing the great value of remote sensing assets in distinct disciplines.

Glossary—List of Acronyms

AQAST	Air Quality Applied Sciences Team
AMSR-E	Advanced Microwave Scanning Radiometer—Earth Observing System
CALIPSO	Cloud-Aerosol Lidar and Infrared Pathfinder Satellite Observations
CDC	Centers for Disease Control and Prevention
CHIRPS	Climate Hazards Group InfraRed Precipitation with Station
CMAP	Climate Prediction Center (CPC) Merged Analysis of Precipitation
CMORPH	CPC MORPHing technique
DoD	Department of Defense
ENACTS	Enhancing National Climate Services
EO	Earth Observation
EPA	Environmental Protection Agency
EVI	Enhanced Vegetation Index
GEO-CAPE	Geostationary Coastal and Air Pollution Events
GHG	Greenhouse Gasses
GPCP	Global Precipitation Climatology Project
GIMMS	Global Inventory Monitoring and Modeling Studies
GPM	Global Precipitation Measurement
GRACE	Gravity Recovery and Climate Experiment
GTS	Global Telecommunication System
HAQ	Health and Air Quality
HyspIRI	Hyperspectral Infrared Imager
LST	Land Surface Temperature
MODIS	Moderate Resolution Imaging Spectroradiometer
MOPITT	Measurement of Pollution in the Troposphere
NAAQS	National Ambient Air Quality Standard
NDVI	Normalized Difference Vegetation Index
NMS	National Meteorological Service
NOAA	National Oceanic and Atmospheric Administration

S. Connor
School of Environmental Sciences University of Liverpool, Liverpool, UK

J. Gunn
Spatial Informatics Group—Natural Assets Laboratory (SIG-NAL), 11 Pond Shore Drive, Cumberland, ME, USA

PACE	Pre-Aerosol, Clouds, and ocean Ecosystems
$PM_{2.5}$	Particulate Matter smaller than 2.5 microns in diameter
SeaWiFS	Sea-Viewing Wide Field of View Sensor
SMAP	Soil Moisture Active/Passive
SMOS	Soil Moisture and Ocean Salinity
SWOT	Surface Water and Ocean Topography
TEMPO	Tropospheric Emissions: Monitoring of Pollution
TRMM	Tropical Rainfall Measuring Mission
VCAP	Vectorial CAPacity model
VL	Visceral Leishmaniasis
VIIRS	Visible Infrared Imager Radiometer Suite
VOC	Volatile Organic Compounds
WHO	World Health Organization

1 Introduction

The 2007 Earth Science Decadal Survey (NRC 2007), in its vision for the future of Earth science and applications, specifically called for "a vision that includes advances in fundamental understanding of the Earth system and increased application of this understanding to serve the nation and the people of the world". The survey goes on to emphasize the role of remote sensing for improving air quality and health conditions: "Further improvements in the application of remote sensing technologies will allow better understanding of disease risk and prediction of disease outbreaks, more rapid detection of environmental changes that affect human health, identification of spatial variability in environmental health risk, targeted interventions to reduce vulnerability to health risks, and enhanced knowledge of human health-environment interactions." Finally, the Survey issues the challenge facing the Earth science community: "Addressing these societal challenges requires that we confront key scientific questions related to … transcontinental air pollution … impacts of climate change on human health, and the occurrence of extreme events, such as severe storms, heat waves, …"

Recognizing the great potential of remote sensing assets for advancing knowledge and applications in air quality and health, in 2001 NASA's Applied Sciences Program defined health and air quality (HAQ) as one of its applications areas to support the use of Earth observations, particularly regarding infectious and vector-borne diseases and environmental health issues. The area also addresses issues of toxic and pathogenic exposure and health-related hazards and their effects for risk characterization and mitigation. HAQ promotes uses of Earth observing data and models regarding implementation of air quality standards, policy, and regulations for economic and human welfare. The area also addresses effects of climate change on public health and air quality to support managers and policy makers in planning and preparations.

Specifically within air quality the key question being addressed by HAQ is:

How will continuing economic development affect the production of air pollutants, and how will these pollutants be transported across oceans and continents?

The HAQ area developed the Air Quality Applied Sciences Team (AQAST) to serve the needs of US air quality management (pollution monitoring and forecasting, quantifying emissions, understanding and monitoring pollutant transport, and understanding climate-air quality interactions) through the use of Earth science satellite observations, models, and latest scientific knowledge. AQAST members complete long-term applications projects but also support short term, quick response projects through 'Tiger Teams'.

The range of Earth observing systems used for air quality applications involves Terra/Aqua Moderate resolution Imaging Spectroradiometer (MODIS: land surface); Landsat (land surface); Tropical Rainfall Measuring Mission/Global Precipitation Measurement (TRMM/GPM: Precipitation); Soil Moisture Active/Passive and Soil Moisture and Ocean Salinity (SMAP/SMOS: soil moisture); and Aura (ozone, air quality and climate). While Centers for Disease Control and Prevention (CDC) represents the key stakeholder agency of the HAQ program, interactions also occur with National Oceanic and Atmospheric Administration (NOAA), Environmental Protection Agency (EPA), Department of Defense (DoD), US Global Change Research Program, nonprofit and for-profit sectors and GEO Health and Environment Community of Practice.

The major challenges related to the use of remote sensing data for HAQ applications include:

- Determining types and vertical distribution of aerosols and particulates;
- Lack of data due to clouds when using optical data;
- Determining human exposure to pollutants at fine spatial scales;
- Availability of real-time air quality and other environmental data for use in warning systems.

The focus of the health program is on understanding the relationships between climate change and human health and on improving predictability of health epidemics. Specific areas of concern are how the following will change with climate:

- Heat-related illnesses and deaths, including cardiovascular disease
- Health impacts of extreme weather events (injuries, mental health issues)
- Air pollution-related health effects (asthma and cardiovascular disease)
- Water- and food-borne diseases and illnesses
- Vector- and rodent-borne diseases and illnesses (malaria, dengue, chikungunya, West Nile, etc.)

1.1 Future Prospects

After several decades of research and development to create the capabilities to control diseases and monitor air quality using remote sensing technologies, the pieces are falling into place to support global implementation of such technologies. Comprehensive and integrated early warning systems are rapidly improving and are available to minimize the impact of deadly diseases and environmental hazards such as floods and extreme heat (Hossain 2015), and the barriers to implementation, namely cost and data management capabilities, are being torn down. Data and good intentions alone, however, are not sufficient. Capacity-building efforts are needed in many countries to improve technology transfer, and to better structure national information systems and decision-making processes, if these nations are to derive full benefit from this powerful technology.

In spite of these major advances, the global scientific community is only beginning to scratch the surface of potential societal applications of remote sensing data. Based on current plans of NASA and other national space agencies, Earth observation data will be abundant in the coming decades, and much of the data will be well suited for applications in health and air quality. In order to bring the benefits of these data to society, a well-designed strategy for effective use of these data to monitor, forecast, and warn the public is critical. This strategy must engage multiple factions within the Earth science community—remote sensing scientists for developing useful products, developers to create analysis and visualization tools, and social scientists, educators and public officials to devise effective messaging to convey critical information to the public in a way that promotes action.

There are several existing sensors and missions that are well-suited to provide niche data sets for health and air quality applications, such as:

- Measurement of Pollution in the Troposphere (MOPITT, http://terra.nasa.gov/about/terra-instruments/mopitt)
- Cloud-Aerosol Lidar and Infrared Pathfinder Satellite Observations (CALIPSO, http://science.nasa.gov/missions/calipso/)
- SMAP (http://science.nasa.gov/missions/smap/)
- SMOS (http://www.esa.int/Our_Activities/Observing_the_Earth/SMOS)
- Aura (http://science.nasa.gov/missions/aura/).

Even more useful inputs to HAQ monitoring and forecasting applications will be provided by sensors that are still in the pre-formulation phase, including:

- Pre-Aerosol, Clouds, and Ocean Ecosystems (PACE; http://decadal.gsfc.nasa.gov/PACE.html);
- Tropospheric Emissions: Monitoring of Pollution (TEMPO; http://science.nasa.gov/missions/tempo/);
- Hyperspectral Infrared Imager (HyspIRI; http://hyspiri.jpl.nasa.gov/);
- Geostationary Coastal and Air Pollution Events (GEO-CAPE; http://geo-cape.larc.nasa.gov/)

Other satellite and sub-orbital sensors, not yet to the pre-formulation phase, may be key elements of a future ($\sim$2030) HAQ system. For example, nanosatellites, including CubeSats (http://www.nasa.gov/mission_pages/cubesats/index.html) flying in constellations may play an important role in a future HAQ observational system. Tropospheric and stratospheric environmental observations from small unmanned aerial vehicles may be commonplace within this time frame. Also, crowd sourced environmental measurements and health outcomes, the 'complementary system of observations of human activities' recommended in the Decadal Survey (NRC 2007), may be extremely useful for calibration and validation of HAQ models, but the path to bring these into the mainstream is not currently clear.

2 Case Studies

2.1 Wildfires and Public Health

With as much as 40 % of terrestrial carbon in the U.S. being stored in forests (Pacala et al. 2007), sequestration of greenhouse gasses (GHGs) in terrestrial biomass has become an important element in strategies aimed at mitigating global climate change. Carbon dioxide emissions from wildfire in the U.S. have been estimated to be equivalent to approximately 4–6 % of anthropogenic sources at the continental scale (Wiedinmyer and Neff 2007).

The Western U.S. has millions of acres of overstocked forestlands at risk of large, uncharacteristically severe wildfire. There are a variety of factors contributing to this risk, including human-induced changes from nearly a century of timber harvest, grazing, and particularly, fire exclusion (North et al. 2015; Miller et al. 2009). For instance, the mid-elevation conifer forests of the Sierra Nevada, California, contain vast areas of high density, relatively homogenous, second growth coniferous forests that are increasingly prone to high-severity fires (Miller et al. 2009). Historic factors contributing to these conditions include the elimination of burning by Native Americans during the mid- to late nineteenth century (Anderson 2005), removal of large trees through the early twentieth century via railroad and selective logging (Stephens 2000), a nearly 100-year policy of fire exclusion (North et al. 2015; Stephens and Ruth 2005), and extensive use of clear-cut harvesting and overstory removals on public lands through the 1980s (Hirt 1996).

State and federal agencies have been very effective at suppressing the vast majority of wildfire ignitions (Dombeck et al. 2004), but the total area burned in the U.S. has increased in recent decades (Westerling et al. 2014). The US Forest Service (USFS) estimates that an average of more than 73,000 wildfires burn about 7.3 million acres each year of private, state, and federal land and more than 2,600 structures (USFS 2015) nationwide. Though the number of wildland fires decreased since 1960, the total acreage burned each year is trending upwards. Large fires covering over 100,000 acres are more frequent and cover more acreage now than in the 1990s (NIFC 2015). Trends of both increased fire sizes and uncharacteristically severe burning have been demonstrated throughout the western U.S. (Miller et al. 2009) and increasing fire sizes are expected to continue under changing climates (Westerling et al. 2006). For decades, scientists and managers have understood that the threat fire would pose to forests in this condition (Biswell 1959, 1989). In the western U.S., however, it was not until the early 1990s that federal land management agencies were given direction to manipulate stands, with the specific objective of modifying landscape level fire behavior.

Fire is a natural disturbance process that has an important role in determining forest species composition, structure, and stand development pathways. Fire has been a regular occurrence in these forests for millennia and continues to be today, despite the changes in spatial and temporal patterns over the last century (Biswell 1959; Kilgore and Taylor 1979; Collins et al. 2011). Most low to moderate intensity fires in western U.S. forests historically included some patches of high severity fire (Arno et al. 2000; Fulé et al. 2003; Beaty and Taylor 2008; Perry et al. 2011). Current wildfire high-severity patch sizes and areas in some forests that once burned frequently with low-moderate intensity fire regimes are well outside historical conditions (Miller et al. 2009).

Fire also drives large-scale transformations that can impede the ability of forests to deliver ecosystem services such as carbon sequestration (Millar and Stephenson 2015). In western U.S. forests, which include some of the globally highest densities of carbon storage, fires can drive the significant loss of carbon stocks from forests to the atmosphere (Gonzalez et al. 2015). High-severity fires also impair the future sequestration of carbon from likely shifts in ecosystem composition, shifting ecosystems from forest to grasslands and shrublands (Savage and Mast 2005; Roccaforte et al. 2012).

The purpose of this western US-based case study is to describe the significance and consequences of high-severity wildfires in terms of GHG emissions and air quality impacts. This chapter describes the current state of science in terms of management options as well as impediments toward lowering GHG emissions from wildfires. We also present a carbon offset methodological framework that accounts for GHG emission reductions from these treatments that could be used to reduce overall treatment costs.

2.1.1 Wildfire Emissions and Air Quality

Fire converts carbon stored in the forest floor materials and live and dead vegetation to atmospheric CO_2, CO, CH_4, other gasses, and particulate matter (Stephens et al. 2007). These other emissions include nitrogen oxides (NO_x), NH_3, SO_2, and fine particulates (Urbanski 2014). Along with the protection of people, property, and atmospheric GHG emissions, concerns of local air quality impacts from smoke have become another important goal of land management (Schweizer and Cisneros 2014).

Wildfires can have a significant adverse impact on regional air quality, depending on fuel load, terrain, and atmospheric conditions. During recent major wildfires, ambient air levels of particulate matter ($PM_{2.5}$) have been measured at over 10 times that of the U.S. EPA National Ambient Air Quality Standard (NAAQS) (MDEQ 2007; CARB 2009). Wildfires have also been directly responsible for exceedances of the ground level ozone NAAQS through generation of NOx and volatile organic compounds (VOC) precursors (SMAQMD 2011).

Air pollutants from wildfires are a well-established threat to public health, particularly increasing the risk for lung and heart disease and respiratory infections.

Exposure to air pollutant emissions from wildfires has been definitively linked to an increase in local hospital admissions, particularly children from ages 1–4 and the elderly over 65, for respiratory related conditions including eye, nose and throat irritation, chest pain, and asthma, bronchitis, pneumonia, and chronic obstructive pulmonary disease (Delfino et al. 2009).

2.1.2 Effects of Fuel Treatments

Effectiveness of Fuel Treatments

It is virtually impossible to exclude fire from most fire-prone landscapes, such as those found across the western U.S., over long periods of time (Reinhardt et al. 2008). During extreme weather conditions, suppression efforts can become overwhelmed and fires can quickly grow to cover very large areas. Suppression efforts can be rendered ineffective under less than extreme weather conditions when fuel and forest structure conditions result in relatively homogenous forests prone to high-intensity burning. This may be particularly true as the effects of climate change are manifested on fire severity trends (Millar et al. 2007; Miller et al. 2009). "No treatment" or "passive management" (Agee 2002; Stephens and Ruth 2005) perpetuates the potential for exacerbated fire behavior in forests (Moghaddas et al. 2010).

Modification of fuel structures and reduction of unnaturally high fuel loads in order to alter fire patterns and behavior are a primary component of planning efforts such as the National Cohesive Wildland Fire Management Strategy (Wildland Fire Leadership Council 2011) and the Sierra Nevada Forest Plan Amendment (USFS 2004). Various methods for fuel modification, collectively termed "fuel treatments," include shredding of understory biomass (mastication), removal of sub-merchantable small diameter trees and understory biomass (e.g., thinning from below), pre-commercial and commercial timber harvest (e.g., whole tree removal), and prescribed fire to remove surface fuels (shrub, grass, and down woody debris) and trees with low branches (ladder fuels). These treatments reduce or alter fire behavior, spatial patterns, effects on ecosystems and GHG emissions mainly by reducing the potential for crown fires and therefore fire severity (Moghaddas and Craggs 2007; Stephens et al. 2009a, b, c, 2012a; Moghaddas et al. 2010). These studies document treatment effects on fire behavior across several treatment types and provide guidance on designing treatments for forest stands.

Efficacy and effects of fuel treatments in real world situations has been demonstrated in real wildfire conditions (Moghaddas and Craggs 2007; Safford et al. 2009), but the majority of scientific evidence for their use comes from modeling efforts (Stephens and Moghaddas 2005; Stephens et al. 2009a; Moghaddas et al. 2010; Collins et al. 2011). Overall, there is clear consensus in the published literature that fuel treatments, specifically those that incorporate thinning from below and treat surface fuels with prescribed fire, reduce potential fire severity under a range of moderate to extreme weather conditions.

If climate change impacts continue even under the most conservative projections, it is likely that coniferous forests in the Sierra Nevada will experience longer fire seasons (Westerling et al. 2006) which are relatively drier and more conducive to high-intensity fire (Miller et al. 2009). To reduce this hazard, there is not one fuel treatment strategy. Rather, a combination of strategies is needed, especially when dealing with complex landscapes and management objectives.

Though now recognized as an important tool for fire protection and ecosystem process restoration, detailed strategies for application of the various techniques in different vegetation types at a landscape scale are still under study (Collins et al. 2010, 2011). Where the management of naturally ignited wildfires for resource benefit is a viable management option, further research is needed to support the integration of fuel treatments and beneficial fire use as a fuel management strategy. At a landscape scale, this is a difficult proposal and will likely challenge many of the current paradigms within management, policy, and regulatory frameworks (Germain et al. 2001; Collins et al. 2010).

Carbon Consequences of Fuel Treatments

Concerns over global climate change seemingly place fuel treatments that reduce near-term forest carbon stocks at odds with long-term carbon sequestration objectives in terrestrial vegetation as a means of climate change mitigation. Though wildfires also combust biomass and can be a significant source of atmospheric carbon emissions in the near-term (Randerson et al. 2006; Ager et al. 2010), they may also act as mechanisms for long-term carbon sequestration in some systems.

Several recent studies have investigated whether the potential future GHG emissions avoided through fuel treatment can offset immediate losses of stored carbon and carbon emitted during operations, and even possibly result in net positive carbon storage over longer time periods (Hurteau and North 2009, 2010; North et al. 2009; Stephens et al. 2009b, 2012a; Ager et al. 2010; Reinhardt and Holsinger 2010; Campbell et al. 2011; North and Hurteau 2011; Stephens et al. 2012a, b; Saah et al. 2012, 2015).

Carbon stored in forests is reduced in the short term through fuel treatments that are designed to reduce fire severity and smoke emissions. Other short term emissions include fossil fuel emissions from machinery used during treatments and processing of biomass and emissions from prescribed fires. While Campbell et al. (2011) found that more carbon is lost to treatments than what would be spared from loss by wildfire, particularly if fire is infrequent, others have concluded that in the long term, fuel treatments may result in overall increases in stored carbon, for example through reduced fire effects, carbon sequestration in large, fire resistant trees (Stephens et al. 2009a, b, c; Hurteau and North 2010), and in the form of durable wood products which might replace other fossil fuel intensive products (product substitution). Whether residues from fuel treatments are used to generate electricity or if it is burned on site in piles can also greatly affect GHG benefits as well as air pollutant balances (Jones et al. 2010; Lee et al. 2010; Springsteen et al. 2011, 2015).

Some carbon pools, such as soil organic matter, are relatively unchanged by wildfire regardless of the fire's severity. In a nationwide study of fire and fire surrogate treatment effects on carbon storage, Boerner et al. (2008) found the network-wide effects of these treatments on soil carbon modest and transient. For above ground carbon pools, Hurteau et al. (2008) pointed out that thinning can be thought of as increasing 'rotation length' by moving more forest carbon into longer residence-time storage. Thinning in Sierra Nevada mixed conifer leads to carbon storage in fewer, but larger, trees which are more representative of pre-settlement forest conditions. Hurteau and North (2009) concluded that the 1865 reconstruction stand structure, in which current stand density was reduced while large, fire resistant pines were retained, may be the best stand structure for achieving high carbon storage while minimizing potential wildfire emissions in fire-prone forests.

Whether or not fuels treatments safeguard enough carbon to offset their carbon cost depends on many factors including initial forest structure and carbon stocks, existing fuel loads, expected wildfire frequency and severity, fuel treatment type and intensity, and the fate of merchantable forest products. In general, overstory thinning plus prescribed fire removes more carbon than other common fuel treatment types while prescribed fire only or understory thinning only removes the least.

2.1.3 Carbon Offset Protocol for Avoided Wildfire Emissions

For western US forests, the paradox around carbon sequestration is that policies are in place that encourages carbon sequestration through afforestation, reforestation, and other silvicultural practices (CAR 2012). In contrast, the treatments that reduce carbon loss to wildfire are poorly acknowledged by current climate-sensitive policy. This is especially relevant when considering that long-term wildfire carbon emissions can be three times the size of their direct carbon emissions during the fire itself (Stephens et al. 2009a, b, c).

As market-based approaches to mitigating global climate change are being considered and implemented, one important emerging strategy for changing the economics of fuels treatments is to sell carbon emission offsets. Offsets are tradable certificates or permits representing the right to emit a designated amount of carbon dioxide or other GHGs. These offsets are generated when projects or actions reduce GHG emissions beyond what is required by permits and rules, and can be traded, leased, banked for future use, or sold to other entities that need to provide emission offsets (Sedjo and Marland 2003). In the case of fuel treatments, carbon emission offsets can theoretically be generated by projects that reduce potential emissions from wildfire, as by modifying the probability of extreme fire behavior for a given portion of land. In 2006, the California legislature enacted Assembly Bill 32: The Global Warming Solutions Act (AB32), setting emissions goals for 2020 and directing the Air Resources Board to develop reduction measures to meet targets (State of California 2006). Forest management (including fuel treatments) is one area that has been targeted for project-based offset development. The EPA and agencies implementing AB32 require that carbon emission offsets be quantifiable, real, permanent, enforceable, verifiable, and surplus.

Development of carbon emission offsets as an effective tool for forest and fire mangers, therefore, requires an integrated approach that considers wildfire probabilities and expected emissions, as well as net expected carbon sequestration or loss over time.

Fuels treatments involve tradeoffs between reducing the risk of carbon loss due to wildfires and increasing carbon emissions due to the fuels treatment themselves. Typically fuel treatments increase carbon emissions in the short term but in specific contexts can reduce long-term carbon emissions when wildfires are considered. A key issue is the probability of fire occurring after treatment implementation. Treatments that are not impacted by wildfire do not mitigate potential emissions, and can therefore be carbon sources as opposed to sinks (Campbell et al. 2011).

The Climate Action Reserve's Forest Project Protocol Version 3.3 (CAR 2012) provides forest owners with a platform to market carbon stored in their forest. By acknowledging the mitigation potential of wildfire-related carbon losses through fuel reduction treatments (Hurteau and North 2010), this protocol does not classify fuel reduction treatments as an immediate emission (Hurteau and North 2009). However, a clear framework that accounts for the net emission savings and endorsed by leading carbon market platforms such as the California Air Resources Board is missing to date. Necessary steps toward development of such a protocol are under way (e.g., Saah et al. 2012). Methodological development will require a consensus on how to approach the following framework elements: (1) delineate and characterize appropriate spatial scales; (2) define acceptable treatment and no treatment (baseline) forest management practices; (3) identification of appropriate forest growth and yield models; (4) definition of forest biomass removals life cycle assessment boundaries; (5) defining mechanisms for determining fuel treatment effectiveness and longevity; (6) quantification methods for determining direct wildfire emissions and indirect treatment effects of fuel treatments on fire within the greater landscape; (7) parameters for determining the probability of fire ignition and fire return interval; (8) standards for quantifying the risk of vegetation conversion after high-severity fires (e.g., forest converted permanently to brush land); and, (9) acceptable practice for prorating total short and long-term emissions.

2.2 Water-Borne Diseases

The World Health Organization (WHO) reports that over 3.5 million people die annually because of various water-related diseases (WHO/UNICEF 2014). Over a billion people in the world still lack access to safe water, and over two billion lack sufficient sanitation facilities. Due to the insufficient safe water and sanitation access in many parts of the world, and lack of knowledge and understanding of the microscale and macroscale processes and transmission pathways of many water-related diseases, the burden of water-borne diseases remains unacceptably high (Akanda et al. 2014).

As the world is rapidly urbanizing, most of the world's emerging megacities will be situated in coastal areas of Asia and Africa (Akanda and Hossain 2012). Many of these regions will experience significant shifts in regional climate patterns and coastal sea level rise. As a result, the relationship between climate, water, and health will become more intimate, existing vulnerabilities will worsen, and natural disasters such as droughts and floods will expose millions to displacement, unsafe living conditions, and water-borne diseases. There is great potential to channel the efforts and skillsets of the satellite remote sensing community for public health benefits to reduce the burden from water-related diseases by providing timely and efficient prediction of water-related disasters, assessment of population vulnerability in regions at risk, and strengthening prevention efforts with operational early warning systems, as well as, education and outreach.

For example, the ongoing seventh pandemic of cholera started in the 1960s and has affected over seven million people in over 50 countries. The WHO estimates that cholera affects over three million people and causes 250–300 thousand deaths every year. Despite the continuous progress on the research on the causative pathogen *V. cholerae* and the development of vaccines, there is limited understanding of the causative pathways and the role of large-scale physical processes and climate phenomena in propagating outbreaks across regions (Akanda et al. 2014). This case study analyzes how satellite remote sensing can be used in detecting appropriate large-scale processes behind endemic and epidemic cholera outbreaks in South Asia, where combination of droughts and floods and coastal ecological conditions create a favorable transmission environment.

The timing of seasonal cholera outbreaks can be anticipated reasonably well in endemic settings such as the Bengal delta region of South Asia or in the Lakes region in Sub-Saharan Africa (Jutla et al. 2015a, b). However, the potential magnitude or the location of outbreaks remains hard to anticipate, and the preemptive positioning of the appropriate level of human and material resources in anticipation of outbreaks have been especially difficult. Predictions based on the larger scale processes that have sufficient system memory can add significant 'lead-time' ahead of impending outbreaks for preemptive preparations. Such predictions can be valuable for endemic regions by adding the potential timing, intensity, and location of outbreaks; and especially for epidemic regions by identifying regions at risk of potential outbreaks and assessing the population vulnerability to cope with such disasters (Akanda et al. 2012).

The bacterium *V. cholerae* can survive, multiply, and proliferate in favorable aquatic and brackish estuarine environments (Colwell and Huq 2001). Primary outbreaks of cholera have been reported in coastal areas of South Asia, Africa, and South America over the last several decades (Huq and Colwell 1996; Griffith et al. 2006). Ever since the first correlative study relating cholera incidence with increased algae in the waters of rural South Asia, various studies have postulated connections between coastal cholera outbreaks and plankton abundance in the marine environment (Cockburn and Cassanos 1960; Huq and Colwell 1996).

The endemic and epidemic outbreaks in South Asia have been linked to a number of environmental and climatic variables, such as precipitation (Longini et al. 2002; Pascual et al. 2002; Hashizume et al. 2008), coastal phytoplankton abundance (Lobitz and Colwell 2000), floods (Koelle et al. 2005), water temperature (Colwell 1996), peak river level (Schwartz et al. 2006), and sea surface temperature (Cash et al. 2008). The link with environmental factors have also been studied for other cholera affected regions in the world, such as Southeast Asia (Emch et al. 2008), Sub-Saharan Africa (Hashizume et al. 2008), southern Africa (Bertuzzo et al. 2008), and South America (Gil et al. 2004).

Cholera incidence in the Bengal Delta region shows distinct seasonal and spatial variations in shape of single annual and biannual peaks, in sharp contrast to the typically seen single peaks in other regions (Akanda et al. 2009). The outbreaks in this region typically propagate from the coastal areas in spring to the inland areas in fall aided by the monsoon season in between. In sum, the transmission process exhibits two distinctly different transmission cycles, pre-monsoon and post-monsoon, influenced by coastal and terrestrial hydroclimatic processes, respectively, revealing a strong association of the space time variability of incidence peaks with seasonal processes and extreme hydroclimatic events (Akanda et al. 2011).

Application of remote sensing to study cholera dynamics has been an emerging research area with the availability of longer and more accurate datasets over the last few decades (Jutla et al. 2015b). A principal motivation for using satellite remote sensing in monitoring the environmental conditions conducive for the cholera pathogen stemmed from the fact that it shows strong association with coastal phytoplankton (Colwell 1996). Although, chlorophyll variations on a daily scale exhibited random variability with very limited memory (Uz and Yoder 2004), earlier studies by Lobitz and Colwell (2000) showed promise in monitoring the coastal growth environment for the bacteria with the help of chlorophyll measurements in coastal Bay of Bengal. In this endemic region, the seasonal freshwater scarcity from upstream regions in the flat deltaic landscape provides a vast growth environment during the dry spring months, which can be monitored by studying the space–time variability of chlorophyll in northern Bay of Bengal and thus linking coastal processes to predicting cholera outbreaks in the region (Akanda et al. 2013).

Jutla et al. (2012) quantified the space–time distribution of chlorophyll in the coastal Bay of Bengal region using data from SeaWiFS (Sea-Viewing Wide Field-of-View Sensor) using 10 years of data (2000–2009). A key finding of the study was that while the variability of chlorophyll at daily scale resembled white noise, chlorophyll values showed distinct seasonality at monthly scales and with increased spatial averaging. The first cholera outbreaks near the coastal areas in spring were found to be correlated with the northward movement of the plankton-rich seawater and increased salinity of the estuarine environment favoring increased growth and abundance of the cholera bacteria in river corridors. In a follow-up study, Jutla et al. (2013) showed that seasonal endemic cholera outbreaks in the Bengal Delta region can be predicted up to three months in advance with a prediction accuracy of over 75 % by using satellite-derived chlorophyll (as a surrogate for coastal plankton and pathogen growth) and air temperature (as a surrogate for upstream snowmelt and freshwater flow) data. A high prediction accuracy was achievable because the two seasonal peaks of cholera in spring and fall were predicted using two separate models representing distinctive seasonal environmental processes. The inter-annual variability of pre-monsoon cholera outbreaks in spring were satisfactorily linked with coastal plankton blooms, and the post-monsoon cholera outbreaks in fall were related to breakdown of sanitary conditions due to monsoon flooding and subsequent inundation in floodplain regions (Akanda et al. 2013; Jutla et al. 2013).

With more than a decade of terrestrial water storage data from the Gravity Recovery and Climate Experiment (GRACE) mission, conditions have recently emerged for predicting cholera occurrence with increased lead-times. The lead–lag relationships between terrestrial water storage in the Ganges–Brahmaputra–Meghna basin of South Asia and endemic cholera in Bangladesh were investigated in a study by Jutla et al. (2015a). Availability of data on water scarcity and abundance in large river basins, a prerequisite for developing cholera forecasting systems, are difficult to obtain. The ubiquitous use of river water for irrigation, sanitation, and consumption purposes in this heavily populated basin region exposes the riverine societies to the infectious disease (Akanda et al. 2013). Open mixing of water bodies and channels with $V.$ $cholerae$ leads to transmission through breakdown in sanitation in heavy monsoon rainfall, and inundated areas are enriched with bacteria already present in the ecosystem. According to Jutla et al. (2015a); water availability showed a strong asymmetrical association with cholera prevalence in spring ($\tau = -0.53$; $P < 0.001$) and in autumn ($\tau = 0.45$; $P < 0.001$) up to 6 months in advance. The study concluded that one unit (centimeter of water) decrease in water availability in the basin increased the odds of spring cholera prevalence above normal by 24 %, while an increase in regional water by one unit due to flooding increased odds of above average fall cholera by 29 %.

Satellite remote sensing data were also used successfully to capture the changes in hydroclimatic conditions related to the infamous cholera outbreak of 2008 in Zimbabwe. The first cases of cholera in Zimbabwe were reported to the World Health Organization in August 2008; but between then and June 2009, a massive outbreak erupted with a total of 98,522 cholera cases and 4,282 deaths (WHO/UNICEF 2009). The case fatality ratio in this epidemic, i.e., the ratio of deaths to total cases reported as a measure of the intensity of the epidemic, were found to be much higher than those reported in other countries or areas where appropriate treatment was available. In a recent study by Jutla et al. (2015b), large-scale hydroclimatic processes estimated using TRMM-based precipitation and gridded air temperature were linked with epidemiological data to assess risk of disease occurrence in a retrospective manner.

Although the precise location of the region of a disease outbreak could not be determined with the existing resolution of the TRMM data, a provincial analysis approach (averaging of all pixels in a particular province) showed strong correlation of precipitation and air temperature (Jutla et al. 2015b). The study postulates that if anomalous temperature in a vulnerable region is followed by heavy precipitation, the risk of a cholera outbreak increases significantly, especially in areas where the drinking water source and sanitation infrastructure are poorly maintained or unavailable. The empirical observations and relationships between satellite sensed precipitation and temperature for the 2008 Zimbabwe cholera outbreaks were expanded for five other countries (Mozambique, South Sudan, Rwanda, Central African Republic, and Cameroon) in the Sub-Saharan Africa region. The study showed that anomalous above average air temperatures, followed by above average precipitation at least 1 month in advance were correlated and statistically significant with the first reporting of cholera outbreaks in all sites.

The results demonstrate that satellite data measurements of climatic and environmental variables such as precipitation, temperature, terrestrial water storage, and coastal chlorophyll can be meaningful predictors over a range of space and time scales and can be effective in monitoring ecological conditions conducive for the pathogen *V. cholerae* and developing a cholera prediction model with several months lead time. Such an approach also shows the potential of newer missions with the increased ability to monitor hydrological and environmental conditions such as river elevation from Surface Water and Ocean Topography (SWOT) and soil moisture from SMAP. The findings and understandings derived from South Asia and Sub-Saharan Africa may serve as the basis for the development of "climate informed" early warning systems for preempting epidemic cholera outbreaks in vulnerable regions and prompting effective means for intervention in vulnerable regions.

2.3 Vector-Borne Diseases

Epidemics of vector-borne diseases still cause millions of deaths every year. Malaria remains a major global health problem, with an estimated three billion people at risk of infection in over 109 countries, 250 million cases annually and one million deaths. Malaria is highly sensitive to climate variations, thus climate information can either be used as a resource, for example in the development of early warning systems (DaSilva et al. 2004), or must be accounted for when estimating the impact of interventions (Aregawi et al. 2014).

In Ethiopia, the determinants of malaria transmission are diverse and localized (Yeshiwondim et al. 2009), but altitude (linked to temperature) is a major limiting factor in the highland plateau region, and rainfall in the semiarid areas. A devastating epidemic caused by unusual weather conditions was documented in 1958, affecting most of the central highlands between 1600 and 2150 m elevation with an estimated three million cases and 150,000 deaths (Fontaine et al. 1961). Subsequently, cyclic epidemics of various dimensions have been reported from other highland areas, with intervals of approximately 5–8 years with the last such epidemic occurring in 2003. Most of these epidemics have been attributed to climatic abnormalities, sometimes associated with El Nino, although other factors such as land-use change may also be important.

Endemic regions of Visceral leishmaniasis (VL) exist within East Africa with a geographic hotspot in the northern states of South Sudan. This region experiences seasonal fluctuations in cases that typically peak during the months of September through January (SONDJ) (WHO 2013; Gerstl et al. 2006; Seaman et al. 1996). In the northern states of South Sudan alone, VL epidemics have recently been observed with a reported 28,512 new cases from 2009 to 2012 (Ministry of Health —Republic of South Sudan 2013). Without proper treatment, mortality in South Sudan is high, with numbers approaching 100,000 during one multi-year epidemic in the late 1980s and early 1990s (Seaman et al. 1996).

The habitat of the VL vector P. orientalis is determined by specific ecological conditions including the presence of specific soils, woodlands, and mean maximum daily temperature (Thomson et al. 1999; Elnaiem et al. 1999; Elnaiem 2011). Research has also documented associations between environmental factors and VL that may contribute to outbreaks of the disease including: relative humidity (Salomon et al. 2012; Elnaiem et al. 1997), precipitation (Gebre-Michael et al. 2004; Hoogstraal and Heyneman 1969), and normalized difference vegetation index (NDVI) (Elnaiem et al. 1997; Rajesh and Sanjay 2013). Additionally, Ashford and Thomson (1991) suggested the possibility of a connection between a prolonged flooding event in the 1960s and the corresponding 10-year drop in VL within the northern states of South Sudan. Although the importance of environmental variables in relation to the transmission dynamics of VL has been established, the lack of in situ data within the study region has led to inconclusive results regarding these relationships. By exploiting the advantages of sustained and controlled Earth monitoring via NASA's Earth observations, the relationship between environmental factors and the spatiotemporal distribution of VL in the northern states has been shown (Sweeney et al. 2014), therefore showing how Earth observations can be used for mapping risks of leishmaniasis.

Recently, methodology has been developed which use remote sensing to monitor climate variability, environmental conditions, and their impacts on the dynamics of infectious diseases. These methods and results are described in the following sections.

2.3.1 Climate and Environmental Factors: How Do They Help?

To date much of the debate has centered on attribution of past changes in disease rates to climate change, and the use of scenario-based models to project future changes in risk for specific diseases. While these can give useful indications, the unavoidable uncertainty in such analyses, and contingency on other socioeconomic and public health determinants in the past or future, limit their utility as decision support tools. For operational health agencies, the most pressing need is the strengthening of current disease control efforts to bring down current disease rates and manage short-term climate risks, which will, in turn, increase resilience to long-term climate change. The WHO and partner agencies are working through a range of programs to (i) ensure political support and financial investment in preventive and curative interventions to reduce current disease burdens; (ii) promote a comprehensive approach to climate risk management; (iii) support applied research,

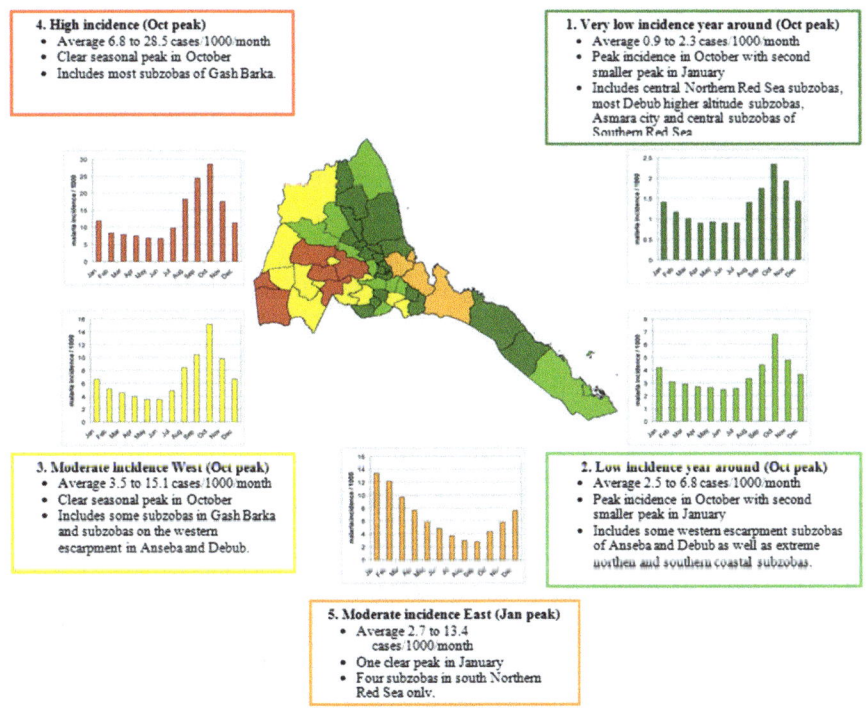

Fig. 1 A climate risk map showing the spatiotemporal stratification of malaria incidences in Eritrea at district level. *Source:* Ceccato et al. (2007)

through definition of global and regional research agendas, and targeted research initiatives on priority diseases and population groups (Campbell-Lendrum et al. 2015).

Risk Maps

A risk map can aid in understanding the relationship between the diseases and the climate. Figure 1 shows how a risk map was produced to understand the relationship between climate and malaria in Eritrea. This analysis indicates that malaria incidence exists across a rather sharp spatial gradient in Eritrea and that this variation is driven by climatic factors. It is clear that malaria is high in the western part of Eritrea and peaks in September–October immediately after the July–August rainy season. Conversely, in the highlands of Eritrea (central green part), malaria incidence is low because of low temperatures at the high altitudes. On the east coast, there are some areas where malaria peaks in January because of rainfall that occurs occasionally in December.

Early Warning System

The WHO has developed a framework for creating an early warning system for malaria (DaSilva et al. 2004), which can be used as a framework for floods, natural disasters, droughts, and other events that have a relationship with climate and environmental factors. The framework is composed by four components:

1. Vulnerability assessment
2. Climate forecasting
3. Monitoring of climate and environmental factors, including precipitation, temperature, presence of vegetation, or water bodies that influence mosquito development
4. Case surveillance

2.3.2 Accessing Quality Data Through Earth Observations

Decision-makers and researchers often face a lack of quality data required for optimal targeting the intervention and surveillance of vector-borne diseases. The decisions are critical as they impact on the lives of many people: "bad data create bad policies". Climate data and information—whether station- or satellite-generated—can increasingly be accessed freely online. Station data, of varying quality, can typically be obtained from a country's National Meteorological Service (NMS). However, station data are not always available. Some of the station data provided by the NMS are freely available through the Global Telecommunication System (GTS) but often lack of the required spatial resolution needed.

Satellites provide raw global data that are continuously archived. In many cases the raw data may be free, but not all interfaces allow free access to their archived data. Sources for satellite-generated climate data are varied, and a selection is provided below. The following are likely to be the most useful of the freely available satellite-based estimates.

Precipitation

- *Global Precipitation Climatology Project (GPCP)* combines satellite and station data. The monthly data extend from 1979 onwards. This product has a low (2.5°) spatial resolution, but is of interest when creating long time series to understand trends in past precipitation.
- *Climate Prediction Center (CPC) Merged Analysis of Precipitation (CMAP)*, similar to GPCP, combines satellite and station data to produce 5-day and monthly aggregations at a 2.5° spatial resolution.
- *CPC MORPHing technique (CMORPH)* provides global precipitation estimates at very high spatial (8 km) and temporal (30 min) resolution. This product is

suitable for real-time monitoring of rainfall, provided a long history is not required, as data are only available from January 1998.
- *Tropical Rainfall Measuring Mission (TRMM)* provides monthly estimates of precipitation in the tropics at 0.25° spatial resolution. They are available from January 1998, with a latency of about a month or more. The TRMM mission ended in 2015.
- *Global Precipitation Measurement (GPM)* provides estimates of precipitation globally. They are available from March 2014 to present at 0.1° spatial resolution. The GPM is an extension of the TRMM rain-sensing package.
- *Enhancing National Climate Services* (*ENACTS*) program combines all available rain gauge data from the National Meteorology Agencies of Ethiopia, Gambia, Ghana, Madagascar, Mali, Rwanda, Tanzania, and Zambia with satellite data for the last 30 years with high spatial resolution (4 km) and a 10-day temporal resolution. ENACTS is expected to expand other countries in Africa.
- *Climate Hazards Group InfraRed Precipitation with Station* (*CHIRPS*) data are produced using a similar technique to that of ENACTS. CHIRPS provides daily rainfall from 1981–present at a 0.05° resolution and covers 50°S–50°N. More details of CHIRPS are given in Sect. 2.3.3.

Temperature

In many parts of the world, the spatial distribution of weather stations is limited and the dissemination of temperature data is variable, therefore limiting their use for real-time applications. Compensation for this paucity of information may be obtained by using satellite-based methods. The estimation of near-surface air temperature (Ta) is useful for tracking vector-borne diseases. It affects the transmission of malaria (Ceccato et al. 2012) in the highlands of East Africa. Air temperature is commonly obtained from synoptic measurements from weather stations. However, the derivation of Ta from satellite-derived land surface temperature (Ts) is far from straightforward. Studies showed that it is possible to retrieve high-resolution Ta data from the MODIS Ts products (Vancutsem et al. 2010; Ceccato et al. 2010). Ts is available from MODIS dating to ∼2000 and from Visible Infrared Imager Radiometer Suite (VIIRS) since 2012 at a 1 km spatial resolution. Separate estimates for daytime and nighttime temperatures are available, from which daily maximum and minimum air temperature estimates can be derived.

Vegetation

Remote sensing can be used to distinguish vegetated areas from other surface covers. Various vegetative properties can be gleaned from indices such as the

NDVI, including, but not limited to, leaf area index, biomass, and greenness. Practitioners can access data on vegetation cover through the following sources:

- *Global NDVI* from legacy NASA sensors is available from NASA's Global Inventory Monitoring and Modeling Studies (GIMMS) for the period from 1981 to date. The dataset has been shown to be valid in representing vegetation patterns in certain regions (but not everywhere) and should be used with caution (Ceccato 2005).
- *MODIS NDVI* and enhanced vegetation index (EVI) are available for 16-day periods from April 2000 at 250 m resolution. The NDVI is an updated extension to the Global NDVI.

2.3.3 Improving Data Quality and Accessibility

To address the spatial and temporal gaps in climate data as well as the lack of quality-controlled data, approaches are being developed based on the idea of merging station data with satellite and modeled data. Some of these methods are developing platforms in which the now-quality-improved data can be accessed, manipulated and integrated into the programs of national-level stakeholders and international partners.

ENACTS Approach

The IRI has been working with national meteorological agencies in Africa (Ethiopia, Gambia, Ghana, Madagascar, Mali, Rwanda, Tanzania, and Zambia) to improve the availability, access, and use of climate information by national decision-makers and their international partners. The approach, ENACTS, is based on three pillars (Dinku et al. 2011, 2014a, b):

1. **Improving data availability**: Availability of climate data is improved by combining quality-controlled station data from the national observation network with satellite estimates for rainfall and elevation maps and reanalysis products for temperature. The final products are datasets with 30 or more years of rainfall and temperature for a 4 km grid across the country.
2. **Enhancing access to climate information**: Access to information products is enhanced by making information products available online.
3. **Promoting the use of climate information**: Use of climate information is promoted by engaging and collaborating directly with potential users.

By integrating ground-based observations with proxy satellite and modeled data, the ENACTS products and services overcome issues of data scarcity and poor quality, introducing quality-assessed and spatially complete data services into national meteorological agencies to serve stakeholder needs. One of the strengths of

ENACTS is that it harnesses all local observational data, incorporating high-definition information that globally produced or modeled products rarely access. The resulting spatially and temporally continuous data sets allow for the characterization of climate risks at a local scale. ENACTS enables analysis of climate data at multiple scales to enhance malaria control and elimination decisions.

CHIRPS Approach

A similar approach has been developed by CHIRPS, a near-global rainfall dataset covering 1981-present (http://pubs.usgs.gov/ds/832/pdf/ds832.pdf). CHIRPS incorporates 0.05° resolution satellite imagery with in situ station data to create gridded rainfall time series for trend analysis and seasonal drought monitoring. Two CHIRPS products are produced operationally: a rapid preliminary version, and a later final version. The preliminary product, which uses only a single station source, GTS, is available for the entire domain 2 days after the end of a pentad. The final CHIRPS product takes advantage of several other stations sources and is complete sometime after the 15th of the following month. Final monthly, dekad, pentad, and daily products are calculated at that time.

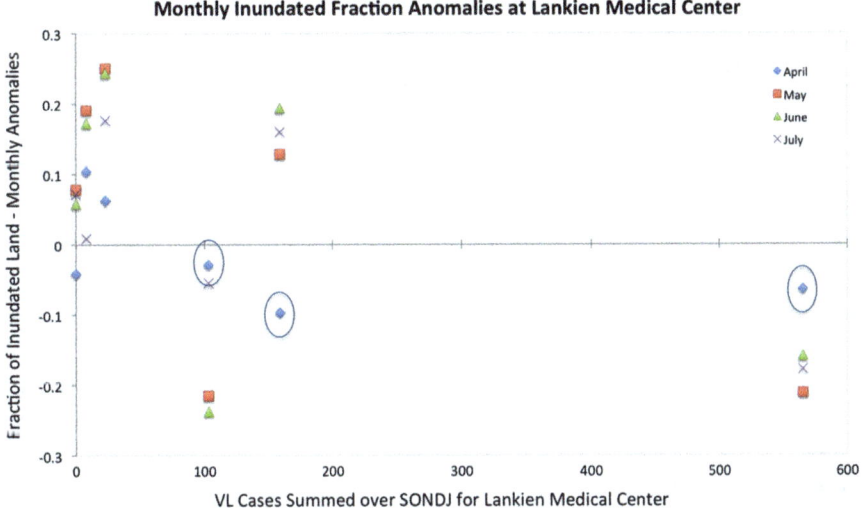

Fig. 2 Monthly inundation anomalies (AMJJ) and VL cases summed over SONDJ for Lankien Medical Center in Jonglei State

2.3.4 Examples of Analysis

Inundation Products for Leishmaniasis

Surface inundation datasets have been developed to examine inundation dynamics (McDonald et al. 2011) using the QuikSCAT and Advanced Microwave Scanning Radiometer—Earth Observing System (AMSR-E) active/passive microwave datasets over the period 2004–2009 for the South Sudan (Schroeder et al. 2010). Three environmental variables (NDVI, precipitation, and inundation) aided in determining whether wet or dry years were more conducive to the transmission of leishmaniasis.

Inundation during April–July also exhibited a strong inverse relationship with VL cases in SONDJ. Results are typified by the Lankien Medical Center analysis

Fig. 3 Landsat image over the Rift Valley in Ethiopia, represented with false color combination of the middle infrared, near-infrared and red. *Green* indicates vegetation, *blue* shows water bodies, and *pink–brown* shows bare soils. *Green dots* represent villages with low malaria transmission, *yellow* and *orange dots* represent villages with medium to medium high malaria transmission, and *red dots* represent villages with high malaria transmission

where below average inundation during April displays an inverse relationship with VL cases in the following SONDJ (Fig. 2).

Water Bodies Products for Malaria

Using Landsat images at 30 m resolution, it is possible to map small water bodies where mosquitoes will breed and transmit diseases such as malaria, dengue fever, chikungunya, West Nile Fever and Zika (Pekel et al. 2011). This is done by combining the middle infrared channel (sensitive to water absorption), the near-infrared channel (sensitive to bare soil and vegetation canopy), and the red channel (sensitive to chlorophyll absorption). In Fig. 3, malaria data are overlaid on a Landsat image for the Rift Valley in Ethiopia. Villages with low malaria transmission are located either in the highland areas where temperature is the limiting

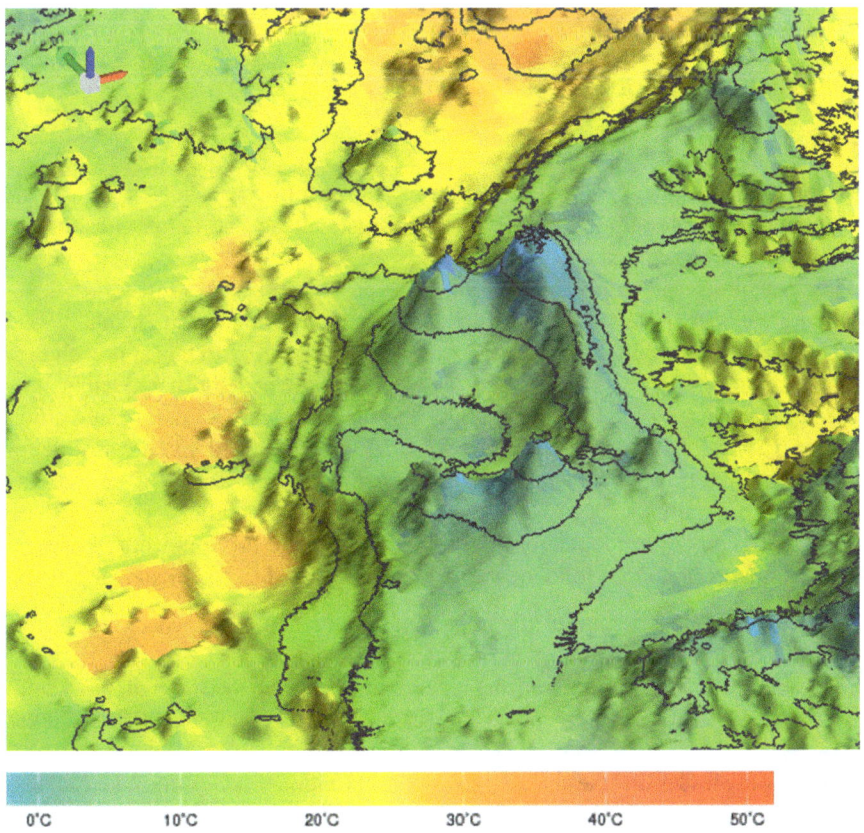

Fig. 4 MODIS nighttime LST showing difference of temperatures according to the elevation in the highlands of Ethiopia

factor or in the dry Rift Valley. Villages with high malaria transmission are located in the dry valley where green vegetation and water bodies are present.

Temperature for Malaria

Nighttime Aqua MODIS Land Surface Temperature (LST) in the highlands of Ethiopia close to the Rift Valley is overlaid in Fig. 4 with altitude lines showing that for a same altitude, temperature can be either favorable for mosquitoes breeding (>16 °C) or unfavorable (<16 °C). This information is important for the Ministry of Health to target control measure to fight malaria.

Time series of MODIS LST can show how the temperature varies in the highlands and therefore impacts the transmission of malaria. Integrating precipitation and temperature into a vectorial capacity model (VCAP), allows the Ministries of Health in Africa to assess the risk of malaria transmission in the epidemic zones of Africa (Ceccato et al. 2012).

3 Conclusion and Way Forward

The challenge issued by the 2007 Decadal Survey (NRC 2007): "Addressing these societal challenges requires that we confront key scientific questions related to … transcontinental air pollution … impacts of climate change on human health, and the occurrence of extreme events, such as severe storms, heat waves, …" frames a powerful approach for building a future in which Earth observation (EO) data are used to greatly improve quality of life globally through advances in air quality and human health monitoring, forecast, and warning systems. Key to this approach is to mobilize, through better EO data and associated analytical tools, modeling, and training, the capacity-building community that is reliant on EO data. The participants in a 2015 'NASA E2 Capacity Building Workshop' noted that it was time for the EO-based capacity-building community to broaden the focus of current EO application programs (such as NASA's Applied Sciences program) to tackle these critical issues through application of EO-based (orbital and non-orbital) science. The participants posed the following key questions related to health and air quality:

- How can we better adapt to the impact of climate change on changing vector- and water-borne disease burden on vulnerable populations?
- If capacity to build EO-based health monitoring improves around the world, how do we measure the societal impact in terms of quality of life and lives saved?
- How can we identify the most impactful intervention strategy for endemic and epidemic diseases in order to design EO-based decision-making tools?

- How can the use of small satellites, unmanned aerial systems, and crowd sourcing programs (citizen science) assist in building and improving more relevant health and air quality monitoring tools that currently use data from conventional satellites?
- What type of disease-relevant and region-specific EO tools is needed to empower the health community?
- Recognizing the inherent nexus between water and water-borne disease, how can we facilitate greater interaction between technical experts in the water and health communities?

The participants also made the following recommendations related to health and air quality:

- A greater focus is needed on understanding how EO systems can best address the impact of climate change on future disease burden.
- Recognizing the strong connections of water resources (availability) with water-borne diseases, water community technical experts that use EO systems and data should partner more effectively with the traditional health community.
- There needs to be greater investment in small satellites and citizen science programs (volunteered geographic information) for health monitoring.
- Programs need to be in place that facilitate clearer communication and trust building between the health stakeholder community and scientists who use EO data for capacity building of health institutions around the world.
- In an effort to build durable capacity of EO systems, space agencies and other national or global organizations should identify strategic partners from philanthropic and private sector organizations with overlapping priorities that rely on monitoring of environmental and Earth science data in their operations.

NASA's HAQ program has mapped a vision statement to define where it aspires to be in 2030 in terms of global capacity. The program will issue the following health 'grand challenges' to the community:

Malaria—Risk characterization models will be deployed regionally. A unified dynamic malaria risk model would be a major achievement for end users worldwide and would provide economic savings by scale and elimination of duplicative models.

Vector-borne diseases: The World Health Organization and other health officials will have access to global risk maps of infectious diseases associated with viral hemorrhagic fevers on a daily basis to aid health officials' preparedness.

Monitoring and forecasting Harmful Cyanobacterial Blooms (HABs): Cyanobacterial harmful algal blooms are a concern for water supplies due to their potential to produce toxins in lakes and estuaries. In cooperation with NOAA, new satellite-derived products will be developed for the issuance of daily HAB bulletins to pertinent US end users.

The air quality 'grand challenges' for 2030 are:

Accurate ground level aerosol and constituent and ozone measurements will routinely be obtained from a combination of remote sensing, in situ observations, and models. Total columnar aerosol and ozone estimates are routinely made from satellite sensors, but these have limited value for determining human exposure to harmful pollutants.

The HAQ program will have established strong relationships with federal, state, local, and international partners to identify unique applications of satellite observations and realize their operational use. These applications will provide critical components for integration with forecast models and decision support systems. NASA's participation in health and air quality applications research and related transition to operations activities currently performed with EPA, NOAA, CDC, and others will fill a significant niche in national capabilities and will be vital components in future domestic and international programs.

In other words, by 2030, the goal is for EO products developed under the HAQ program to be widespread, regularly accessed, and indispensable to the public and end users. The HAQ program will be known as the 'go to' program for information about vector-borne and infectious disease risks, environmental health risks, and dangerous air pollution episodes.

References

Agee, J. K. (2002). The fallacy of passive management. *Conservation in Practice, 3*(1), 18–26. doi:10.1111/j.1526-4629.2002.tb00023.x.

Ager, A. A., Vaillant, N. M., & Finney, M. A. (2010). A comparison of landscape fuel treatment strategies to mitigate wildland fire risk in the urban interface and preserve old forest structure. *Forest Ecology and Management, 259*, 1556–1570.

Akanda, A. S., & Hossain, F. (2012). The climate-water-health nexus in emerging megacities. *EOS Transactions, 93*(37), 353–354.

Akanda, A. S., Jutla, A. S., & Islam, S. (2009). Dual peak cholera transmission in Bengal Delta: A hydroclimatological explanation. *Geophysical Reseach Letters, 36*, L19401.

Akanda, A. S., Jutla, A. S., Alam, M., de Magny, G. C., Siddique, A. K., Sack. R. B., et al. (2011). Hydroclimatic influences on seasonal and spatial cholera transmission cycles: Implications for Public Health Intervention in Bengal Delta. *Water Resources Research, 47*, W00H07.

Akanda, A. S., Jutla, A. S., Gute, D. M., Evans, T., & Islam, S. (2012). Reinforcing cholera intervention through prediction aided prevention. *Bulletin of the World Health Organization, 90*, 243–244.

Akanda, A. S., Jutla, A. S., Gute, D. M., Sack, B. R., Alam, M., Huq, A., et al. (2013). Population Vulnerability to Biannual Cholera Peaks and associated Macro-scale drivers in the Bengal Delta. *American Journal of Tropical Medicine and Hygeine, 89*(5), 950–959.

Akanda, A. S., Jutla, A. S., & Colwell, R. R. (2014). Global diarrhea action plan needs integrated climate-based surveillance. *The Lancet Global Health, 2*(2), e69–e70.

Anderson, M. K. (2005). *Tending the wild: Native American knowledge and the management of California's natural resources.* Berkeley, CA: University of California Press.

Aregawi, M., Lynch, M., Bekele, W., Kebede, H., Jima, D., Taffese, H.S., et al. (2014). Measure of trends in malaria cases and deaths at hospitals, and the effect of antimalarial interventions, 2001–2011, Ethiopia. *PLoS ONE 9*(1), e106359.

Arno, S. F., Parsons, D. J., & Keane, R. E. (2000). Mixed-severity fire regimes in the Northern Rocky Mountains: Consequences of fire exclusion and options for the future. *Proceedings of the Wilderness Science in a Time of Change Conference*. RMRS Proceedings 15 (Vol. 5, pp. 225–232). USDA Forest Service, Rocky Mountain Research.

Ashford, R. W., & Thomson, M. (1991). Visceral Leishmaniasis in Sudan. A delayed development disaster. *Annals of Tropical Medicine and Parasitology, 85*, 571–572.

Beaty, R. M., & Taylor, A. H. (2008). Fire history and the structure and dynamics of a mixed conifer forest landscape in the northern Sierra Nevada, Lake Tahoe Basin, CA. *Forest Ecology and Management, 255*, 707–719.

Bertuzzo, E., Azaele, S., et al. (2008). On the space-time evolution of a cholera epidemic. *Water Resources Research, 44*, W01424.

Biswell, H. (1959). Man and fire in ponderosa pine in the Sierra Nevada of California. *Sierra Club Bulletin, 44*, 44–53.

Biswell, H. H. (1989). *Prescribed burning in California wildland vegetation management*. Berkeley, CA: University of California Press.

Boerner, R. E. J., Huang, J., & Hart, S. C. (2008). Fire, thinning, and the carbon economy: Effects of fire and fire surrogate treatments on estimated carbon storage and sequestration rate. *Forest Ecology and Management, 255*, 3081–3097.

Campbell, J. L., Harmon, M. E., & Mitchell, S. R. (2011). Can fuel-reduction treatments really increase forest carbon storage in the western U.S. by reducing future fire emissions? *Frontiers in Ecology and the Environment*. doi:10.1890/110057.

Campbell-Lendrum, D., Manga, L., Bagayoko, M., & Sommerfeld, J. (2015). Climate change and vector-borne diseases: What are the implications for public health research and policy? *Philosophical Transactions of the Royal Society B, 370*, 20130552. doi:10.1098/rstb.2013.0552.

CARB (California Air Resources Board), Summer Wildfire PM Exceptional Event Demonstration. (2009). http://www2.epa.gov/air-quality-analysis/exceptional-events-submissions-table#PM10.

Cash, B. A., Rodo, X., & Kinter III, J. L (2008). Links between Tropical Pacific SST and cholera incidence in Bangladesh: Role of the Eastern and Central Tropical Pacific. *Journal of Climate, 21*, 4647–4663.

Ceccato, P. (2005). Operational Early Warning System Using SPOT-VGT and TERRA-MODIS to Predict Desert Locust Outbreaks. *Proceedings of the 2nd VEGETATION International Users Conference*, March 24–26, 2004. Antwerpen. In: F. Veroustraete, E. Bartholome, W. W. Verstraeten, (Eds.), Luxembourg: Luxembourg: Office for Official Publication of the European Communities, ISBN 92-894-9004-7, EUR 21552 EN, 475 p.

Ceccato, P., Ghebremeskel, T., Jaiteh, M., Graves, P. M., Levy, M., Ghebreselassie, S., et al. (2007). Malaria stratification, climate and epidemic early warning in Eritrea. *American Journal of Tropical Medicine and Hygiene, 77*, 61–68.

Ceccato, P., Vancutsem, C., & Temimi, M. (2010). Monitoring air and land surface temperatures from remotely sensed data for climate-human health applications. *International Geoscience and Remote Sensing Symposium (IGARSS)*, 178–180.

Ceccato, P., Vancutsem, C., Klaver, R., Rowland, J., & Connor, S. J. (2012). A vectorial capacity product to monitor changing malaria transmission potential in epidemic regions of Africa. *Journal of Tropical Medicine, 2012*, Article ID 595948. doi:10.1155/2012/595948, 6 pp.

Climate Action Reserve (CAR). (2012). *Climate Action Reserve's Forest Project Protocol Version 3.3*. Retrieved October 29, 2015 from http://www.climateactionreserve.org/how/protocols/forest/dev/version-3-3/.

Cockburn, T. A., & Casanos, J. G. (1960). Epidemiology of Endemic Cholera. *Public Health Reports*, Vol. 75, No. 9.

Collins, B. M., Stephens, S. L., Moghaddas, J. J., & Battles, J. (2010). Challenges and approaches in planning fuel treatments across fire-excluded forested landscapes. *Journal of Forestry, 108* (1), 24–31.

Collins, B. M., Everett, R. G., & Stephens, S. L. (2011). Impacts of fire exclusion and managed fire on forest structure in an old growth Sierra Nevada mixed-conifer forest. *Ecosphere, 2*, art 51.

Colwell, R. (1996). Global climate and infectious disease: The cholera paradigm. *Science, 274* (5295).

Colwell, R., & Huq, A. (2001). Marine ecosystems and cholera. *Hydrobiologia, 460*.

DaSilva, J., Garanganga, B., Teveredzi, V., Marx, S. M., Mason, S. J., & Connor, S. J. (2004). Improving epidemic malaria planning, preparedness and response in Southern Africa. *Malaria Journal, 3*, 37.

Delfino, R. J., Brummel, S., Wu, J., Stern, H., Ostro, B., Lipsett, M., et al. (2009). The relationship of respiratory and cardiovascular hospital admissions to the southern California wildfires of 2003. *Occupational and Environmental Medicine, 66*, 189–197.

Dinku, T., Ceccato, P., & Connor, S. J. (2011). Challenges to satellite rainfall estimation over mountainous and arid parts of East Africa. *International Journal of Remote Sensing, 32*(21), 5965–5979.

Dinku, T., Block, P., Sharoff, J., & Thomson, M. (2014a). Bridging critical gaps in climate services and applications in Africa. *Earth Perspectives, 1*(15).

Dinku, T., Kanemba, A., Platzer, B., & Thomson, M. C. (2014b). Leveraging the climate for improved malaria control in Tanzania. *Earthzine*. http://www.earthzine.org/2014/02/15/.

Dombeck, M. P., Williams, J. E., & Wood, C. A. (2004). Wildfire policy and public lands: Integrating scientific understanding with social concerns across landscapes. *Conservation Biology, 18*, 883–889.

Elnaiem, D. A. (2011). Ecology and control of the sand fly vectors of Leishmania donovani in East Africa, with special emphasis on Phlebotomus orientalis. *The Journal of Vector Ecology, 36* (s1), S23–S31.

Elnaiem, D. A., Hassan, H. K., & Ward, R. D. (1997). Phlebotomine sandflies in a focus of visceral leishmaniasis in a border area of eastern Sudan. *Annals of Tropical Medicine and Parasitology, 91*(3), 307–318.

Elnaiem, D. A., Hassan, H. K., & Ward, R. D. (1999). Associations of Phlebotomus orientalis and other sandflies with vegetation types in eastern Sudan focus of kala-azar. *Medical and Veterinary Entomology, 13*, 198–203.

Emch, M., Feldacker, C., Yunus, M., Streatfield, P. K., DinhThiem, V., Canh do, G., et al. (2008). Local environmental predictors of cholera in Bangladesh and Vietnam. *The American Journal of Tropical Medicine and Hygiene, 78*(5), 823–832.

Fontaine, R. E., Najjar, A. E., & Prince, J. S. (1961). The 1958 malaria epidemic in Ethiopia. *American Journal of Tropical Medicine and Hygiene, 10*, 795–803.

Fulé, P. Z., Crouse, J. E., Heinlein, T. A., Moore, M. M., Covington, W. W., & Verkamp, G. (2003). Mixed severity fire regime in a high-elevation forest: Grand Canyon, Arizona. *Landscape Ecology, 18*, 465–486.

Gebre-Michael, T., Malone, J. B., Balkew, M., Ali, A., Berhe, N., Hailu, A., et al. (2004). Mapping the potential distribution of Phlebotomus martini and P. orientalis (Diptera: Psychodidae), vectors of kala-azar in East Africa by use of geographic information systems. *Acta Tropica, 90*, 73–86.

Germain, R. H., Floyd, D. W., & Stehman, S. V. (2001). Public perceptions of the USDA Forest Service public participation process. *Forest Policy and Economics, 3*, 113–124.

Gerstl, S., Amsalu, R., & Ritmeijer, K. (2006). Accessibility of diagnostic and treatment centres for visceral leishmaniasis in Gedaref State, northern Sudan. *Tropical Medicine and International Health, 11*(2), 167–175.

Gil, A. I., Louis, V. R., Rivera, I. N. G., Lipp, E., Huq, A, Lanata, C.F., et al. (2004). Occurrence and distribution of Vibrio cholerae in the coastal environment of Peru. *Environmental Microbiology, 6*(7), 699–706.

Gonzalez, P., Battles, J. J., Collins, B. M., Robards, T., & Saah, D. S. (2015). Aboveground live carbon stock changes of California wildland ecosystems, 2001–2010. *Forest Ecology and Management*. doi:10.1016/j.foreco.2015.03.040.

Griffith, D. C., Kelly-Hope, L. A., & Miller, M. A. (2006). Review of reported cholera outbreaks worldwide, 1995–2005. *The American Journal of Tropical Medicine and Hygiene, 75*(5), 973–977.

Hashizume, M., Armstrong, B., Hajat, S., Wagatsuma, Y., Faruque, A. S., Hayashi, T., et al. (2008). The Effect of Rainfall on the Prevalence of Cholera in Bangladesh. *Epidemiology, 19* (1).

Hirt, P. W. (1996). *A conspiracy of optimism: Management of the national forests since World War Two*. Lincoln, Neb: University of Nebraska Press.

Hoogstraal, H., & Heyneman, D. (1969). Leishmaniasis in the Sudan Republic. *The American Journal of Tropical Medicine and Hygiene, 18*(6). International Soil Reference and Information Centre (ISRIC)—World Soil Information, 2013. Soil property maps of Africa at 1 km. www.isric.org.

Hossain, F. (2015). Data for all: Using satellite observations for social good. *Eos, 96*. doi:10.1029/2015EO037319. Retrieved October 14 2015.

Huq, A., & Colwell, R. R. (1996). Vibrios in the marine and estuarine environment: Tracking Vibrio cholerae. *Ecosystem Health, 13*(9).

Hurteau, M., & North, M. (2009). Fuel treatment effects on tree-based forest carbon storage and emissions under modeled wildfire scenarios. *Frontiers in Ecology and the Environment, 7*, 409–414.

Hurteau, M., & North, M. (2010). Carbon recovery rates following different wildfire risk mitigation treatments. *Forest Ecology and Management, 260*, 930–937.

Hurteau, M. D., Koch, G. W., & Hungate, B. A. (2008). Carbon protection and fire risk reduction: Toward a full accounting of forest carbon offsets. *Frontiers in Ecology and the Environment, 6* (9), 493–498. doi:10.1890/070187.

Jones, G., Loeffler, D., Calkin, D., & Chung, W. (2010). Forest treatment residues for thermal energy compared with disposal by onsite burning: Emissions and energy return. *Biomass Bioenergy, 34*, 737–746.

Jutla, A. S., Akanda, A. S., & Islam, S. (2012). Satellite remote sensing of space-time plankton variability in the Bay of Bengal: Connections to cholera outbreaks. *Remote Sensing of Environment, 123*, 196–206.

Jutla, A. S., Akanda, A. S., & Islam, S. (2013). A framework for predicting endemic cholera using satellite derived environmental determinants. *Environ Modeling and Software, 47*, 148–158.

Jutla, A., Unnikrishnan, A., Akanda, A. S., Huq, H., & Colwell, R. R. (2015a). Predictive time series analysis linking Bengal cholera with terrestrial water storage measured from GRACE sensors. *American Journal of Tropical Medicine and Hygiene* (Accepted).

Jutla, A., Aldaach, H., Billian, H., Akanda, A. S., Huq, A., & Colwell, R. R. (2015b). Satellite based assessment of hydroclimatic conditions related to cholera in Zimbabwe. *PLoS-ONE*. doi:10.1371/journal.pone.0137828.

Kilgore, B. M., & Taylor, D. (1979). Fire history of a sequoia-mixed conifer forest. *Ecology, 60*, 129–142.

Koelle, K., Rodo, X., Pascual, M., Yunus, M., & Mostafa, G. (2005). Refractory periods and climate forcing in cholera dynamics. *Nature, 436*, 4.

Lee, C., Erickson, P., Lazarus, M., & Smith, G. (2010). *Greenhouse gas and air pollutant emissions of alternatives for woody biomass residues*. Seattle, WA: Stockholm Environmental Institute. http://data.orcaa.org/reports/all-reports-entries/woody-biomass-emissions-study/.

Lobitz, B., Beck, L. Huq, A. Wood, B. Faruque A., & Colwell, R. (2000). Climate and infectious disease: Use of remote sensing for detection of vibrio cholerae. *PNAS, 97*(4).

Longini, M., Zaman, K., Yunus, M., & Siddique, A. K. (2002). Epidemic and Endemic Cholera Trends over a 33-Year Period in Bangladesh. *J. Infec. Dis., 186*, 246–251.

McDonald, K. C., Chapman, B., Podest, E., Schroeder, R., Flores, S., Willacy, K., et al. (2011). Monitoring inundated wetlands ecosystems with satellite microwave remote sensing in support of earth system science research. Conference Paper, *34th International Symposium on Remote Sensing of Environment—The GEOSS Era: Towards Operational Environmental Monitoring*, 4p.

MDEQ (Montana Department of Environmental Quality), Exceptional Events Demonstration for PM10 and PM2.5 Data for 2007 Wildfires (2007). http://www2.epa.gov/sites/production/files/2015-05/documents/mt_deq_demonstration_121407.pdf.

Millar, C. I., & Stephenson, N. L. (2015). Temperate forest health in an era of emerging mega disturbance. *Science, 349*(6250), 823–826.

Millar, C. I., Stephenson, N. L., & Stephens, S. L. (2007). Climate change and forests of the future: Managing in the face of uncertainty. *Ecological Applications, 17*(8), 2145–2151. doi:10.1890/06-1715.1.

Miller, J. D., Safford, H. D., Crimmins, M., & Thode, A. E. (2009). Quantitative evidence for increasing forest fire severity in the Sierra Nevada and southern Cascade Mountains, California and Nevada, USA. *Ecosystems, 12*, 16–32.

Moghaddas, J. J., & Craggs, L. (2007). A fuel treatment reduces fire severity and increases suppression efficiency in a mixed conifer forest. *International Journal of Wildland Fire, 16*(6), 673–678. doi:10.1071/WF06066.

Moghaddas, J. J., Collins, B. M., Menning, K., Moghaddas, E. E. Y., & Stephens, S. L. (2010). Fuel treatment effects on modeled landscape-level fire behavior in the northern Sierra Nevada. *Canadian Journal of Forest Research, 40*, 1751–1765.

National Fire Interagency Fire Center (NIFC). (2015). *1997–2014 large fires*. Retrieved 10 October, 2015 from https://www.nifc.gov/fireInfo/fireInfo_stats_lgFires.html.

National Research Council (NRC). (2007). *Earth science and Applications from space: National imperatives for the next decade and beyond*. Washington, DC: The National Academies Press. doi:10.17226/11820.

North, M. P., & Hurteau, M. D. (2011). High-severity wildfire effects on carbon stocks and emissions in fuels treated and untreated forest. *Forest Ecology and Management, 261*, 1115–1120.

North, M. P., Stine, P., O'Hara, K., Zielinski, W., & Stephens, S. (2009). *An ecosystems management strategy for Sierra mixed-conifer forests*. PSW GTR-220 with addendum. Albany, CA, USA: USDA Forest Service, Pacific Southwest Research Station.

North, M. P., Stephens, S. L., Collins, B. M., Agee, J. K., Aplet, G., Franklin, J. F., et al. (2015). Reform forest fire management. *Science, 349*(6254), 1280–1281.

Pacala, S., Birdsey, R. A., Bridgham, S. D., Conant, R. T., Davis, K., Hales, B., et al. (2007). The North American carbon budget past and present. In: A. W. King, L. Dilling, G. P. Zimmerman, D. M. Fairman, R. A. Houghton, G. Marland (Eds.), The First State of the Carbon Cycle Report (SOCCR): *The North American Carbon budget and implications for the global carbon cycle* (Vol. 29–36, pp. 167–170). Asheville, NC: National Oceanic and Atmospheric Administration, National Climatic Data Center.

Pascual, M., Bouma, M., & Dobson, A. (2002). Cholera and climate: Revisiting the quantitative evidence. *Microbes and Infection, 4*(2).

Pekel, J. F., Ceccato, P., Vancutsem, C., Cressman, K., Vanbogaert, E., & Defourny, P. (2011). Development and application of multi-temporal colorimetric transformation to monitor vegetation in the Desert Locust Habitat. *IEEE Journal of Selected Topics in Applied Earth Observations and Remote Sensing, 4*(2), 318–326.

Perry, D. A., Hessburg, P. F., Skinner, C. N., Spies, T. A., Stephens, S. L., Taylor, A. H., et al. (2011). The ecology of mixed severity fire regimes in Washington, Oregon, and Northern California. *Forest Ecology and Management, 262*, 703–717.

Rajesh, K., & Sanjay, K. 2013. Change in global climate and Prevalence of Visceral Leishmaniasis. *International Journal of Scientific and Research Publications, 3*(1).

Randerson, J., Liu, H., Flanner, M. Chambers, S., Jin, Y., Hess, P., et al. (2006). The impact of boreal forest fire on climate warming. *Science, 314,* 1130–1132.

Reinhardt, E., & Holsinger, L. (2010). Effects of fuel treatments on carbon-disturbance relationships in forests of the northern Rocky Mountains. *Forest Ecology and Management, 259,* 1427–1435.

Reinhardt, E. D., Keane, R. E., Calkin, D. E., & Cohen, J. D. (2008). Objectives and considerations for wildland fuel treatment in forested ecosystems of the interior western United States. *Forest Ecology and Management, 256*(12), 1997–2006. doi:10.1016/j.foreco.2008.09.016.

Roccaforte, J. P., Fulé, P. Z., Chancellor, W. W., & Laughlin, D. C. (2012). Woody debris and tree regeneration dynamics following severe wildfires in Arizona ponderosa pine forests. *Canadian Journal of Forest Research, 42,* 593–604.

Saah, D., Robards, T., Moody, T., O'Neil-Dune, J., Moritz, M., Hurteau, M., et al. (2012). *Developing an Analytical Framework for Quantifying Greenhouse Gas Emission Reductions from Forest Fuel Treatment Projects in Placer County*, California. Prepared for: United States Forest Service: Pacific Southwest Research Station, 130p.

Saah, D., Schmidt, D., Roller, G., Moody, T. J., Moghaddas, J. J., & Freed, T. (2015). *Carbon storage and mass balances: Characteristics of forest carbon and the relationship between fire severity and emissions in the Sierra Nevada Mountains, California*, USA. Prepared for: California Energy Commission CEC-600-10-006, 95p.

Safford, H. D., Schmidt, D. A., & Carlson, C. H. (2009). Effects of fuel treatments on fire severity in an area of wildland–urban interface, Angora Fire, Lake Tahoe Basin, California. *Forest Ecology and Management, 258,* 773–787. doi:10.1016/j.foreco.2009.05.024.

Salomon, O. D., Quintana, M. G., Mastrangelo, A. V., & Fernandez, M. S. (2012). Leishmaniasis and climate change—case study: Argentina. *Journal of Tropical Medicine.* doi:10.1155/2012/601242.

Savage, M., & Mast, N. J. (2005). How resilient are southwestern ponderosa pine forests after crown fires? *Canadian Journal of Forest Research, 35,* 967–977.

Schroeder, R. M., Rawlins, A., McDonald, K. C., Podest, E., Zimmermann, R., & Kueppers, M. (2010). Satellite microwave remote sensing of North Eurasian inundation dynamics: Development of coarse-resolution products and comparison with high-resolution synthetic aperture radar data. *Environmental Research Letters*, special issue on Northern Hemisphere high latitude climate and environmental change, *5,* 015003 (7 pp). doi:10.1088/1748-9326/5/1/015003.

Schwartz, B. S., Harris, J. B., Khan, A. I., Larocque, R. C., Sack, D. A., Malek, M. A., et al. (2006). Diarrheal epidemics in Dhaka, Bangladesh, during three consecutive floods: 1988, 1998, 2004. *The American Journal of Tropical Medicine and Hygiene, 74*(6).

Schweizer, D., & Cisneros, R. (2014). Wildland fire management and air quality in the southern Sierra Nevada: Using the Lion Fire as a case study with a multi-year perspective on PM 2.5 impacts and fire policy. *Journal of Environmental Management, 144,* 265–278.

Seaman, J., Mercer, A. J., & Sondorp, E. (1996). The epidemic of visceral leishmaniasis in western Upper Nile, southern Sudan: course and impact from 1984 to 1994. *International Journal of Epidemiology, 25*(4), 862–871.

Sedjo, R. A., & Marland, G. (2003). Inter-trading permanent emissions credits and rented temporary carbon emissions offsets: Some issues and alternatives. *Climate Policy, 3,* 435–444

SMAQMD (Sacramento Metropolitan Air Quality Management District). (2011). Exceptional Events Demonstration for 1-Hour Ozone Exceedance in the Sacramento Regional Nonattainment Area due to 2008 Wildfires.

Springsteen, B., Christofk, T., Eubanks, S., Mason, T., Clavin, C., & Storey, B. (2011). Emission reductions from woody biomass waste for energy as an alternative to open burning. *Journal of the Air and Waste Management Association, 61,* 63–68.

Springsteen, B., Christofk, T., York, R. A., Mason, T., Baker, S., Lincoln, E., et al. (2015). Forest biomass diversion in the Sierra Nevada: Energy, economics and emissions. *California Agriculture, 69*(3), 142–149.

State of California. (2006). Assembly Bill 32: The California Global Warming Solutions Act of 2006. Nunez, F. September 27, 2006.

Stephens, S. L. (2000). Mixed conifer and upper montane forest structure and uses in 1899 from the central and northern Sierra Nevada, CA. *Madrono, 47*, 43–52.

Stephens, S. L., & Moghaddas, J. J. (2005). Silvicultural and reserve impacts on potential fire behavior and forest conservation: Twenty-five years of experience from Sierra Nevada mixed conifer forests. *Biological Conservation, 125*(3), 369–379. doi:10.1016/j.biocon.2005.04.007.

Stephens, S. L., & Ruth, L. W. (2005). Federal forest-fire policy in the United States. *Ecological Applications, 15*(2), 532–542. doi:10.1890/04-0545.

Stephens, S. L., Martin, R. E., & Clinton, N. D. (2007). Prehistoric fire area and emissions from California's forests, woodlands, shrublands and grasslands. *Forest Ecology and Management, 251*, 205–216.

Stephens, S. L., Moghaddas, J., Hartsough, B., Moghaddas, E., & Clinton, N. E. (2009a). Fuel treatment effects on stand level carbon pools, treatment related emissions, and fire risk in a Sierran mixed conifer forest. *Canadian Journal of Forest Research, 39*, 1538–1547.

Stephens, S. L., Moghaddas, J., Edminster, C., Fiedler, C., Hasse, S., Harrington, M., et al. (2009b). Fire treatment effects on vegetation structure, fuels, and potential fire severity in western U.S. forests. *Ecological Applications, 19*, 305–320.

Stephens, S. L., Moghaddas, J. J., Edminster, C., Fiedler, C. E., Haase, S., Harrington, M., et al. (2009c). Fire treatment effects on vegetation structure, fuels, and potential fire severity in western U.S. forests. *Ecological Applications, 19*(2), 305–320. doi:10.1890/07-1755.1. PMID: 19323192.

Stephens, S. L., McIver, J. D., Boerner, R. E. J., Fettig, C. J., Fontaine, J. B., Hartsough, B. R., et al. (2012a). Effects of forest fuel reduction treatments in the United States. *BioScience, 62*, 549–560.

Stephens, S. L., Boerner, R. E. J., Moghaddas, J. J., Moghaddas, E. E. Y., Collins, B. M., Dow, C. B., et al. (2012b). Fuel treatment impacts on estimated wildfire carbon loss from forests in Montana, Oregon, California, and Arizona. *Ecosphere, 3*(5), 38. http://dx.doi.org/10.1890/ES11-00289.1.

Sweeney, A., Kruczkiewicz, A., Reid, C., Seaman, J., Abubakar, A., Ritmeijer, K., et al. (2014). Utilizing NASA Earth observations to explore the relationship between environmental factors and Visceral Leishmaniasis in the Northern States of the Republic of South Sudan. *Earthzine IEEE*.

Thomson, M. C., Elnaiem, D. A., Ashford, R. W., & Connor, S. J. (1999). Towards a kala azar risk map for Sudan: Mapping the potential distribution of Phlebotomus orientalis using digital data of environmental variables. *Tropical Medicine and International Health, 4*(2), 105–113.

Urbanski, S. (2014). Wildland fire emissions, carbon, and climate: Emission factors. *Forest Ecology and Management, 317*, 51–60.

US Forest Service (USFS). (2004). Sierra Nevada Forest Plan Amendment (SNFPA) Final Supplemental Environmental Impact Statement (FEIS) and Record of Decision (ROD). January 2004.

US Forest Service (USFS). (2015). Managing Wildfires. http://www.fs.fed.us/managing-land/fire.

Uz, B. M., & Yoder, J. A. (2004). High frequency and mesoscale variability in SeaWiFS chlorophyll imagery and its relation to other remotely sensed oceanographic variables. *Deep Sea Research II, 51*(10–11), 1001–1017. doi:10.1016/j.dsr2.2004.03.003.

Vancutsem, C., Ceccato, P., Dinku, T., & Connor, S. J. (2010). Evaluation of MODIS Land surface temperature data to estimate air temperature in different ecosystems over Africa. *Remote Sensing of Environment, 114*(2), 449–465.

Westerling, A. L., Hidalgo, H. G., Cayan, D. R., & Swetnam, T. W. (2006). Warming and earlier spring increase western U.S. forest wildfire activity. *Science, 313*(5789), 940–943. doi:10.1126/science.1128834. PMID:16825536.

Westerling, A., Brown, T., Schoennagel, T. Swetnam, T., Turner, M., & Veblen, T. (2014). Briefing: Climate and wildfire in western U.S. forests. In: V. A. Sample; R. P. Bixler (Eds.), *Forest conservation and management in the Anthropocene: Conference proceedings.*

Proceedings. RMRS-P-71. Fort Collins, CO: US Department of Agriculture, Forest Service. Rocky Mountain Research Station. pp. 81–102.

WHO/UNICEF. (2009). *Diarrhea: Why children are still dying and what can be done*, Geneva, Switzerland.

WHO/UNICEF. (2013). *Ending Preventable Child Deaths from Pneumonia and Diarrhea by 2025. The Integrated Global Action Plan for Pneumonia and Diarrhea (GAPPD)*. Retrieved from www.unicef.org/media/files/Final_GAPPD_main Report-_EN-8_April_2013 pdf.

WHO/UNICEF. (2014). *Neglected Tropical Diseases: The 17 neglected tropical diseases*. Retrieved from http://www.who.int/neglected_diseases/diseases/en/.

Wiedinmyer, C., & Neff, J. C. (2007). Estimates of CO_2 from fires in the United States: Implications for carbon management. *Carbon Balance and Management, 2*(1), 1–12.

Wildland Fire Leadership Council. (2011). *A National Cohesive Wildland Fire Strategy*. http://www.forestsandrangelands.gov/strategy/documents/reports/1_CohesiveStrategy03172011.pdf. Accessed 10/29/2015.

Yeshiwondim, A. K., Gopal, S., Hailemariam, A. T., Dengela, D. O., & Patel, H. P. (2009). Spatial analysis of malaria incidence at the village level in areas with unstable transmission in Ethiopia. *International Journal of Health Geographics, 8*, 5. doi:10.1186/1476-072X-8-5.

Satellite Remote Sensing in Support of Fisheries Management in Global Oceans

Dovi Kacev and Rebecca L. Lewison

Abstract The world's oceans are dynamic: environmental conditions and ecosystems in marine environments fluctuate spatially and temporally on multiple scales. Spatially, the ocean varies with water depth, ocean currents, and oceanic fronts. This abiotic and biotic variability makes managing resources in the dynamic ocean environment extremely difficult, and as a result, fisheries management often serves as one of the textbook examples of an unstructured or 'wicked' environmental problem. This chapter provides an overview of the role satellite remote sensing can play in ocean and fisheries management. Currently, there are very few applications available that enable managers to use satellite earth observations in a scientifically robust but straightforward manner. The chapter recommends collaboration between researchers, scientists, data analysts and conservation practitioners to develop accessible tools, all the while ensuring such approaches are scientifically robust and defensible and are directly meeting the needs of the management community. Continued support and resources for satellite earth observations, distribution, integration, and management-relevant science are needed to maximize the return on this investment in support of sustainable fisheries.

1 Fisheries Management in a Dynamic Ocean: Overview

The world's oceans are dynamic: environmental conditions and ecosystems in marine environments fluctuate spatially and temporally on multiple scales (Hazen et al. 2013). For example, sea surface temperature fluctuates temporally daily, seasonally, and in multi-year cycles such as the El Nino, Southern Oscillation (ENSO) cycle. Spatially, the ocean varies with water depth, ocean currents, and oceanic fronts. Other abiotic conditions like currents, eddy structure, and salinity thresholds also fluctuate in both predictable and non-predictable ways. Biotic conditions in the ocean are just as complex (Haury et al. 1978). Habitat associations

D. Kacev · R.L. Lewison (✉)
Department of Biology, San Diego State University, 5500 Campanile Drive, San Diego, CA 92182-4614, USA
e-mail: rlewison@mail.sdsu.edu

of species or assemblages vary in space and time, as ocean conditions change, and biotic patterns, such as migration or persistent movement pathways, i.e., corridors, are far more dynamic than in terrestrial systems (Hazen et al. 2012). This abiotic and biotic variability makes managing resources in the dynamic ocean environment extremely difficult, and as a result, fisheries management often serves as one of the textbook examples of an unstructured or 'wicked' environmental problem (Balint et al. 2011; Hisschemöller and Hoppe 1995).

In the US, ocean fisheries management is legislated by several legislative instruments with the Magnuson Stevens Act as the primary law governing marine fisheries. The Magnuson–Stevens Act was originally enacted as the Fishery Conservation and Management Act of 1976. Two major recent sets of amendments to the law were the Sustainable Fisheries Act of 1996 and the Magnuson–Stevens Fishery Conservation and Management Reauthorization Act (MSFCMA) of 2006. The MSFCMA and other relevant policies mandate a management approach that balances the often conflicting objectives of economic viability with ecological sustainability, i.e., fisheries managers are tasked with protecting target and nontarget species and stocks, protecting essential habitats as well as supporting viable fishing economies.

Globally, fisheries management is often a part of a larger ecosystem-based management framework; a transition from single-species, stock assessments, or management plan to ecosystem approaches which requires the ability to monitor the ocean conditions and variability that drive many biotic and abiotic processes (Scheurell and Williams 2005; Wells et al. 2008). The vast temporal and spatial scales of the global oceans, the limited ability to collect ocean data directly, and the move to manage ocean systems rather than ocean species has led to the reliance on advancing technology in satellite remote sensing (SRS) data (Collie et al. 2013). Hereafter, the term SRS will be used to signify Earth observation (EO) in the previous chapters.

2 Current Status of Ocean SRS: The Fisheries Management Context

Traditional approaches to collect oceanographic data relied on ship board equipment and observer driven measurements. However, since the early 1970s, fisheries scientists have implemented airborne remote sensing (ARS) using balloons, helicopters, and airplanes to study oceans from above (reviewed in Santos 2000; Klemas 2013). In some areas, ARS relies on naked-eye observations from flown transects over ocean waters. Although the human eye and airborne sensors have a wide spatial range and high resolution, ARS is limited by human skill and endurance as well as by water depth and both water and atmospheric visibility (Klemas 2013). The addition of airborne sensors like cameras, low-light-televisions, infrared cameras, and LiDAR (Light Detection and Ranging) has made ARS more reliable, but they are still limited temporally and spatially (Klemas 2013; Rose et al. 2000).

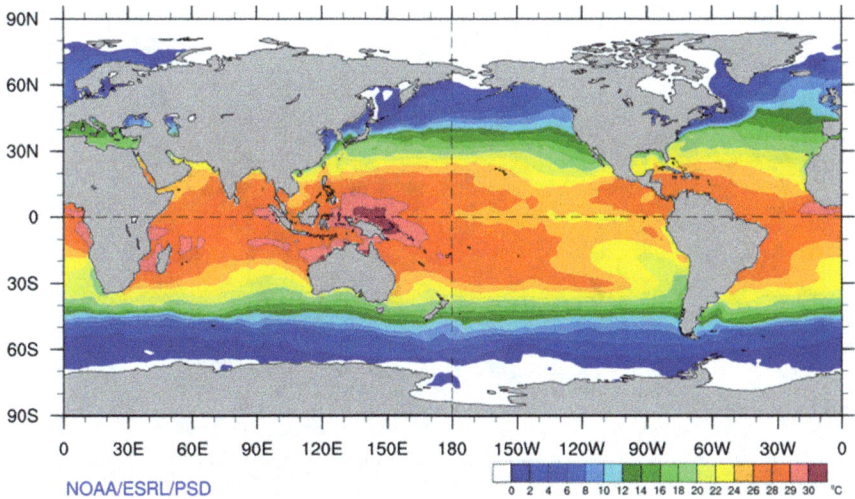

Fig. 1 This image shows the global weekly average SSTs, January 30 2011–February 5 2011. *Image credit* Earth System Research Laboratory Physical Sciences Division. Public domain

With rapid increases in satellite technology and data storage in recent years, SRS has become an essential element of fisheries management due to its ability to collect data at regular intervals for long time periods across the entire globe (Chassot et al. 2011; Pettorelli et al. 2014). The scale and scope of SRS data allow for the implementation of near-real-time ecosystem and forecast models that can be applied to ocean management. To contextualize the application of SRS technology to fisheries management, we describe some of the most common SRS data used in fisheries.

Sea surface temperature: Sea surface temperature (SST) is a measure of the energy from the molecules at the top layer of the ocean and was one of the first satellite-sensed ocean variables applied to fisheries. Thermal infrared sensors that measure SST, when there is no cloud cover have been deployed for over four decades (Klemas 2013). SST can be used to detect oceanic fronts and large-scale climatic fluctuations. Maps of SST are directly relevant to fisheries because of the temperature preferences of individual fish species. Satellite platforms use different bands sensitive to the thermal infrared, allowing precise, daily measurement of SST over the world's oceans (Fig. 1). Currently, the most common data source for SST is the NASA-deployed MODIS sensor.

Ocean color: The color in most of the world's oceans is in the visible light region (wavelengths of 400–700 nm) and varies with the concentration of chlorophyll and primary ocean producers, phytoplankton. When more phytoplankton are present, the concentration of plant pigments increases and the greener the water becomes. Subtle changes in ocean color signify various types and quantities of marine phytoplankton. Changes in ocean color signify changes in chlorophyll concentrations and can be used to infer biological productivity (Fig. 2). Ocean color

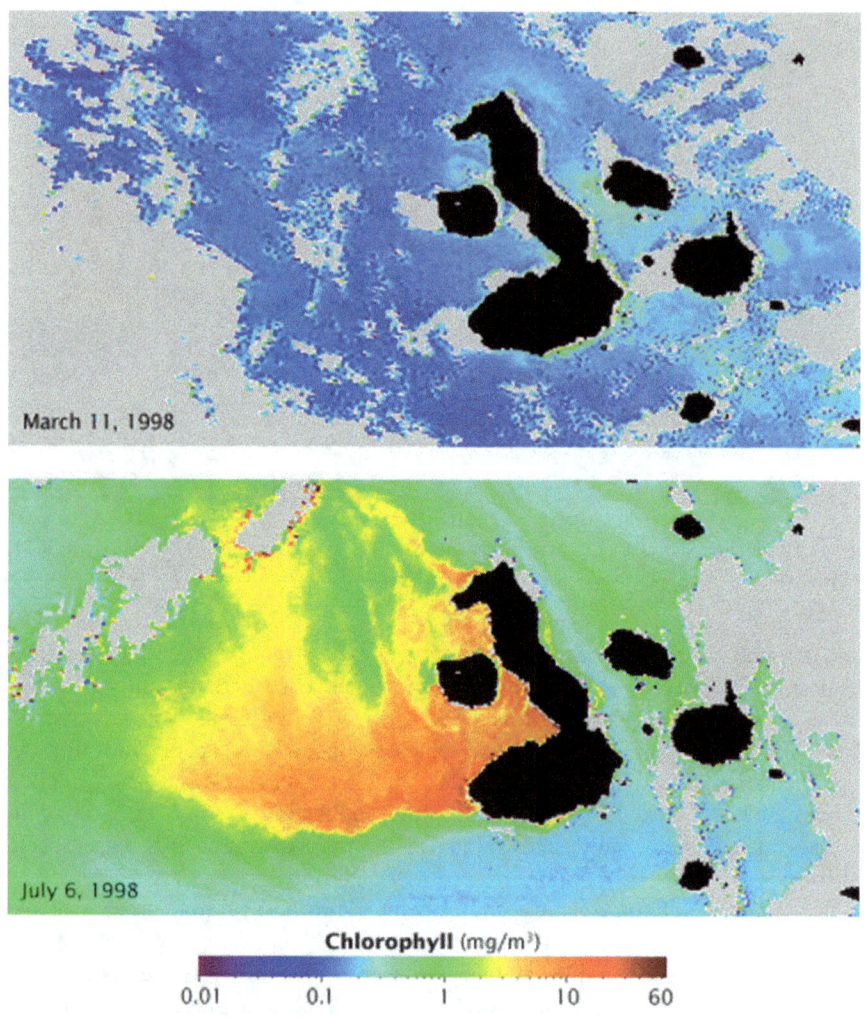

Fig. 2 Chlorophyll concentration during El Niño (*top*) and La Niña (*lower*) where *blue* represents low concentrations, while *yellow*, *orange* and *red* indicate high concentrations. *Image credit* GSFC Ocean Color team and GeoEye, NASA Earth Observatory. Public domain

data is also useful for detecting terrestrial runoff and tracking oil spills (Wilson, IOCCG). Currently, the most commonly used satellite sensors for ocean color include Nimbus-7, SeaWiFS, and MODIS.

Sea surface salinity: Salinity in the ocean is defined as the grams of salt per 1000 grams of water. Since salt dissolves in sea water creating increased reflectivity and decreased emissivity. Salinity can be determined using microwave radiometers when accounting for SST. Salinity varies due to evaporation and precipitation over the ocean as well as river runoff and ice melt. Along with SST, it is a major driver of

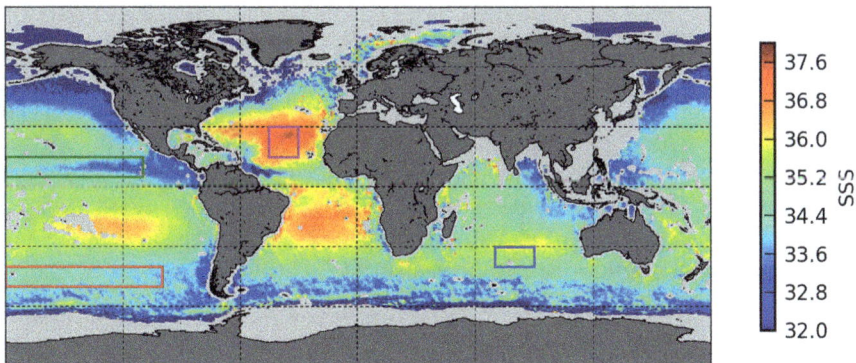

Fig. 3 A map of surface salinity levels recorded between August and September, 2010 and 2011. *Image credit* SMOS GLOSCAL Cal/Val project. Public domain

abiotic processes like ocean circulation (Fig. 3). The most commonly used salinity sensors are the European SMOS and the NASA-launched Aquarius/SAC-D.

Ocean currents: Because of the complexity, the range of temporal and spatial scales and the dependence on local conditions, satellite remote sensing is an ideal tool to measure ocean currents. Currents are responsible for water exchange between different parts of the ocean. Satellite sensors are not capable of measuring ocean currents directly. However, remotely sensed data are used to assess current velocity with a variety of methods and products including Ocean Surface Current Analyses Realtime (OSCAR; http://www.oscar.noaa.gov, http://podaac.jpl.nasa.gov), as well as Mercator/SURCOUF (Larnicol et al. 2006; http://www.mercator-ocean.fr), and the Centre de Topographie des Océans et de l'Hydrosphère (CTOH; Sudre and Morrow 2008; http://ctoh.legos.obs-mip.fr). These products provide global surface current data directly calculated from satellite altimetry and ocean vector winds. The current sensors for ocean currents include Jason-2/OSTM and QuikSCAT.

Sea surface height: Just as bathymetry measures the relief or topography of the ocean floor, sea surface height measures the topography of the ocean surface. In the context of fisheries, sea surface height is often used to calculate anomalies, which are related to eddy structure and strength and other important features in understanding fish species and community distributions (Fig. 3). Sea surface height data is available from Topex/Poseidon and Jason-1 and currently from Jason-2, Ocean Surface Topography Mission.

3 Utility of SRS in Fisheries Management

The formidable challenges of fisheries management, i.e., the concurrent objectives of harvesting and protecting living marine resources have been a joint focus of the National Oceanic and Atmospheric Administration (NOAA) and the National

Aeronautics and Space Administration (NASA) for decades. For example, NOAA and NASA sponsored a 2006 workshop, "Integrating Satellite Data Into Ecosystem-Based Management of Living Marine Resources," to identify specific ways to incorporate Earth science satellite observations, data, and associated models into NOAA fisheries management (EOS 2006). A similar workshop was convened in 1996 (Boehlert and Schumaker 1997). The interest in capitalizing on SRS to support fisheries management and the larger goals of ecosystem-based management continues to expand and develop. Here, we describe some of the key areas of SRS application in fisheries management.

3.1 Identifying Ocean Features and Regions

Ocean features such as ocean currents, eddies, and fronts often serve as aggregating features for fisheries target species. The spatial and temporal scale of many of these features makes them difficult to study with traditional, shipboard, or ARS methods. SRS, in contrast, has the capacity to monitor these dynamic ocean features on broad scales as it can combine thousands of daily, global satellite images using a number of different data integration methods.

Detecting ocean fronts is often central to identifying important fishing areas and there are several methods that use SRS data to do so including gradient measurement and histogram-based methods (Chassot et al. 2011). The gradient method can detect fronts on a scale >4 km using time-averaged data although it can be limited when the SRS data are noisy. A more widely used approach is the histogram-based method (Kahru et al. 1995; Chassot et al. 2011), which uses single image analyses and can detect mesoscale circulation patterns such as eddies, which are often important for fish recruitment processes and tracking marine pollutants (Klemas 2012a).

Studying large-scale ocean currents is also vital to studying large marine ecosystems. SRS, like infrared SST, is used to track currents such as the Gulf Stream Kuroshio currents which moves warm water and creates warm water eddies along the current boundaries (Klemas 2012b). Another SRS approach to studying large marine ecosystems is to use "feature mapping," which uses SRS imagery to track the movement of ocean abnormalities such as chlorophyll plumes (ocean color) or temperature anomalies (SST). By tracking the displacement of these features relative to fixed and coastal landmarks, current speeds can be estimated from SRS images.

3.2 Identifying Species-Specific Habitat Features

One of the fundamental goals of fishery management is to support fisheries catches or yield. Beyond measuring general ocean features, directly mapping the distribution and ocean habitat for target species is often a central objective of fisheries management applications. The use of fishery logs, fishery-independent surveys, and fish tagging combined with SRS data is common to study and predict fish distributions, where concurrent SRS ocean data are analyzed to understand current and projected future habitat associations for target species. Analyses of in situ biological data paired with SRS data has been identified as one of most powerful means to determine habitat preferences on large spatial scales (Cushing 1982; Bakun 1996; Chassot et al. 2011).

One oceanographic feature often associated with fish distribution is SST (Ramos et al. 1996; Klemas 2013). An early application of SRS on sardine (Sardina pilchardus), an economically important fisheries target species in the Atlantic, used SST to map sardine distribution (Santos and Fiuza 1992). These authors found that sardines were clustered around upwelling areas, ocean regions where cold, nutrient rich water from depth displaces warmer, surface waters. Similar patterns have been found with numerous other fish species, e.g., some species of tuna tend to be found in upwelling regions, particularly near thermal fronts (Santos 2000), areas where cold and warm water masses meet. In some cases, thermal limits for target species can be determined by comparing fish distribution to SST data (Klemas 2013), even relatively small changes in temperature as some fish are sensitive to water changes on the magnitude of 0.1 °C. Fish species distribution can also be influenced by regional productivity as measured by chlorophyll concentration (ocean color). A general trend indicates that viable fishery populations require a chlorophyll concentration; at least 0.2 mg/m^3 (FAO 2003). Particularly in pelagic habitats, which are typically characterized by low biological productivity, fish tend to be found along ocean fronts with high chlorophyll levels (e.g., Royer et al. 2004; Santos 2000).

Species-specific habitat models have become a key element of many fisheries stock assessments, which are population models or evaluations that are used to set harvest levels. Zwolinski et al. (2011) used a combination of predictive habitat modeling and previous catch data to demonstrate how SRS data can be used to improve fishery survey design for Pacific sardine (*Sardinops sagax*), the target species for an economically important fishery in the Northwest United States. In the 1930s, excessive fishing pressure along with environmental fluctuations led to a stock collapse in the species (Radovich 1982) but the stock recovered sufficiently for the fishery to resume in the 1990s (Smith and Moser 2003). Despite the large spatial range of sardine, several recent studies have suggested that sardine distribution is limited temporally by SST ($\sim$ 12–16 °C; e.g., Checkley et al. 2000; Lynn 2003; Weber and McClatchie 2010). To build a Pacific sardine habitat suitability model, Zwolinski et al. (2011) used 12 years of sardine presence data and SRS data, including SST, chlorophyll *a* (ocean color), and SSH. In order to avoid future stock collapse, the Pacific Fishery Management Council has set harvest rates based on annual stock assessment for the species, which requires SRS data.

3.3 Identifying Movement or Migration Pathways

Long distance movements of both fisheries target and nontarget populations challenge fisheries management (Rose et al. 2000) and understanding the ecological drivers of movement patterns on both spatial and temporal scales is essential for management of highly migratory target and nontarget species. The field of biotelemetry has expanded rapidly over the past two decades with developments in satellite tracking technology (Bograd et al. 2010; Hammerschlag et al. 2011). These technological developments have led to an increased understanding in the spatial and temporal scale of animal movements, and combining SRS data with these new types of movement data have broadened the understanding of the drivers and mechanisms behind these movements (Chassot et al. 2011).

Over the past decade, pop-up satellite archival tags (PSATs) have been developed that allow scientists to track species at depth with high frequency temperature-depth-light sensors. PSATs provide previously unavailable observations of fish species that rarely use the surface water over relatively large ocean areas. Pairing data from PSATs and SRS has also led to new insights into migration and movement patterns of difficult-to-monitor highly migratory fish species. Studying an economically important species, the porbeagle sharks (Lamna nasus), in the southwest Pacific Ocean, Francis et al. (2015) used SRS and PSAT data to quantify the horizontal and vertical migration of porbeagles, revealing that, seasonally, porbeagles make horizontal movements of hundreds to thousands of kilometers but also capturing the large daily vertical migrations that this species makes. By integrating a merged SRS product and data from PSAT deployments, Luo et al. (2015) identify new insights into migration pathways for yellowfin tuna (*Thunnus albacares*), blue marlin (*Makaira nigricans*), white marlin (*Tetrapturus albidus*), and sailfish (*Istiophorus platypterus*) in the Caribbean Sea and Gulf of Mexico, where these species were found to track highly specific ocean front and eddy features.

3.4 Forecasting Fishing Hotspots

Marine fisheries are an important global food source, accounting for at least 15 % of the world's animal protein consumption (FAO 2009). The costs associated with commercial fishing, including fuel and search costs, can be substantial (Santos 2000). As such, commercial fishers rely on SRS data to help find productive fishing grounds in an efficient manner. The application of SRS to identifying productive fishing grounds started as early as the 1960s with the availability of SST and ocean color data (Chassot et al. 2011). Since then, many countries have allocated public funds to provide SRS data to fishermen, a step aimed at reducing fishing costs and increasing productivity (Santos 2000). SRS data has been shown to reduce the search time by 50 % for US commercial fishing fleets (Santos 2000) resulting in statistically higher catch rates (Wright et al. 1976).

In the US, NOAA has provided SRS data on ocean fronts and SST on a regular basis since 1980 (reviewed in Santos 2000). Even in these early years, economic analyses demonstrated that SRS support saved the Northwest salmon and albacore fisheries upwards; $500,000 per year (Breaker 1981). Similar analysis showed that individual New England fishers saved thousands of dollars in fuel costs per year (Cornillon et al. 1986). The Japan Fisheries Information Service Centre, often described as the most well organized commercial fishery support service (Santos 2000), also synthesizes SRS data to share with commercial fishers to increase in fishery efficiency and reduce seafood prices for consumers (Yamanaka 1988). Similar SRS data is collected by governments around the globe which is shared with their respective fishing industries (reviewed in Santos 2000).

3.5 Identifying Bycatch Hotspots

Fisheries bycatch, the incidental catch of unused or unmanaged species (Davies et al. 2009) poses a substantial impediment to fisheries sustainability (Hall et al. 2000; Kelleher 2005). Bycatch of juvenile target species, foundational species, like sponges or corals, as well as long-lived marine megafauna has both direct and indirect ecological effects that are challenging to mitigate and manage (Lewison et al. 2004). SRS data are a critical element to many applications that have been developed to avoid or minimize bycatch. Characterization of bycatch spatial patterns (Gardner et al. 2008) and hotspots (Lewison et al. 2009) in conjunction with research on animal movement patterns has identified relationships between a wide range of SRS products like SST, chlorophyll, and SSH and other ocean, fishery, or gear characteristics and bycatch (James et al. 2005; Polovina et al. 2006; Sims et al. 2008; Zydelis et al. 2011; Briscoe et al. 2014). For example, using SRS data to characterize oceanographic conditions and habitat features associated with loggerhead and leatherback turtles off Hawaii, the NOAA-led program Turtlewatch creates weekly maps for fishermen depicting ocean areas where bycatch of both species is more likely (Howell et al. 2008, 2015). SRS data paired with sophisticated predictive ocean models integrating oceanographic and telemetry data have also been developed to forecast the presence of high-risk bycatch species like bluefin tuna in eastern Australia, guiding dynamic spatiotemporal fisheries restriction decisions (Hobday et al. 2010).

3.6 Marine Reserve Identification and Monitoring

The development of marine protected areas (MPAs) has been recognized as one important tool in maintaining marine biodiversity and supporting fisheries yields (Brown et al. 2015). The Convention on Biological Diversity set a target of conserving 10 % of available habitats by the year 2020 (https://www.cbd.int/sp/targets/).

As MPA designation continues to reach the 2020 target, the challenge of protected area selection is central to planning activities. SRS data provides planners and managers with quantifiable information on biotic and abiotic conditions which is a key element for marine reserve design (Kachelriess et al. 2014). In the context of MPAs implementation, SRS data are also vital to monitor the environmental correlates of biodiversity (Soykan and Lewison 2015), monitor MPA-protected systems, and assess the impacts of human threats to biodiversity (Kechelriess et al. 2014).

Two of the major impediments to MPA implementation are characterizing how to designate protected areas that promote biodiversity and how to monitor human impacts on MPA efficacy. SRS data is vital for optimally designating and managing protected areas (Pettorelli et al. 2014). Kechelriess et al. (2014) defined three discrete ways that SRS data can improve MPA design and management: (1) monitoring environmental correlates to biodiversity, (2) monitoring habitats within MPAs, and (3) assessing anthropogenic impacts and threats. SRS-based habit designation is routinely incorporated into terrestrial reserve design (Lengyel et al. 2008) and a similar approach has been proposed for pelagic systems (e.g., Hobday et al. 2011). Currently, SST, ocean color, and current patterns have all been used to characterize pelagic habitats, and incorporating a range in each of those environmental variables can help build MPAs that are more representative of the ocean at large. In coastal, marine systems, SRS can be used to identify structural biota and substrate such as mangroves, seagrass beds, kelp forests, and coral reefs (reviewed in Kachelriess et al. 2014). The ability to pinpoint these biomes at large scale can help managers incorporate each into networks of MPAs.

One of the primary criticisms of MPAs is the resources required to enforce them. SRS data can aid in MPA monitoring at large spatial and temporal scales without the need for expensive ship time. Ocean color sensors can detect suspended particles in the water column indicative of terrestrial runoff (Kachelreiss et al. 2014). Monitoring the presence of these suspended particles helps assess the interaction of land use and marine ecosystem health. Ocean color sensors can also detect the presence of and spreading of oil spills in protected areas. Both large and small oil spills can have devastating impacts on marine habitats. SRS allows them to be detected early to allow for rapid response. Finally, SRS allows for the detection of illegal fishing vessels within MPA boundaries. Large, heavily fished species are often the conservation targets of MPAs and illegal or underreported fishing can have detrimental population and community level results. The ability of remote optical sensors to pick out boats in remote reserves can minimize the impact of illegal fishing. As satellite technology and data precision and accuracy increases and MPAs get implemented, SRS data are likely to continue to serve as an instrumental tool in MPA design and monitoring.

3.7 Dynamic Ocean Management

Given the variable nature of marine systems, recent papers have posited that the inherent multi-objective nature of fisheries (i.e., economic profitability and viability coupled with ecological sustainability) requires a management framework that can integrate biotic and abiotic complexity and ocean dynamics (Hobday et al. 2014; Maxwell et al. 2015; Lewison et al. 2015; Dunn et al. 2016). The term dynamic ocean management refers to methods that explicitly account for dynamic oceanographic and fishing conditions and dynamic management approaches are increasingly gaining traction worldwide, typically building on the foundation of existing SRS applications. Dynamic ocean management combines technological advances to utilize and share near-real-time environmental and biological data to promptly communicate fisheries conditions to a broad stakeholder community (managers, industry, fishermen, and conservation organizations). These types of approaches can allow managers and fishermen to rapidly adjust fishing practices, efforts, and locations in response to changing conditions. Importantly, the development and effective application of these dynamic tools is reliant on robust high-resolution ocean data, making continued collection of SRS critical to the innovation of fisheries and other sectors of ocean management.

3.8 SRS in Fisheries Management: The Road Ahead

There have been numerous papers, workshops and evaluations of the integration of SRS into fisheries management and the conclusion reached in these documents continues to hold true today: while there is a substantial amount of SRS-generated oceanographic information, there remains a limited ability to apply these data. This conclusion is predicated on a critical assumption: SRS ocean data will continue to be collected and that new SRS technologies will be developed that improve accuracy and precision of the ocean data collected. This assumption is paramount as the continued collection of high-resolution SRS ocean data is central to sustainable fisheries management now and in the future. Without continued collection of high-resolution SRS data, sustainable fisheries management, within the larger ecosystem management context, will not be possible.

Even with decades of exciting innovations of SRS applications in fisheries and ocean management, there are several gaps that must be addressed to support the continued use and development of SRS data in fisheries management. These gaps include the ongoing disconnect between data acquisition and fisheries applications, and the need for wider SRS data integration and application. However, the most critical gap to address may be the limited support for actionable, need-driven science that uses SRS (and other) data to support fisheries management.

Both NASA and NOAA have developed many accessible online platforms to promote the use and distribution of SRS data. For example, NASA's Physical

Oceanography Distributed Active Archive Center (PO.DAAC), an element of the Earth Observing System Data and Information System (EOSDIS) provides SRS data to a wide community of users (https://podaac.jpl.nasa.gov/). Similarly, NOAA's ERDDAP web-based data server (http://coastwatch.pfeg.noaa.gov/erddap/information.html) provides a simple, consistent way to combine and download multiple sets of SRS data in common file formats. These and many similar efforts have made tremendous advances in delivering SRS data to the community of potential users. However, these efforts have not permeated the fisheries management community and there remains untapped potential in the development of SRS applications in direct service of fisheries management.

Robust fisheries management, of course, relies, on robust oceanographic data. However, in many cases, SRS and other oceanographic data are necessary but not sufficient, for robust fisheries management. Fisheries management also requires the integration of data on policy, social structure, culture, economics, and other related fields (Hall et al. 2007). Integrated frameworks that merge natural and social science data have been developed and applied to identify the factors affecting ecosystem management (Chan et al. 2012; Schultz et al. 2015), however, these types of integrated frameworks that merge SRS data with socioeconomic, policy and other data are still rare in a fisheries management context.

The final issue to be addressed is the critical need for actionable, management-driven science that uses SRS and other data to support fisheries management. Despite the growing, and in many areas of the world desperate, need for practical and implementable solutions to fisheries management problems, there are still comparatively fewer resources focused on actionable, applied science that can support innovative management solutions. This, in part, stems from a disconnect between the scientific and management communities. Analyzing SRS data is fundamentally complex and often requires highly specialized analytic tools and skills. For some fisheries managers and decision makers, this can be prohibitive. To interpret and apply SRS data or merge it with other in situ biological or other data also requires tools to apply the data into decision-making. There are very few applications available that enable managers to use SRS data in a scientifically robust but straightforward manner. This integration requires collaboration between researchers, scientists, data analysts and conservation practitioners to develop accessible tools, all the while ensuring such approaches are scientifically robust and defensible and are directly meeting the needs of the management community. Even with the remarkable advances in SRS sensors and platforms, the applications of SRS data in support of fisheries management have not yet fully capitalized on the rich SRS data sets. Given the substantial global investment in SRS technology and the ever-increasing pressure for sustainable fisheries worldwide, continued support and resources for SRS data distribution, integration, and management-relevant science are needed to maximize the return on this investment in support of sustainable fisheries.

References

Bakun, A. (1996). *Patterns in the ocean: Ocean processes and marine population dynamics* (p. 323). University of California Sea Grant, San Diego, California, USA, in cooperation with Centro de Investigaciones Biologicas de Noroeste, La Paz, Baja California Sur, Mexico.

Balint, P. J., Stewart, R. E., Desai, A., & Walters, L. C. (2011). *Wicked environmental problems: Managing uncertainty and conflict.* Washington, DC: Island Press.

Boehlert, G. W., & Schumacher, J. D. (1997). Changing oceans and changing fisheries: Environmental data for fisheries research and management. *NOAA Technical Memorandum NMFS.* NOAA-TM-NMFS-SWFSC-23.

Bograd, S. J., Block, B. A., Costa, D. P., & Godley, B. J. (2010). Biologging technologies: New tools for conservation. *Introduction Endangered Species Research, 10,* 1–7. doi:10.3354/esr00269.

Breaker, L. C. (1981). The applications of satellite remote sensing to West Coast fisheries. *Marine Technology Society Journal, 15,* 32–40.

Briscoe, D., Hiatt, S., Lewison, R., & Hines, E. (2014). Modeling habitat and bycatch risk for dugongs in Sabah, Malaysia. *Endanger Species Research, 24,* 237–247. doi:10.3354/esr00600.

Brown, C. J., White, C., Beger, M., Grantham H. S., Halpern, B. S., Klein, C. J., et al. (2015). Fisheries and biodiversity benefits of using static versus dynamic models for designing marine reserve networks. *Ecosphere, 6* (10). Article Number: 182.

Chan, K. M. A., Guerry, A. D., Balvanera, P., Klain, S., Satterfield, T., Basurto, X., et al. (2012). Where are cultural and social in ecosystem services? A framework for constructive engagement. *BioScience, 62,* 744–756. doi:10.1525/bio.2012.62.8.7.

Chassot, E., Bonhommeau, S., Reygondeau, G., Nieto, K., Polovina, J. J., Huret, M., et al. (2011). Satellite remote sensing for an ecosystem approach to fisheries management. *ICES Journal of Marine Science, 68*(4), 651–666. doi:10.1093/icesjms/fsq195/.

Checkley, D. M., Dotson, R. C., & Griffith, D. A. (2000). Continuous, underway sampling of eggs of Pacific sardine (Sardinops sagax) and northern anchovy (Engraulis mordax) in spring 1996 and 1997 off southern and central California. *Deep-Sea Research Part II-Topical Studies In Oceanography, 47*(5–6), 1139–1155.

Collie, J. S., Adamowicz, W. L., Beck, M. W., et al. (2013). Marine spatial planning in practice by: Estuarine Coastal And Shelf. *Science, 117,* 1–11.

Cornillon, P., et al. (1986). Sea surface temperature charts for the southern New England fishing community. *The Marine Technology Society Journal, 20*(2), 57–65.

Costa, D. P., Breed, G. A., & Robinson, P. W. (2012). New insights into Pelagic migrations: Implications for ecology and conservation. D.J. Futuyma (Eds.). *Annual Review of Ecology Evolution and Systematics, 43,* 73–96.

Cushing, D. H. (1982). *Detection of fish* (p. 200). London: Pergamon Press.

Davies, R. W. D., Cripps, S. J., Nickson, A., & Porter, G. (2009). Defining and estimating global marine fisheries bycatch. *Marine Policy, 33,* 661–672. doi:10.1016/j.marpol.2009.01.003.

Dunn, D. C., Maxwell, S. M., Boustany, A. M., & Halpin, P. N. (2016). Dynamic ocean management increases the efficiency and efficacy of fisheries management. *Proceedings of the National Academy of Sciences, 113*(3), 668–673. doi:10.1073/pnas.1513626113.

EOS. (2006). Using satellite data products to manage living marine resources. *Eos, 87*(41), 437–438.

FAO. (2003). The application of remote sensing technology to marine fisheries: An introductory manual (Section 7). *Food and Agriculture Organization Corporate Document Repository.*

FAO. (2009). *The state of world fisheries and aquaculture 2008.* Rome, Italy: FAO Documentation Group. 176 p.

Francis, M. P., Holdsworth, J. C., & Block, B. A. (2015). Life in the open ocean: Seasonal migration and diel diving behaviour of Southern Hemisphere porbeagle sharks (Lamna nasus). *Marine Biology, 162,* 2305–2323. doi:10.1007/s00227-015-2756-z.

Gardner, B., Sullivan, P. J., Morreale, S. J., et al. (2008). Spatial and temporal statistical analysis of bycatch data: Patterns of sea turtle bycatch in the North Atlantic. *Canadian Journal of Fisheries and Aquatic Sciences, 65*(11), 2461–2470.

Hall, M. A., Alverson, D. L., & Metuzals, K. I. (2000). By-catch: Problems and solutions. *Marine Pollution Bulletin, 41*, 204–219. doi:10.1016/S0025-326X(00)00111-9.

Hall, M., Nakano, H., Clarke, S., Thomas, S., Molloy, J., Peckham, S., et al. (2007). Working with fishers to reduce bycatches. In S. Kennelly (Ed.), *Bycatch reduction in the world's fisheries* (pp. 235–288). Dordrecht: Springer.

Hammerschlag, N., Gallagher, A. J., Lazarre, D. M., et al. (2011). Range extension of the Endangered great hammerhead shark Sphyrna mokarran in the Northwest Atlantic: Preliminary data and significance for conservation. *Endangered Species Research, 12*(2), 111–116.

Haury, L. R., McGowan, J. A., & Wiebe, P. H. (1978). Patterns and processes in the time-space scales of plankton distributions. In J.H. Steele (Ed.), *Spatial patterns in plankton communities* (pp. 277–327). Plenum Press.

Hazen, E. L., Maxwell, S. M., Bailey, H., Bograd, S. J., Hamann, M., Gaspar, P., et al. (2012). Ontogeny in marine tagging and tracking science: Technologies and data gaps. *Marine Ecology Progress Series, 457*, 221–240.

Hazen, E. L., Jorgensen, S., Rykaczewski, R. R., Bograd, S. J., Foley, D. G., Jonsen, I. D., et al. (2013). Predicted habitat shifts of Pacific top predators in a changing climate. *Nature Climate Change, 3*, 234–238.

Hisschemoller, M., & Hoppe, R. (1995). Coping with intractable controversies: The case for problem structuring in policy design and analysis. *Knowledge, Technology and Policy, 8*(4), 40–60. doi:10.1007/bf02832229.

Hobday, A. J., Hartog, J. R., Timmis, T., & Fielding, J. (2010). Dynamic spatial zoning to manage southern bluefin tuna capture in a multi-species longline fishery. *Fisheries Oceanography, 19*, 243–253. doi:10.1111/j.1365-2419.2010.00540.x.

Hobday, A. J., Smith, A. D. M., Stobutzki, I. C., Bulman, C., Daley, R., Dambacher, J. M., et al. (2011). Ecological risk assessment for the effects of fishing. *Fisheries Research, 108*, 372–384. doi:10.1016/j.fishres.2011.01.013.

Hobday, A. J., Maxwell, S. M., Forgie, J., Mcdonald, J., Darby, M., Seto, K., et al. (2014). Dynamic ocean management: Integrating scientific and technological capacity with law, policy and management. *Stanford Environmental Law Journal, 33*, 125–165. https://journals.law.stanford.edu/sites/default/files/stanford-environmental-law-journal-selj/print/2014/03/i_hobday_final.pdf.

Howell, E. A., Hoover, A., Benson, S. R., Bailey, H., Polovina, J. J., Seminoff, J. A., et al. (2015). Enhancing the TurtleWatch product for leatherback sea turtles, a dynamic habitat model for ecosystem-based management. *Fisheries Oceanography, 24*, 57–68. doi:10.1111/fog.12092.

Howell, E., Kobayashi, D., Parker, D., Balazs, G., & Polovina, J. (2008). TurtleWatch: A tool to aid in the by-catch reduction of loggerhead turtles Caretta caretta in the Hawaii-based pelagic longline fishery. *Endanger Species Research, 5*, 267–278. doi:10.3354/esr00096.

James, M., Ottensmeyer, C., & Myers, R. (2005). Identification of high-use habitat and threats to leatherback sea turtles in northern waters: New directions for conservation. *Ecology Letters, 8*, 195–201. doi:10.1111/j.1461-0248.2004.00710.x.

Kachelriess, D., Wegmann, M., Gollock, M., et al. (2014). The application of remote sensing for marine protected area management. *Ecological Indicators, 36*, 169–177.

Kahru, M., Hakansson, B., & Rud, O. (1995). Distributions of the sea-surface temperature fronts in the baltic sea as derived from satellite imagery. *Continental Shelf Research, 15*(6), 663–679. Published: MAY 1995.

Kelleher, K. (2005). *Discards in the World's Marine Fisheries: An Update*. Technical Paper. No. 470. Rome: FAO Fisheries. p. 131.

Klemas, V. (2012a). Remote sensing of coastal plumes and ocean fronts: Overview and case study. *Journal Of Coastal Research 28*(1A_S), 1–7.

Klemas, V. (2012b). Remote sensing of environmental indicators of potential fish aggregation: An overview. *Baltica, 25*(2), 99–112.

Klemas, V. (2013). Fisheries applications of remote sensing: An overview. *Fisheries Research, 148*, 124–136.

Larnicol, G., Guinehut, S., Rio, M.-H., Drevillon, M., Faugere, Y., & Nicolas, G. (2006). The global observed ocean products of the French mercator project. In *Proceedings of 15 Years of progress in radar altimetry Symposium*, ESA Special Publication, pp. 614.

Lengyel, S., Déri, E., Varga, Z., Horváth, R., Tóthmérész, B., Henry, P.-Y., et al. (2008). Habitat monitoring in Europe: A description of current practices. *Biodiversity and Conservation, 17*, 3327–3339. doi:10.1007/s10531-008-9395-3.

Lewison, R., Crowder, L., Read, A., & Freeman, S. (2004). Understanding impacts of fisheries bycatch on marine megafauna. *Trends in Ecology and Evolution, 19*, 598–604. doi:10.1016/j.tree.2004.09.004.

Lewison, R. L., Soykan, C. U., & Franklin, J. (2009). Mapping the bycatch seascape: Multispecies and multi-scale spatial patterns of fisheries bycatch. *Ecological Applications, 19*, 920–930. doi:10.1890/08-0623.1.

Lewison, R., Hobday, A., Maxwell, S., Hazen, E., Hartog, J., Dunn, D., et al. (2015). Dynamic ocean management: Identifying the critical ingredients of dynamic approaches to ocean resource management. *BioScience, 65*, 486–498. doi:10.1093/biosci/biv018.

Luo, J., Ault, J. S., Shay, L. K., Hoolihan, J. P., Prince, E. D., Brown, C. A., et al. (2015). Ocean heat content reveals secrets of fish migrations. *PLoS ONE, 10*(10): e0141101. doi:10.1371/journal.pone.0141101.

Lynn, R. J. (2003). Variability in the spawning habitat of Pacific sardine (Sardinops sagax) off southern and central California. *Fisheries Oceanography, 12*(6), 541–553.

Maxwell, S. M., Hazen, E. L., Lewison, R. L., Dunn, D. C., Bailey, H., Bograd, S. J., et al. (2015). Dynamic ocean management: Defining and conceptualizing real-time management of the ocean. *Mar. Policy, 58*, 42–50. doi:10.1016/j.marpol.2015.03.014.

Pettorelli, N., Nagendra, H., Willians, R., Rocchini, D., & Fleishman, E. (2014). A new platform to support research at the interface of remote sensing, ecology and conservation. *Remote Sensing in Ecology and Conservation, 1*, 1–3. doi:10.1002/rse2.1.

Polovina, J., Uchida, I., Balazs, G., Howell, E., Parker, D., & Dutton, P. (2006). The Kuroshio extension bifurcation region: A pelagic hotspot for juvenile loggerhead sea turtles. *Deep Sea Research II, 53*, 326–339. doi:10.1016/j.dsr2.2006.01.006.

Radovich, J. (1982). The collapse of the California sardine fishery. *What have we learned*, pp. 56–77.

Ramos, A. G., Santiago, J., Sangra, P., & Canton, P. (1996). An application of satellite-derived sea surface temperature data to the skipjack and albacore tuna fisheries in the north-east Atlantic. *International Journal of Remote Sensing, 17*, 749–759. doi:10.1080/01431169608949042.

Rose, G. A., deYoung, B., Kulka, D. W., Goddard, S. V., & Fletcher, G. L. (2000). Distribution shifts and overfishing the northern cod (Gadus morhua): A view from the ocean. *Canadian Journal of Fisheries and Aquatic Sciences, 57*, 644–663. doi:10.1139/f00-004.

Royer, F., Fromentin, J.-M., & Gaspar, P. (2004). Association between bluefin tuna schools and oceanic features in the western Mediterranean. *Marine Ecology Progress Series, 269*, 249–263.

Santos, A. M. P. (2000). Fisheries oceanography using satellite and airborne remote sensing methods: A review. *Fisheries Research, 49*(1), 1–20.

Santos, A. M. P, & Fiúza, A. F. G. (1992). Supporting the Portuguese fisheries with satellites. In *ESA ISY-1 (ESA SP-341)*, pp. 663–668.

Scheuerell, M. D., & Williams, J. G. (2005). Forecasting climate-induced changes in the survival of Snake River spring/summer Chinook salmon (Oncorhynchus tshawytscha). *Fisheries Oceanography, 14*(6), 448–457.

Schultz, L., Folke, C., Österblom, H., & Olsson, P. (2015). Adaptive governance, ecosystem management, and natural capital. *Proceedings of the National Academy of Sciences, 112*(24), 7369–7374.

Sims, M., Cox, T., & Lewison, R. (2008). Modeling spatial patterns in fisheries bycatch: Improving bycatch maps to aid fisheries management. *Ecological Applications, 18*, 649–661. doi:10.1890/07-0685.1.

Smith, P. E., & Moser, H. G. (2003). Long-term trends and variability in the larvae of Pacific sardine and associated fish species of the California Current region. *Deep Sea Research Part II: Topical Studies in Oceanography, 50*(14), 2519–2536.

Soykan, C. U., & Lewison, R. L. (2015). Using community-level metrics to monitor the effects of marine protected areas on biodiversity. *Conservation Biology.* doi:10.1111/cobi.12445.

Sudre, J., & Morrow, R. (2008). Global surface currents: A new product for investigating ocean dynamics. *Ocean Dynamics, 58*(2):101–118.

Weber, E. D., & McClatchie, S. (2010). Predictive models of northern anchovy Engraulis mordax and Pacific sardine Sardinops sagax spawning habitat in the California Current. *Marine Ecology Progress Series, 406,* 251–263.

Wright, D. J., Woodworth, B. M., & O'Brien, J. J. (1976). A system for monitoring the location of harvestable coho salmon stocks. *Marine Fisheries Review, 38,* 1–7.

Wells, B. K., Field, J. C., Thayer, J. A., Grimes, C. B., Bograd, S. J., Sydeman, W. J., et al. (2008). Untangling the relationships among climate, prey and top predators in an ocean ecosystem. *Marine Ecology Progress Series, 364,* 15–29.

Yamanaka, I. (1988). *The fisheries forecasting system in Japan for coastal pelagic fish,* (No. 301). Food and Agriculture Organisation.

Zwolinski, J. P., Emmett, R. L., & Demer, D. A. (2011). Predicting habitat to optimize sampling of Pacific sardine (Sardinops sagax). *ICES Journal of Marine Science: Journal du Conseil, 68*(5), 867–879.

Zydelis, R., Lewison, R. L., Shaffer, S., Moore, J., Boustany, A., Roberts, J., et al. (2011). Dynamic habitat models: Using telemetry data to project fisheries bycatch. *Proceedings of the Royal Society B: Biological Sciences, 282,* 1–10. doi:10.1098/rspb.2011.0330.

Improving NASA's Earth Observation Systems and Data Programs Through the Engagement of Mission Early Adopters

Vanessa M. Escobar, Margaret Srinivasan and Sabrina Delgado Arias

Abstract This chapter provides an overview of the NASA Early Adopter Program from the perspective of three new and planned satellite earth observing missions—SMAP, ICESat-2 and SWOT. The level of activity and engagement of the mission's applications is directly related to the maturity of the mission products and algorithms. Early Adopters were introduced to NASA through the SMAP Mission Applications Program in 2010 and it quickly became adopted by other missions (ICESat-2 2013 and SWOT in 2014). Early Adopters work with the mission scientists to test the thematic uses and applications of mission products, while providing feedback on the accomplishments and challenges of the data.

1 Overview

This chapter will give an overview of the NASA Early Adopter Program from the perspective of three missions, SMAP, ICESat-2, and SWOT. The development of a NASA Applications Program is broken up into stages that complement the development of science product development. The level of activity and engagement of the mission's applications is directly related to the maturity of the mission products and algorithms. Early Adopters were introduced to NASA through the SMAP Mission Applications Program in 2010 and it quickly became adopted by other missions (ICESat-2 2013 and SWOT in 2014). Early Adopters work with the mission scientists to test the thematic uses and applications of mission products, while providing feedback on the accomplishments and challenges of the data. Early Adopters are a subset of the mission's community of users that volunteer their efforts and resources to use prelaunch mission data in their organization. The Early

---

V.M. Escobar (✉) · S.D. Arias
NASA GSFC, Greenbelt, USA
e-mail: Vanessa.m.escobar@nasa.gov

M. Srinivasan (✉)
NASA JPL, Pasadena, USA
e-mail: Margaret.srinivasa@jpl.nasa.gov

© Springer International Publishing Switzerland 2016
F. Hossain (ed.), *Earth Science Satellite Applications*,
Springer Remote Sensing/Photogrammetry, DOI 10.1007/978-3-319-33438-7_9

adopter feedback, experience, and recommendations are incorporated into the mission development through strategic communication activities and outreach that help the mission scientist gain an outside perspective of how the science can be used by society. In return, the early adopters are provided guidance and mentorship from the science team in the development of their research and they are given the opportunity to present their results to NASA before any mission data become publically available, accelerating the early adopters research results, and applications after launch. NASA embraced the effort of including Early Adopters into the mission life cycle as a new standard for future missions during the SMAP mission life cycle. Understanding *how* NASA data are used in different types of applications will continue to inform future science for NASA. Incorporating user feedback during mission development and fostering Early Adopters in the development stages of the mission life cycle is also a way of accelerating product knowledge and relevance after launch. The Early adopter program is part of NASA's new approach for making science an integral part of society. Through the Early Adopters, missions incorporate feedback and leverage relationship in an effort to make science applications more impactful and relevant to the everyday users. Application Programs and Early Adopters are redefining how NASA develops science and shares knowledge with the public. The new decadal survey missions now require Early Adopters to be a part of the project architecture from conception through the end of the mission's life cycle, making sure science is not only unique and informative but useful and impactful.

2 Introduction

Satellite data provide continuous information for most Earth systems. These data are valuable for developing scientific products that explain the how earth's systems are integrated. These data can be integrated into models and decision support systems that enhance our knowledge and enable improved natural resource management, disaster prevention and response, and other benefits to society. However, the value is best understood through thematic presentations and explicit communications describing how the data impacts our everyday lives. Providing science data in a user-friendly fashion (format, language, scale, etc.) to decision-makers is a priority for NASA, particularly for the issues that surround climate and climate change. Although satellite data and products prove to be useful, it is still necessary to describe the utility of those product and link them to societal relevant topics so the science resonates with the everyday user and benefactor of the information. Asking, 'how can we identify the policies and decisions that benefit from integrating satellite data products?' is how NASA is addressing applications for future missions. One of NASA's Applied Sciences Program (ASP) strategic goals for science applications is to "ensure that NASA's flight missions plan for and support applications goals in conjunction with their science goals, starting with project planning and extending through the project life cycle" (NASA 2009). An action item related to this goal is to "evaluate the potential for current and planned NASA missions to meet societal needs through applied sciences participation in mission science teams."

United States launched its first Earth satellite on January 31, 1958, when *Explorer 1* documented the existence of radiation zones encircling the Earth. With over 20 satellites orbiting in operations today, NASA remains a leading force in scientific research. The objectives of the existing missions are strategic, identified in decadal surveys from the National Academy of Sciences, and created to meet national or agency objectives. Others are competed missions selected in response to open solicitations. All missions are challenging endeavors that seek to understand science at different capacities and applications.

These missions take years to evolve and they progress in phases (Pre Phase A through Phase F). Each phase of a mission has specific science objectives and goals it must reach before "graduating" to the next phase. Phases Pre-Phase A through D are the "pre-launch" stages of a mission where the algorithms for the science are in development and the calibration and validation of field data against those algorithms are in full effect. As the mission progresses from Phase A to Phase D, a science definition (SDT) team is selected to design the science products for the mission. During this process is when the applications of the science products are defined. Pre-launch can take anywhere from 5 to 8 years to complete before reaching launch (Phase E). Phase E is the "operational" stage of the mission life cycle, beginning at launch, Phase E lasts until the satellite is decommissioned. Phase E is where the satellite observations are developed into actual science products promised by the mission. Leading up to Phase E, the SDT also is recompleted as a new science team (ST), no longer in development of algorithms. The focus of the science team is now to calibrate and validate the actual satellite observations with the algorithms developed in the prelaunch phases of the mission's life cycle. Phase E is also the phase were the public has access to the data and can start using the mission products in their research. Because NASA is a research organization, the operational stages of a mission are designed to last 3–5 years. Some mission lasts much longer than the 3–5 year life expectancy (like MODIS) and continue to operate until it is no longer operational. However, the intent is to have 3–5 years collected and then handed off to an operational user (NOAA, USGA, DoD, etc.) for its continued use. Phase E, traditionally, is when society learns most about the applications of the science products. However, learning about applications after the design and development mission have all been made, makes it an expensive challenge for implementing feedback or making improvements to the science products. Early Adopters were introduced to NASA by the Soil Moisture Active Passive (SMAP) mission in an effort to implement feedback into the design of mission product development. The introduction of Early Adopters to NASA will forever change the way science is developed.

In 2007, the National Research Council (NRC) released the first Decadal Survey for Earth Science (National Research Council 2007). The Decadal Survey outlined research initiatives of national importance for the next decade and identified the development of applications of satellite data as a priority for all future missions.

The Applied Sciences Division responded to the NRC guidance by 'identifying applications leads for all Tier 1 Decadal Survey missions and making applications an integral part of all decadal survey missions'. The intent of this response is for those leading application for missions to engage with the project scientists during the prelaunch development process and to facilitate interaction with the relevant communities of users to ensure that mission data products will achieve maximum value. Because satellite missions can take a decade or more to develop, engaging with potential data users during the early stages of mission development is the best way to ensure that new observations meet the most urgent demands of users and have the broadest impact. Through early engagement in the prelaunch development stage of a project, each individual mission's applications program generates feedback loops that feed into the development of products making more relevant and user-friendly data products at launch. This feedback helps connects end users with science data product developers and distributors (such as the Distributed Active Archive Centers (DAACs)), increasing the visibility of the products and enabling the maximum benefit to society from satellite missions.

The 2007 Decadal Survey served as the catalyst for the development of a formalized Applications Program for NASA missions (National Research Council 2007). The NASA Applied Science Program redefined the way it conducts applications through the NRC guidance and is quickly providing examples of socioeconomic benefits of satellite data through project-specific Early Adopter Programs. The Early Adopter program is a component within each NASA Mission's Applications Program (Moran et al. 2015).

3 The Early Adopter Program

Early Adopters are an inclusive community of users, representing science, commercial, private, academic, and federal organizations that are willing to work alongside mission scientists and provide feedback to the mission scientists related to how the science product(s) are used in their field of expertise. The goal of the Early Adopter Program is to facilitate feedback on mission products prelaunch, and to accelerate the use of these products *post-launch*, by providing specific guidance and early mission data access to Early Adopters who commit to engage in applied research. The Early Adopter program is a non-funded activity that leverages existing resources, relationships, and opportunities. The program provides a framework for building a broad and well-defined user community during the prelaunch phases of the mission to maximize the use of data products after launch and to provide early insight into the range of potential uses of the mission data (Moran et al. 2014).

NASA's SMAP Mission was the first to implement an Early Adopter Program into the mission development structure during prelaunch stage of the mission and continued working with Early Adopters through operations. The overarching objective of the program was to broaden and strengthen the knowledge of the community to the science of SMAP, improve the knowledge of the applications for the science team and demonstrate the value of the data uses in ways that facilitate the science to help society. The Early Adopter program will identify communities of users for the Earth Observation (EO) missions at NASA that will communicate lessons learned for how the mission data are used in areas of society that have an impact on health, natural resources, and the economy. Early Adopters promote applications research to provide a fundamental understanding of how the mission data products can be scaled and integrated into an organization's policy, business or management activities to improve decision-making efforts (Escobar and Arias 2015). Through their use of the Early Adopter program, NASA Earth Observer (EO) missions described in this chapter create the necessary feedback loops between the mission and user communities for an understanding of the utility of mission data products within different decision-making contexts. As a result of SMAP's success with Early Adopters, the program was highlighted as a best practice in the 2012 National Academies Midterm Assessment report (http://www.nap.edu/catalog/13405/earth-science-and-applications-from-space-a-midterm-assessment-of).

All Early Adopters are unique because the conduct applied research at a variety of spatial and temporal scales, aimed towards answering targeted questions that clearly link how science is helping our everyday lives. The benefit of participating in an Early Adopter program is not only the access to early mission data but also the relationships that develop within science team and other Early Adopters (Fig. 1). Mission scientists benefit greatly from Early Adopter feedback, gaining perspective on the utility and challenges of the mission data, while they are still in development. Using feedback early in mission development allows for corrections and adjustments to data products can be made without an increased cost to the mission. These adjustments address things like latency, format and resolution, optimizing cost, and improving the mission understanding of uses of products after launch.

Feedback provides opportunity for improvements to data products not only through mission product design but also data delivery. One example of this improvement comes from the 2012 community assessment of the SMAP community (Brown and Escobar 2014). Here the goal of the assessment was to solicit data requirements, accuracy needs and current understanding of the SMAP mission by the user community. The SMAP user community was asked to describe accuracy needs for spatial, temporal, and latency characteristics of their data use and applications. The results of the assessment identified gaps in the fields of scientific research, operations, and those involved in public policy and decision-making. The results showed 80 % of the reviewers were involved in carrying out soil moisture related research, and 60 % of those reviewers stated that they were not involved policy or decision-making applications (Brown and Escobar 2014). By making the Google Earth file format, (KMZ), a standard format for SMAP data product, the mission enabled policy and decision-making communities who use satellite data, to use a more familiar format for applying SMAP to their efforts. Feedback such as this continues to bring value to NASA and the applications of mission data.

Participants in the Early Adopter program not only gain exclusive access to early mission data but they get increased engagement with the mission science team. The

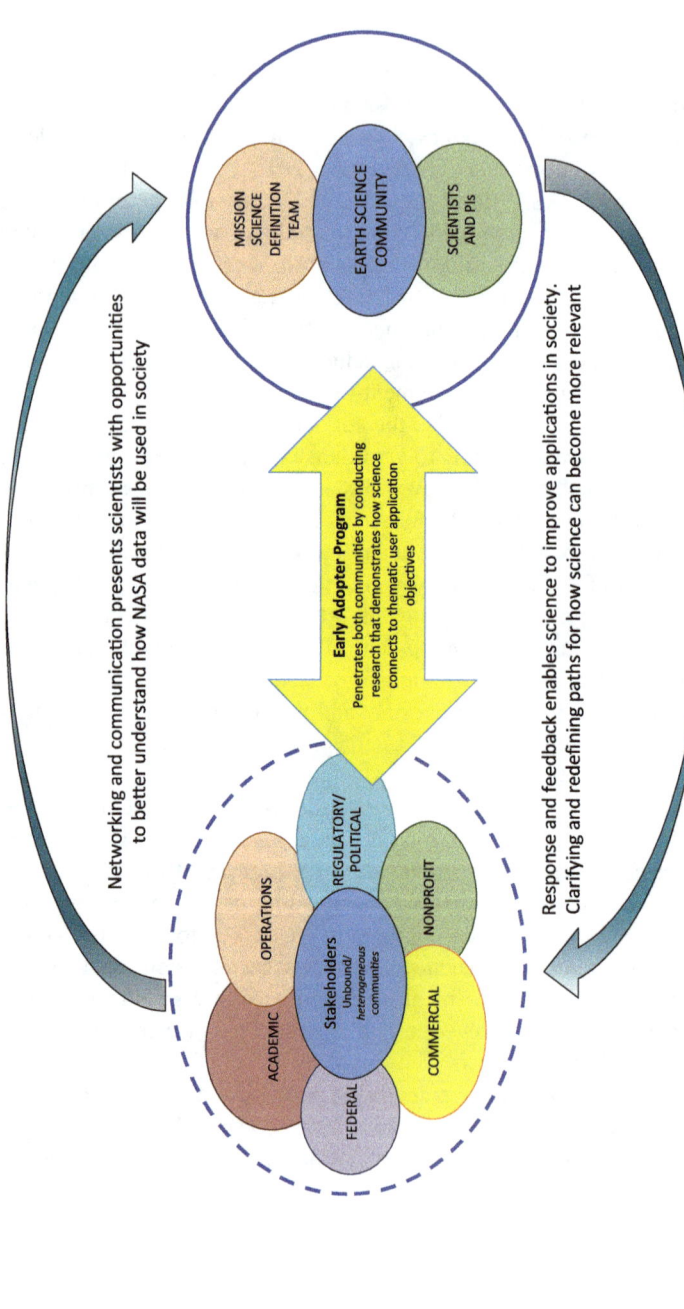

Fig. 1 Early Adopter Program Schematic. Early Adopters penetrate stakeholder groups and the Earth Science community through research that demonstrates how science connects to thematic user application objectives

early mission data provides an advanced into what the mission will produce after launch. This gives the Early Adopter time to prepare and develop methods for ingesting the data; cutting back on time and resources needed had waited to adopt the data after launch. Early Adopters benefit from early mission science and technology to improve their operating or decision-making systems. Both the mission scientists and the Early Adopters gain the unique opportunity to develop relationships that help identify the impact of science and implement lessons learned that reveal products that inform societally relevant applications.

How an Early Adopter Program is implemented and how it evolves is unique to each mission and the science objectives of that mission. NASA's SMAP Mission, the Ice Cloud and Elevation Satellite (ICESat-2) Mission and Surface Water and Ocean Topography (SWOT) Mission have each designed their Early Adopter Program in accordance with the mission science objectives and a targeted user community. Throughout the life of the program, the applications of the mission science and the targeted community expand to include new applications and users organically. The SMAP, ICESAt-2, and SWOT projects are currently at different stages in the project life cycle, therefore each project has different activities they are conducting with their Early Adopter community. All three missions use their Early Adopter program as bridges between the mission science and the end users providing thematically specific sample of how the mission science will benefit from their research objective, which is directly linked to a policy or decision process that is relevant to the user community. The objectives of this chapter are to provide you an understanding of NASA's Early Adopter Program, demonstrate how the program works for three different NASA missions, and give examples of the type of research Early Adopters are conducting for missions at different phases of the mission life cycle.

4 Examples of Mission Early Adopters and Case Studies

The Early Adopter programs described below are a learning experience that provide direction for how mission planning is conducted at NASA and setting some context for the next NRC decadal survey in 2017.

5 NASA's SMAP Mission

One of the first missions recommended by the NRC decadal survey was the SMAP mission (Fig. 2), launched on January 31 2015. The SMAP mission is designed to provide global soil moisture and freeze–thaw measurements from space for applications in fields of weather, climate, drought, flood, agricultural production, human health, and national security. The SMAP mission provides high quality; daily, global soil moisture data, which have both high science and applications value. SMAP measurements are used to enhance the understanding of processes that link

Fig. 2 SMAP Satellite Image
Credit: NASA JPL

the water, energy, and carbon cycles, and to extend the capabilities of weather and climate prediction models (Brown et al. 2012). The satellite orbit provides an 8-day revisit with optimum coverage of global land area at 3-day average intervals.

The range of applications in weather, flooding, drought, agriculture, health, and national security for SMAP are defined partly by the spatial resolution and latency of the SMAP products, ranging from approximately 1–3 km for the synthetic aperture radar (SAR) data to the 39 km × 47 km radiometer data, nominally referred to as 40-km resolution (Moran et al. 2015). The high resolution of the radar is critical for accurate determination of freeze–thaw state, and the low-resolution radiometer measurements are key to deriving the soil moisture product. The radar and radiometer measurements are combined to generate a mid-resolution product that optimizes the resolution and accuracy attributes of the radar and radiometer. Sadly, after 10 weeks of postlaunch measurements the SMAP radar stopped transmitting on July 7 2015. In response, the SMAP Early Adopters conducted investigations of products other than soil moisture and freeze/thaw to increase the science and applications value of SMAP radiometer data stream. The Early Adopters are conducting short-term studies with the 10-week SMAP radar/radiometer product stream to report on the value of the unique SMAP radar/radiometer data for applications.

Some Early Adopters and the SMAP community are assisting to develop and test SMAP science recovery of higher resolution soil moisture and FT. The SMAP Applications Team is engaging with Early Adopters to provide feedback to the mission. This unexpected, yet important, effort will provide lessons learned to NASA Headquarters.

In 2009, SMAP Project funded the first formal Applications Program that engaged with data users (science and commercial) early enough in the mission design to have impact on the developed mission products. The SMAP Early Adopter Program was initiated in 2010 to support prelaunch applied research, which provided fundamental knowledge of how SMAP data products could be scaled and integrated into users' policy, business, and management activities to improve decision-making efforts. Since its inception, the SMAP Early Adopter Program has been recognized by NASA Headquarters as the most valuable component of the SMAP Applications Program. SMAP Early adopters are formally defined as those users who have a direct or clearly defined need for SMAP-like data and who are planning to apply their own resources to demonstrate the utility of SMAP data for their particular system or model.

The goal of the SMAP Early Adopter is to provide quantitative metrics designed to benefit the users, as well as the SMAP Project. With a total of 55 Early Adopters (Fig. 3 and Appendix 1), SMAP Early Adopters conduct research in unprecedented prelaunch preparation for SMAP applications, provided critical feedback to the mission to improve product specifications, are helping improve the distribution of products for postlaunch applications, and provide perspective to the impact SMAP data products have on societal areas of interest. After launch in early 2015, Early Adopters continued to work on their research effort with the use of real observations, each one achieving unique milestone as they progressed. Because Early Adopter feedback was so important to the mission, Phase E case studies were proposed as part of the post-launch applications efforts. Case studies are example projects that demonstrate scientific and societal impact through the efforts of the research. For SMAP Phase E case studies, select Early Adopters demonstrate how SMAP science data are (1) ingested and (2) used technically by their organization, (3) while providing feedback about any challenges, changes, or improvements to their system processes. The Phase E case studies for SMAP show the data life cycle and scientific knowledge gain from the products. Through case studies, SMAP learns about how the use of the data affects the research hypothesis of the Early Adopters but also understand if the science has the potential to impact an application or a decision. The SMAP mission plans to qualify the results for six individual case studies and publish their results after 2018. In this next section, the most recent SMAP Early Adopter case study preliminary results are presented.

Fig. 3 The logos of 55 SMAP Early Adopters

6 SMAP Early Adopter Examples

SMAP Early Adopters are a diverse and international group of scientist, engineers, manages, and entrepreneurs representing applications in a broad range of areas (Fig. 4). The examples of SMAP Early Adopters that follow provide an overview of the program and the areas of SMAP applications. The examples below as a sample of the contributions that continue to be made to the project and demonstrate the value of early mission engagement and stakeholder relationships. The project is enormously grateful for the effort of our Early Adopter community. More results, updates on Early Adopters, and biographies can be found on the SMAP Mission Applications website at http://smap.jpl.nasa.gov/science/applications/.

6.1 Weather and Climate Forecasting

Soil moisture information is important for temperature and precipitation forecasting. Initialization of numerical weather prediction and seasonal climate models with accurate soil moisture information enhances their prediction skills and extends their skillful lead times. Improved seasonal climate predictions benefit climate-sensitive socioeconomic activities, including water management, agriculture, fire, flood, and drought hazards monitoring. SMAP data products are used to constrain land surface models. NOAA researchers Kyle McDonald, Michael Ek, Marouane Temimi, Xiwu Zhan, and Weizhong Zhen are working together on the development of proxies for

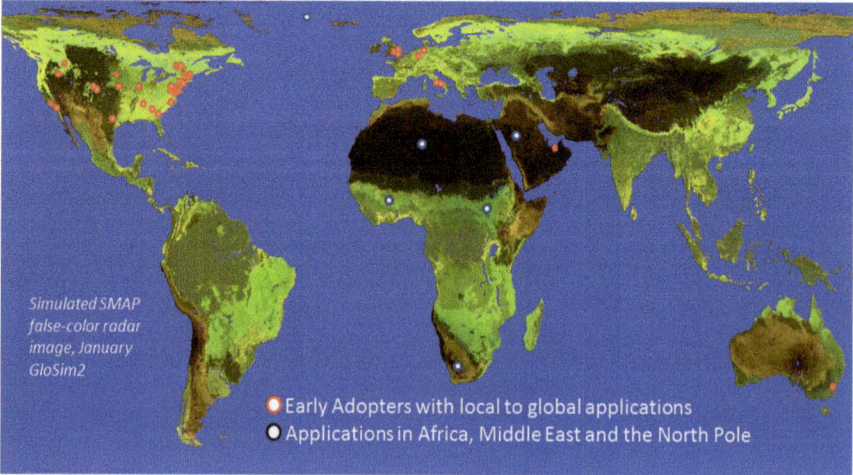

Fig. 4 A simulated image of SMAP radar data scaled backscatter triplets in the order VV, HH, and HV to a 24-bit color map and overlain with the locations of 37 of 55 early adopters conducting applied research over local-to-global domains, and specific applications in Africa, the Middle East, and the North Pole (Moran et al. 2015)

Product	Description	Gridding (Resolution)	Latency**	
L1A_Radiometer	Radiometer Data in Time-Order	-	12 hrs	Instrument Data
L1A_Radar	Radar Data in Time-Order	-	12 hrs	
L1B_TB	Radiometer T_B in Time-Order	(36×47 km)	12 hrs	
L1B_S0_LoRes	Low-Resolution Radar σ_o in Time-Order	(5×30 km)	12 hrs	
L1C_S0_HiRes	High-Resolution Radar σ_o in Half-Orbits	1 km (1–3 km)#	12 hrs	
L1C_TB	Radiometer T_B in Half-Orbits	36 km	12 hrs	
L2_SM_A	Soil Moisture (Radar)	3 km	24 hrs	Science Data (Half-Orbit)
L2_SM_P*	Soil Moisture (Radiometer)	36 km	24 hrs	
L2_SM_AP*	Soil Moisture (Radar + Radiometer)	9 km	24 hrs	
L3_FT_A*	Freeze/Thaw State (Radar)	3 km	50 hrs	Science Data (Daily Composite)
L3_SM_A	Soil Moisture (Radar)	3 km	50 hrs	
L3_SM_P*	Soil Moisture (Radiometer)	36 km	50 hrs	
L3_SM_AP*	Soil Moisture (Radar + Radiometer)	9 km	50 hrs	
L4_SM	Soil Moisture (Surface and Root Zone)	9 km	7 days	Science Value-Added
L4_C	Carbon Net Ecosystem Exchange (NEE)	9 km	14 days	

Fig. 5 SMAP data product table

SMAP freeze/thaw data product. They plan to use SMAP data to assess the improvement in the performance of the NCEP Noah land surface model and assess the assimilation of SMAP freeze/thaw product on the performance NCEP weather forecast models. NOAA Soil Moisture Product System (SMOPS) will ingest available satellite soil moisture observations to NOAA's National Weather Prediction and other operations and research. Preliminary test shows that SMOPS could retrieve soil moisture products from L1B TB (Fig. 5) data with shorter latency, ultimately helping improve operational weather predictions.

SMAP has a total of seven Early Adopters focused on weather and climate forecasting. Table 1 lists SMAP Early Adopters who address weather related applications.

6.2 Droughts and Wildfires

Soil moisture strongly affects plant growth and hence agricultural productivity, especially during conditions of water shortage and drought. Soil moisture information can be used to predict wildfires, determine prescribed burning conditions, and estimate smoldering combustion potential of organic soils. For wildfires, SMAP soil moisture products provide more useful and accurate data on toxic air-quality events and smoke whiteouts (thus increasing transportation safety) and can inform prescribed fire activities (increasing efficiency). Early Adopters focused on drought will use data assimilation to assimilate SMAP data into land surface models. These models can be run at any scale, and the SMAP retrievals act as a constraint on the model, so while the loss of the radar does mean that there is a reduced quality of information, it is not a fundamental limitation.

Table 1 Weather and climate forecasting early adopters

Weather and climate forecasting	
Researcher	Organization
* Stephane Bélair	Meteorological Research Division, Environment Canada (EC)
* Lars Isaksen and Patricia de Rosnay	European Centre for Medium-Range Weather Forecasts (ECMWF)
* Xiwu Zhan, Michael Ek, John Simko and Weizhong Zheng	NOAA National Centers for Environmental Prediction (NCEP), NOAA National Environmental Satellite Data and Information Service (NOAA-NESDIS)
* Michael Ek, Marouane Temimi, Xiwu Zhan and Weizhong Zheng	NOAA National Centers for Environmental Prediction (NCEP), NOAA National Environmental Satellite Data and Information Service (NOAA-NESDIS), City College of New York (CUNY)
* John Galantowicz	Atmospheric and Environmental Research, Inc. (AER)
◊ Jonathan Case, Clay Blankenship and Bradley Zavodsky	NASA Short-term Prediction Research and Transition (SPoRT) Center
◊ Steven Quiring	Texas A&M University

* Early Adopters selected in 2011–2012. ◊ Early Adopters selected from 2013 forward

Famine Early Warning Systems Network (FEWS NET), was created in 1985 by the US Agency for International Development (USAID) after devastating famines in East and West Africa, FEWS NET is a leading provider of early warning and analysis on acute food insecurity and provides objective, evidence-based analysis to help government decision-makers and relief agencies plan for and respond to humanitarian crises in Central Asia and North Africa. Analysts and specialists in 22 field offices work with US government science agencies, national government ministries, international agencies, and NGOs to produce forward-looking reports on more than 36 of the world's most food-insecure countries.

The network of researchers and analysts provide monthly reports and maps detailing current and projected food insecurity, timely alerts on emerging or likely crises and specialized reports on weather and climate, markets and trade, agricultural production, livelihoods, nutrition, and food assistance.

Improved seasonal soil moisture forecasts using SMAP data will benefit famine early warning systems particularly in sub-Saharan Africa and South Asia, where hunger remains a major human health factor and the population harvests its food from rain-fed agriculture in highly monsoonal (seasonal) conditions. Improving the ability to forecast agricultural drought has the potential to improve forecasts of food-related economic stresses in local areas within this region, and this will improve both medium-term and long-term decision-making among several agencies about where, when and how to distribute food aid. In order for SMAP soil moisture data to be informative to the FEWS-NET analysis, the data need to show clear value of SMAP data for estimating agricultural drought, and relevance to forecasting local

Table 2 Drought and wildfires early adopters

Droughts and wildfires	
Researcher	Organization
* Jim Reardon and Gary Curcio	US Forest Service (USFS)
* Chris Funk, Amy McNally and James Verdin	USGS & UC Santa Barbara
◊ Brian Wardlow and Mark Svoboda	Center for Advanced Land Management Technologies (CALMIT), National Drought Mitigation Center (NDMC)
◊ Uma Shankar	The University of North Carolina at Chapel Hill—Institute for the Environment
◊ Javier Fochesatto	University of Alaska
◊ Amir AghaKouchak	University of California, Irvine
◊ Renato D'Auria	ALTEC S.p.A
◊ Rong Fu	University of Texas; SMAP contact: Randy Koster

* Early Adopters selected in 2011–2012. ◊ Early Adopters selected from 2013 forward

food availability and economic stress. Primarily, this is expected to happen by demonstrating that SMAP can facilitate improvements to monthly rainfall estimates used by FEWS-Net analysts. The primary challenge is if the value of SMAP-derived drought analysis products is not clearly demonstrated in case studies, the analysts will not incorporate this information into their routine efforts. So far, this SMAP Early Adopter project has demonstrated that SMAP generally quantitatively and qualitatively agrees with precipitation-forced model simulations of soil moisture drought. This shows that SMAP can be *expected* to do no harm. The "no-harm" analysis must be expanded to cover some very clear and specific significant past events, rather than just statistics calculated over the entire historical record. For more information about FEWS NET, please visit www.fews.net.

Table 2 lists Early Adopters who are working with SMAP data products to address applications focused on drought and wildfires.

6.3 Flood and Landslide

Floods affect more people globally than any other type of natural hazard, causing some of the largest economic, social, and humanitarian losses (Zurich Reinsurance 2015). Soil moisture is a key variable in water-related natural hazards including floods and landslides. High-resolution observations of soil moisture and landscape freeze/thaw status will lead to improved flood forecasts, especially for intermediate to large watersheds where most flood damage occurs. The use of SMAP data will help improve the communication and response of government agencies and emergency managers to a full range of emergencies and disasters. SMAP Early

Adopter, Dr. Luca Brocca from the Research Institute for Geo-Hydrological Protection in Italy is working with SMAP data to address flood prediction estimates to better inform the Umbria Region Civil Protection Centre and the National Department of Civil Protection. There are hosts of ways in which research can enhance flood resilience. Understanding potential flood threats to help aid in planning resources for preparedness and flood response is one way Dr. Brocca is focused on improving. Dr. Brocca's SMAP EA project aims is to understanding the capability of using SMAP soil moisture for operational hydrological applications throughout Europe. Specifically, his Early Adopter project uses soil moisture data for (1) assimilation into continuous rainfall runoff modeling to demonstrate the impact on runoff prediction with emphasis on extremes, and (2) to address the estimation of rainfall from soil moisture data. The work on data assimilation will be implemented by the Umbria Region Civil Protection Centre for improving the early warning system for flood and landslide forecasting in Umbria, Italy (Fig. 6). The use of soil moisture data to correct and/or replace precipitation estimates are applied to the entire Italian territory for the hydrometeorological and hydraulic risk assessment at the Italian Department of Civil Protection. Combining this applications research with the operational user's guide to improve communications of the flood risk to come will help people make better choice and take mitigating actions.

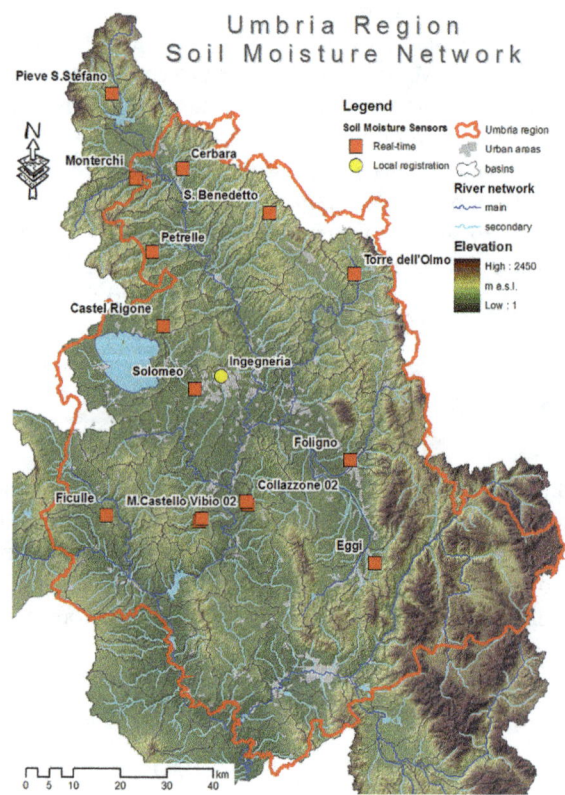

Fig. 6 Umbria Region Soil Moisture Network

Fig. 7 Soil Moisture correlation between SMAP and in situ data from Umbria Region Soil Moisture Network

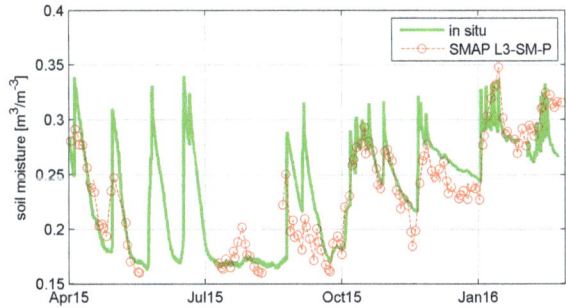

As an Early Adopter, during the prelaunch stages of SMAP, Dr. Brocca worked with the mission to apply simulated data as a "fit" for his early model runs and prepare for using real SMAP observations. Using proxy data (European soil moisture data sets SMOS, (Soil Moisture Ocean Salinity) Mission), and prelaunch preparations with SMAP, Dr. Brocca was able to quickly demonstrate how the ground observations in the Umbria Region Soil Moisture Network correlated very well with the SMAP soil moisture passive data through the real-time monitoring network operating (Fig. 7). Through Dr. Brocca's work, the Umbria Region Civil Protection Centre will have a 6–8 month lead time for implementing the most up to date satellite data to inform flood planning enabling better preparedness and community planning for early warning systems that save lives and get people to safety faster. The SMAP project intends to assess how this Early Adopter research will potentially impact the decisions made by flood management and emergency response personnel at the Italian Department of Civil Protection and provides that feedback to NASA.

Table 3 lists all SMAP Early Adopters working with SMAP data products to address flood and landslide applications.

Table 3 Flood and landslides

Floods and landslides	
Researcher	Organization
* Fiona Shaw	Willis, Global Analytics
* Kashif Rashid and Emily Niebuhr	UN World Food Programme
◇ Konstantine Georgakakos	Hydrologic Research Center
◇ Luca Brocca	Research Institute for Geo-Hydrological Protection, Italian Dept. of Civil Protection
◇ Jennifer Jacobs	University of New Hampshire
◇ Huan Wu, Xiwu Zhan, and Robert F. Adler	University of Maryland, NASA Jet Propulsion Laboratory (JPL), and NOAA/NESDIS/STAR
◇ G. Robert Brakenridge	Dartmouth Flood Observatory, University of Colorado

* Early Adopters selected in 2011–2012. ◇ Early Adopters selected from 2013 forward

6.4 Agriculture Productivity

According to UN projections, the world's population could reach 9.15 billion by 2050, creating a 60 % increase in demand for food. Research on global agricultural productivity focuses on quantifying comparable productivity growth measures for countries and regions worldwide. SMAP data will provide information on water availability and environmental stress for estimating plant productivity and potential yield. The availability of direct observations of soil moisture status and the timing and extent of potential frost damage from SMAP will enable significant improvements in operational crop productivity and water stress information systems by providing realistic soil moisture and freeze/thaw observations as inputs for agricultural prediction models. Improved models will provide crucial information for decision-makers managing water and other resources, especially in data-sparse regions. Another Early Adopter for the SMAP mission is the USDA National Agricultural Statistics Service (NASS). The USDA publishes weekly national and state-level reports of crop progress, crop conditions, and soil moisture during the growing season (Bolten et al. 2009). These reports require accurate, spatially and spectrally detailed, timely satellite data and products of soil moisture, inundation, drought, etc., for monitoring and assessment of agriculture. The USDA's National Agricultural Statistics Service conducts hundreds of surveys every year and prepares reports covering virtually every aspect of U.S. agriculture. SMAP Early Adopters Zhengwei Yang and Rick Mueller efforts focus are to study the feasibility of using SMAP mission results to support US national crop condition monitoring and other NASS operational data needs, such as crop yield modeling needs, to improve NASS cropland soil moisture monitoring results in consistency, reliability, objectivity, and efficiency, and to reduce survey burden and cost.

Production and supplies of food and fiber, prices paid and received by farmers, farm labor and wages, farm finances, chemical use, and changes in the demographics of US producers. NASS currently relies heavily on 4,000 "enumerators" around the country to produce its assessments of crop conditions and soil moisture. SMAP data is a gold mine of objective, reliable and accurate moisture data for the agency. It promises to sharpen NASS's reporting and possibly replace part of that enumerator network, which would be a huge cost savings to the agency, according to Rick Mueller, head of spatial analysis research at NASS. NASS is also preparing to use SMAP data to improve its state-by-state assessments of soil moisture released in USDA's weekly crop progress reports.

Below in Table 4 are the Early Adopters working with SMAP data products to address agricultural productivity.

Table 4 Agricultural productivity

Agricultural productivity	
Researcher	Organization
* Catherine Champagne	Agriculture and Agri-Food Canada (AAFC)
* Zhengwei Yang and Rick Mueller	USDA National Agricultural Statistical Service (NASS)
* Amor Ines and Stephen Zebiak	International Research Institute for Climate and Society (IRI) Columbia University
* Jingfeng Wang, Rafael Bras, Aris Georgakakos and Husayn El Sharif	Georgia Institute of Technology (GT)
* Curt Reynolds	USDA Foreign Agricultural Service (FAS)
◊ Alejandro Flores	Boise State University
◊ Barbara S. Minsker	University of Illinois and sponsored by John Deere Inc.
◊ Lynn J. Torak	U.S. Geological Survey, Georgia Water Science Center
◊ Kamal Labbassi	Faculty of Sciences, MARSE, El Jadida, Morocco
◊ Shibendu Ray	Mahalanobis National Crop Forecast Centre, New Delhi, India
◊ Niladri Gupta	Tocklai Tea Research Institute

* Early Adopters selected in 2011–2012. ◊ Early Adopters selected from 2013 forward

6.5 Human Health

Soil moisture has a direct effect on dust generation and air quality in desert and arid environments. Improved seasonal soil moisture forecasts using SMAP data help inform early warning systems for famine in agriculture/food production, it will benefit environmental risk models and early warning systems related to the potential expansion of many disease vectors that are constrained by the timing and duration of seasonal frozen temperatures. Indirect benefits are realized, as SMAP data will enable better weather forecasts that lead to improved predictions of heat stress and virus spreading rates. SMAP Early Adopter, Dr. Hosni Ghedira, Dr. Imen Gherboudj, and Dr. Naseema from the Masdar Institute in the United Arab Emirates are working on health applications using SMAP data. The team is quantifying the effects of soil moisture on the total dust emission over the Middle East and North Africa (MENA) region (as illustrated in Fig. 8) using a parameterization scheme that requires satellite/reanalysis data, such as soil moisture, wind speed, soil texture, grain size distribution, soil erodibility, aeolian surface roughness, and vegetation cover, all either measured or estimated. This Early Adopter research is expected to improve the accuracy of estimating dust emissions. To date, Dr. Ghedira and his

Fig. 8 Dust storm in Abu Dhabi. *Source* http://forum.skyscraperpage.com

team have run an analysis of their research comparing maps from the European Space Agency's Soil Moisture and Ocean Salinity (SMOS) mission and the Advanced Microwave Scanning Radiometer—Earth Observing System (AMSR-E). The simulated dust flux shows better correlation with the SMOS soil moisture estimates than the AMSR-E soil moisture estimates, especially in the seasonal scale. Results demonstrated that when the soil moisture increases by 10 %, the dust fluxes decrease by 20–45 % over the Arabian Peninsula desert and less by 20 % over North Africa deserts. Information about increases or decreases dust fluxes can later be shared with public health organizations as an early warning of potential dust storms that could impact respiratory conditions. This Early Adopter research continues to be modified as new information from SMAP becomes available.

Table 5 shows all SMAP Early Adopters focused on health applications.

Table 5 Human health early adopters

Human health	
Researcher	Organization
* Hosni Ghedira	Masdar Institute, UAE
* Kyle McDonald and Don Pierson	City College of New York (CUNY) and CREST Institute, New York City Dept. of Environmental Protection
◊ James Kitson, Andrew Walker and Cameron Hamilton	Yorkshire Water, UK
◊ Luigi Renzullo	Commonwealth Scientific and Industrial Research Organisation (CSIRO), Australia
◊ David DuBois	New Mexico State University

* Early Adopters selected in 2011–2012. ◊ Early Adopters selected from 2013 forward

6.6 National Security

National security includes mobility of soldiers and equipment through data scarce remote regions around the world. Information on surface soil moisture and freeze/thaw is critical to evaluating ground mobility of military vehicles through these areas. The integration of soil moisture has been determined to be the single most critical parameter in state-of-the-ground mobility models Soil moisture and freeze/thaw data are key to a broad array of military and civil works capabilities including road and bridge building, dam and levee assessment/construction, mining, forestry, and tactical decision aid design and development. The application of SMAP data to unmanned vehicle applications is essential to effective autonomous cross-country route selection as well as route recommendations to the users of these vehicles.

SMAP Early Adopter, Mr. Derek Ward, Lockheed Martin, is working in collaboration with all U.S. Department of Defense (DoD) Early Adopters to develop comprehensive feedback for worldwide route invariant mobility predictions for friendly, neutral, and threat vehicles across the globe. The objective is to use SMAP data paired with high-resolution soil type data to provide enhanced manned/unmanned ground vehicle mobility mapping. The group examined the use of SMAP data for assessing the movement of a variety of manned and unmanned vehicles operating in a cross-country mode in the Pacific Rim. Preliminary result from this DoD effort demonstrated success in using SMAP radar data (when available) over the mapped terrain; areas that previously looked dark (saturated and unmaneuverable in terms of vehicle weight, tire pressure and soils strength) appeared bright and "useable" and able to sustain the weight, tire pressure of specified vehicles without risk of getting stuck (Fig. 9). This detailed information was only made possible through the use of the observed soil moisture content measured by the SMAP radar paired with existing high resolution-soil maps. Since the loss of the SMAP radar, the passive product from SMAP (radiometer) is used to derive a radar analog by pairing the lower resolution radiometer soil moisture measurements with existing 1 km resolution soil type maps. Knowing the underlying soil type/saturation behaviors, elevation data, reported local weather, as well as the measured radiometer soil saturation value allows an inference to be made of the current soil type saturation state at the native resolution of the soil map (1 km). This methodology scales easily to instrument resolution, and so enables a complete exploration of the instrument resolution space, while providing breakpoints and desired specifications for future passive or active instruments. Direct comparison of the radar data to the radiometer/soil type high resolution-inference, is also possible. Vehicle specific mobility maps generated for this effort helped assess the value and potential impact of this instrument, but more importantly provided NASA with a more detailed understanding of how such soil saturation data is used operationally—defining a path for DoD, NATO, ESA, and NASA to collaborate in the future with regards to instrument specification, design, and most importantly joint funding on dual use missions. Table 6 lists the Early Adopters working in the area of National Security.

Fig. 9 Canadian Leopard Tank in Afghanistan, stuck in transit. Courtesy of Derek ward

Table 6 National security early adopters

National security	
Researcher	Organization
* John Eylander and Susan Frankenstein	U.S. Army Engineer Research and Development Center (ERDC) Cold Regions Research and Engineering Laboratory (CRREL)
* Gary McWilliams, George Mason, Li, Andrew Jones and Maria Stevens	Army Research Laboratory (ARL); U.S. Army Engineer Research and Development Center (ERDC) Geotechnical and Structures Laboratory (GSL); Naval Research Laboratory (NRL); and Colorado State University (CSU)
◊ Kyle McDonald	City College of New York (CUNY)
◊ Georg Heygster	Institute of Environmental Physics, University of Bremen, Germany
◊ Lars Kaleschke	Institute of Oceanography, University of Hamburg
◊ Jerry Wegiel	Headquarters Air Force Weather Agency
◊ Matthew Arkett	Canadian Ice Service
◊ Derek Ward	Lockheed Martin Missiles and Fire Control
◊ David Keith	U.S. National Ice Center, Naval Research Laboratory
◊ Christopher Jackson and Frank Monaldo	NOAA NESDIS Center for Satellite Applications and Research (STAR)

* Early Adopters selected in 2011–2012. ◊ Early Adopters selected from 2013 forward

In addition to the application topics presented above, SMAP has Early Adopters in areas of decision services such as technology-communication, software, and policy. These commercial applications will help improve technology used for communicating information with regards to farming, drought, emergency response to weather, and carbon emissions. Table 7 lists SMAP Early Adopters in other general application for communicating and expanding.

Table 7 General application early adopters

General	
Researcher	Organization
◊ Srini Sundaram	Agrisolum Limited, UK
* Rafael Ameller	StormCenter Communications, Inc.
◊ Joey Griebel	Exelis Visual Information Solutions
◊ Kimberly Peng	Africa Soil Information Service (AfSIS) and Center for International Earth Science Network (CIESIN)
◊ Tyler Erickson and Rebecca Moore	Google Earth Engine, Google, Inc
◊ Jeff Morisette	USGS and DOI North Central Climate Science Center, the National Drought Mitigation Center
◊ Benjamin White	Integra, LLC

* Early Adopters selected in 2011–2012. ◊ Early Adopters selected from 2013 forward

The unexpected loss of the SMAP Radar has increased the importance of feedback from Early Adopters. An additional effort was recently added to the SMAP Mission Phase E Applications Plan that include using the existing 10 weeks of Radar data SMAP collected. Because of the ongoing relationships with Early Adopters, the mission has been able to communicate this need to the Early Adopter community openly in an effort to collect examples of the value of SMAP radar data to inform the next Earth Science Decadal Survey.

The SMAP Applications Program expects to see improvements across many applications but especially those focused on weather forecasting, drought and flood prediction, and agriculture productivity. Early Adopters have pushed the envelope of science applications and provided a unique and valuable perspective to the uses of SMAP products that would not have been available to the mission scientists if not for the Early Adopter engagement. The SMAP program anticipates that the Early Adopters will make discoveries that the SMAP science team did not expect, that there will be new science questions that develop from Early Adopter efforts, and that there will be unique and innovative ways people use SMAP data because of the feedback from and lessons learned by the Early Adopters.

More details on SMAP Applications, the SMAP Early Adopters and the Case Study progress can be found be going to the SMAP Applications website: http://smap.jpl.nasa.gov/science/applications/.

7 The Ice Cloud and Elevation Satellite (ICESat-2) Mission

The ICESat-2 mission (Fig. 10) is the second decadal survey mission to implement a formal Early Adopter program. The ICESat-2 Applications program follows the SMAP model and works in conjunction with the ICESat-2 Project Science Office to

Fig. 10 ICESat-2 Satellite Image Credit: Orbital; Earth image illustrating AMSR-E sea ice; *Image Credit* NASA Scientific Visualization Studio

provide a framework for building a broad and well-defined user community during the prelaunch phases of the mission with the goal to maximize the use of the data products after launch in 2017. The application program seeks to identify applications related to the mission science objectives driving the instrument design—studies of ice sheets, sea ice, and vegetation—and those related to areas of interest to the mission for which its developing data products, but for which it does not have mandated science requirements: atmosphere, inland water, and ocean.

The ICESat-2 Applications Team seeks to *achieve the greatest possible utility of NASA's Investment in Earth Science and observations by enabling its application to practical societal needs*. The program is structured to address three overarching goals: (1) enhance applications research, (2) increase collaboration, and (3) accelerate applications. The main objective for enhancing applications research is to show the expected and actual value that the satellite data has for different end users.

In the prelaunch period of the mission, the applications program is identifying and collaborating with a variety of organizations and individuals in order to understand and clarify their specific requirements and needs for ICESat-2-like data. The following are some of the questions that the program seeks to elucidate through engagement with different stakeholders:

- **WHY?** What key decisions need to be addressed?
- **WHAT?** What are current data sources and existing data requirements?
- **WHEN?** What are latency needs?
- **HOW?** How do we translate science to end use and decision support? What methods are needed?

In 2013, the ICESat-2 mission accepted its first Early Adopter and continues to attract individuals and groups from all over the world who volunteer to test the utility of ICESat-2 for a specific decision process or operation. One year prior to its expected launch date of October 2017, the Early Adopter program now hosts 16 early adopters who are providing insight into a range of potential uses for the mission, as described below.

7.1 ICESat-2 Applications

ICESat-2 Early Adopters have identified various applications their research can potentially inform with the use of the new satellite observations. The next section describes the Early Adopter research that supports the primary objectives of the mission or those related to sea ice, ice sheets, and vegetation.

7.1.1 Sea Ice

ICESat-2 has a science objective to estimate sea ice thickness to examine ice/ocean/atmosphere exchanges of energy, mass, and moisture. It will produce monthly maps of sea ice above-water height, or freeboard, for the Arctic and Southern Oceans. Four Early Adopters (Table 8) are exploring the use of ICESat-2 for prediction of the changing ice environment. The following snapshots of their work where developed from their Early Adopter proposals.

Investigators from the U.S. Naval Research Laboratory have proposed to use ICESat-2 within the Navy's Arctic Cap NOWCAST Forecast System to improve ice edge forecasts in the northern hemisphere. Successful implementation is expected will translate to improved quality of the U.S. Navy, NOAA, and National Guard's National Ice Center's ice analysis and tactical sea ice forecasts which are of essence for navigation and arctic shipping (Fig. 11), as well as contribute to increasing the public's understanding of recent changing conditions in the Arctic (Posey et al. 2013).

Andrew Roberts and the researchers at the Naval Postgraduate School, University of Colorado at Boulder, and Los Alamos National Laboratory are working together to develop an ICESat-2 emulator that can be used in earth system models to sample simulated sea ice freeboard (sea ice above-water height) and snow cover. The emulator is expected to allow an apple-to-apple comparison of sea ice state statistics (e.g., ice concentration, ice thickness, and snow cover) with the upcoming ICESat-2 observations that would not be possible with the current sampling regime used within the models. Successful implementation of the

Table 8 Sea ice application early adopters

Prediction of changing ice environment—sea ice	
Early adopter PI(s)	Organization
* Pamela G. Posey	Naval Research Laboratory at John C. Stennis Space Center
◊ Andrew Roberts, Alexandra Jahn, Adrian Turner	Naval Postgraduate School, University of Colorado at Boulder & Los Alamos National Laboratory
◊ Andy Mahoney	Geophysical Institute, University of Alaska Fairbanks
◊ Stephen Howell	Environment Canada, Climate Research Division

* Early Adopters selected in 2013. ◊ Early Adopters selected from 2014 forward

Fig. 11 U.S. Coast Guard navigating through Sea Ice in the Arctic

ICESat-2 emulator is expected to improve the U.S. Department of Defense (DoD) (an ICESat-2 Early Adopter end user) capability and capacity to operate in the Arctic in changing climate and geostrategic situations (Roberts et al. 2014).

Lack of a reliable means for mapping landfast ice in an automated fashion from remote sensing data is one of the motivations for the Early Adopter research being conducted at the Geophysical Institute at the University of Alaska Fairbanks. Operational sea ice analysts, such as those from the NOAA National Weather Service Ice Desk, rely on their skilled interpretation to delineate landfast ice in operational ice charts. This work is considered laborious and subject to a level of objectivity that makes it challenging to analyze trends and interpret events. Use of ICESat-2 data to automate the process of deriving landfast ice extent may improve scheduling coastal deliveries and the identification of potential places of refuge for vessels needing safe anchorage in protected bays, among other applications (Mahoney 2014).

Environment Canada is exploring the use of ICESat-2 ice thickness estimates for use in a variety of research activities including seasonal and decadal model validation and optimization, ice operations, and understanding sea ice variability and change. The accuracy of sea ice thickness data is considered crucial for mid-range forecasts and correctly simulating short-term events such as ice internal pressure buildups, which are also considered very important for navigation purposes. Environment Canada will collaborate with the Canadian Ice Service to provide operationally up to date and accurate sea ice information to mariners (Howell et al. 2014).

7.1.2 Ice Sheets

ICESat-2 has two science objectives for ice sheets:

1. Quantify polar ice-sheet contributions to current and recent sea level change and the linkages to climate conditions; and

2. Quantify regional signatures of ice-sheet changes to assess mechanisms driving those changes and improve predictive ice sheet models.

Proposed Early Adopter research (Table 9) using ICESat-2 ice surface elevation products include work by the Florida Atlantic University (FAU) and the State University of New York (SUNY) at Buffalo. The following snapshots of their work were developed from their Early Adopter proposals. Figure 12 shows a typical polar ice sheet.

Early Adopter research proposed by researchers at FAU involves using the new ICESat-2 elevation information to compute a more precise past/current ice volume discharge for automatically generating an improved Dynamic DEM (DDEM). The DDEM is expected will allow for improved monitoring of sudden increases in sea level for coastal communities (Nagarajan et al. 2013).

Improved volcanic deformation monitoring and eruption forecasting are two motivations for the current Early Adopter research being conducted at the SUNY at Buffalo. ICESat-2 observations may allow for the generation of high spatial and temporal resolution Digital Elevation Models (DEMs) needed as model inputs for the prediction of evolving pyroclastic density currents. These hot avalanches of volcanic rock and gas pose a hazard during ongoing eruptive crises when topographic changes make in-filled valleys less able to contain them. Another focus of this research involves determining surface change histories related to subglacial

Table 9 Land ice application early adopters

Hazard monitoring and forecasting—land ice	
Early adopter PI(s)	Organization
* Greg Babonis	State University of New York at Buffalo
* Sudhagar Nagarajan	Florida Atlantic University

* Early Adopters selected in 2013. ◊ Early Adopters selected from 2014 forward

Fig. 12 Polar ice sheet. *Image Credit:* Christy Hansen, NASA, 2012

volcanic events in areas such as Antarctica, Iceland, and in southern Andean ice fields (Babonis et al. 2013). Table 9 lists Hazard Monitoring and Forecasting for Land Ice Early Adopters.

7.1.3 Vegetation

ICESat-2 has an objective to measure vegetation canopy height as a basis for estimating large-scale biomass and biomass change. The mission will develop high-precision (centimeter-level) elevation measurements over land with a track density of 2 km along the equator after 2 years through off-nadir pointing.

7.1.4 ICESat-2 Data for SemiArid Vegetation Mapping

Researchers at Boise Center Aerospace Laboratory are exploring the use of ICESat-2, both independently and leveraging synergistic remote sensing data, for quantifying the composition and structural variability of short-statured semiarid vegetation like that in Fig. 13. Having estimates of semiarid ecosystem vegetation biomass is expected to allow for quantification of carbon and fuel loads, and to monitor change in regions such as the Great Basin in the western US This is of interest to various end users including the U.S. Forest Service, USDA Agricultural Research Service, and U.S. Geological Survey for long-term land management and understanding ecosystem vulnerability (Glenn et al. 2013).

Estimating above-ground biomass, tracking forest growth over time, as well as monitoring forest-related harvesting and land use are the primary motivations behind the Early Adopter research being conducted at Virginia Polytechnic Institute and State University (Virginia Tech). The Early Adopter group at Virginia Tech is developing algorithms and software to identify ground, top-of-canopy, and canopy height using simulated ICESat-2 data in preparation for the postlaunch observations. Successful processing of ICESat-2 will translate into improved accuracy of

Fig. 13 *Image Credit:* U.S. Fish and Wildlife Service, 2013

canopy height estimates for both of the government and private industry end users identified for this research: the USDA Forest Service and the American Forest Management (Wynne et al. 2013).

Early Adopter research being conducted at the Hunter College of the City University of New York proposes using ICESat-2 data within an existing web service architecture that already fuses GLAS data with Landsat-scale datasets on demand. ICESat-2 is expected to provide additional vegetation structure information at high spatial resolution that may improve the accuracies of the current vegetation products generated by a web service, including vegetation heights and maps of vertical foliage profiles (Ni-Meister 2014).

Of expressed interest to the USDA Forest Service Pacific Northwest Station Forest Inventory is the Early Adopter research proposed by the Geophysical Institute at the University of Alaska Fairbanks, which consists of analyzing the ICESat-2 prelaunch data to differentiate high latitude vegetation architecture type, landscape gradients and stands, and improve classification in two scenarios: Arctic Tundra and Boreal Forest. Successful post-launch processing of ICESat-2 data will benefit the continuous monitoring of vast regions of responsibility by the USDA Forest Service (Fochesatto and Huettmann 2015) Table 10 lists these Early Adopters, and Table 11 lists those for Inland Water and Atmosphere applications.

The table in Appendix 2 shows the work proposed by all of the ICESat-2 Early adopters. It provides an initial look at how the ICESat-2 observations can be integrated into applications of benefit to society. An ongoing goal of the Applications program is to discover new applications for the mission and identify gaps in the communities engaged. The following table shows the Early Adopters that have proposed research supporting applications relating to inland water and atmosphere.

The full ICESat-2 Applications plan and strategy can be found on the ICESat-2 applications website at http://icesat.gsfc.nasa.gov/icesat2/apps-ov.php.

Table 10 Vegetation application early adopters

Monitoring and planning for land management—vegetation	
Early adopter PI(s)	Organization
* Nancy F. Glenn	Boise Center Aerospace Laboratory, Boise State University
* Lynn Abbott & Randy Wynne	Virginia Polytechnic Institute and State University
* Wenge Ni-Meister	Hunter College of The City University of New York
◊ G. Javier Fochesatto, Falk Huettmann	University of Alaska Fairbanks

* Early Adopters selected in 2013. ◊ Early Adopters selected from 2014 forward

Table 11 Inland water and atmosphere application early adopters

Hazard monitoring and forecasting—inland water and atmosphere	
Early adopter PI(s)	Organization
* Lucia Mona	Institute of Methodologies for Environmental Analysis of the National Research Council of Italy
◊ Birgit Peterson	U.S. Geological Survey
◊ Charon Birkett	Earth System Science Interdisciplinary Center, University of Maryland
◊ Guy J-P. Schumann	Joint Institute for Regional Earth System Science and Engineering, University of California
◊ Kuo-Hsin Tseng	Center for Space and Remote Sensing Research, National Central University, Taiwan
◊ Rodrigo C.D. Paiva	Hydraulic Research Institute at Federal University of Rio Grande do Sul, Brazil

* Early Adopters selected in 2013. ◊ Early Adopters selected from 2014 forward

8 The Surface Water and Ocean Topography (SWOT) Mission

NASA and the French space agency, CNES, with contributions from the Canadian Space Agency (CSA) and United Kingdom Space Agency (UKSA) are developing new wide swath altimetry technology that will measure the height of surface water (i.e., lakes, rivers, wetlands) and the global ocean with unprecedented spatial coverage, temporal sampling, and spatial resolution compared to existing technologies. The data and information products from the Surface Water and Ocean Topography (SWOT) mission (Fig. 14), along with the airborne concept-validation project, AirSWOT, will benefit society in two critical areas; freshwater on land, and the oceans' role in climate change. It will provide important observations of the amount and variability of water stored in global lakes, reservoirs, wetlands, and river channels and will support derived estimates of river discharge. SWOT will also provide critical information necessary for water management, particularly in international hydrological basins, and will be useful for monitoring the hydrologic cycle, flooding, and characterizing human impacts on a changing environment. The SWOT mission is in development and preparing for launch in 2020.

The applied science community is a key element in the success of the SWOT mission in demonstrating the high value of the science and information products for addressing societal issues and needs. The SWOT applications framework includes a network of scientists, applications specialists, academics, and SWOT Project members to promote applications research and engage a broad community of potential SWOT data users. A defined plan and a guide describing the SWOT Early Adopters program includes providing proxies for SWOT data, including sophisticated ocean and hydrology simulators, AirSWOT data sets, and other satellite datasets as cornerstones for the program (Srinivasan et al. 2015).

Fig. 14 Artist rendition of the SWOT spacecraft Credit: NASA/JPL

The anticipated science and engineering advances that SWOT will provide may be transformed into valuable services to decision-makers, operational users, and governing organizations focused on addressing global disaster risk reduction initiatives and potential science-based mitigation activities for water resources challenges of the future. A broad range of applications can inform inland and coastal managers and marine operators of terrestrial and oceanic phenomena relevant to their operations.

SWOT will extend existing space-based altimetry, a mature remote sensing technique with over 23 years of continuous ocean observations from the Jason-series of satellites and other international platforms. The suite of operating altimetry satellites provide highly accurate satellite-to-surface measurements, allowing for global and long-term monitoring of ocean circulation up to mesoscale (200 km and larger) processes. Many applications in scientific and operational areas can be realized. With continued altimetry observations, it has been possible to observe phenomena occurring over long time periods; the rise in global mean sea level and climatic events such as El Niño, La Niña, and the Pacific Decadal Oscillation. Moreover, as data are made available within a few hours, the measurements can be used in ocean, meteorological, and climate forecast models.

The next section describes potential SWOT Early Adopters and the best-suited EA application topics for SWOT data products.

9 SWOT Early Adopters and Potential Applications

Satellite altimetry revolutionized the study of the global oceans over two decades. Altimetry measurements have also been applied to the study of river and lake water levels but are limited by the coarse resolution of the data. Key hydrological questions on the storage of water on land and its discharge remain unanswered (Fu et al. 2009). The repeated high-resolution elevation measurements projected for SWOT would directly support science findings and operational uses by hydrologists and oceanographers who are focused on applications in the areas of ocean circulation, coastal management, and water management (Fu et al. 2009). Aside from science investigators, applications users would include representatives from the U.S. National Oceanographic and Atmospheric Administration (NOAA), the Navy, the U.S. Geological Survey (USGS), and international weather/climate organizations.

Similar to SMAP and ICESat-2, the primary goals of the SWOT applications program is to (1) promote the use of SWOT data products for the benefit of society and (2) to educate the community of potential SWOT end users and decision-makers so that they understand the mission capabilities for their specific application. In order to achieve these objectives, NASA's ASP, the mission international partners, and the SWOT applications team will facilitate the development of accessible data and information products that may be used by end users, known as the community of practice. From this community of users, SWOT Early Adopters are identified. The SWOT Early Adopter Program is modeled after the successful NASA SMAP program and aims to provide similar feedback mechanisms between the mission scientists and the community. Once the mission is launched in 2020, the SWOT Early Adopters will contribute to the mission successes by demonstrating the benefits of SWOT data to society (Srinivasan et al. 2012, 2014).

The SWOT science focus has traditionally divided the mission science objectives into the categories of hydrology and oceanography. The following are the primary areas of application for the SWOT mission:

- Hydrology: developing world water problems, food security (flooding and drought)
- Oceanography: coastal applications (circulation, impacts), marine operations support/open ocean issues
- Climate: regional capabilities, coastal and agricultural impacts

In addition to science focus areas, the operational applications and societal benefits for the SWOT mission may include;

- Water management: reservoirs, floods, ecology,
- International rivers: flood and drought management,
- Insurance: hydrodynamics and flood risks,
- Transportation: shipping, barges,
- Agriculture: water management to support irrigation,
- Energy: water availability in new regions,

- Spills and pollution: mapping of potential spill,
- Ocean and coastal circulation models,
- Climate studies: ocean circulation, heat content, regional sea level studies.

By the time of the SWOT launch in 2020, the applications team will have engaged a core group of potential users capable of incorporating the measurements and information products from SWOT in their operations. Methods for engagement and communication will be modeled after the predecessors, and then modified to fit the specific needs of SWOT. Below are three potential activities that may be common to SWOT application communities. Grouping these users into the activity areas identified can facilitate collaboration and the development of tools and models to better utilize potential SWOT data products (Srinivasan et al. 2013).

Forecasting applications may use improved models developed from potential SWOT data to support forecasts of a quantity or system status, for example, reservoir storage next month or an ocean state in two weeks. These forecasts could then be used in decision-making contexts. Important considerations for this perspective include; data latency, observations, and uncertainties used by other models.

- A potential Early Adopter for this application will be the U.S. National Weather Service (NWS) regional flood forecasting center. SWOT measurements will be used to update their system-wide storage and flow databases and to aid in operations. Other applications may include large-scale river basin monitoring to facilitate sharing of water resources across international borders, and more accurate weather and climate forecasting. Existing altimeter data can be assimilated into flood models for SWOT, and synergistic use of other satellite missions (Landsat, SMAP, SMOS, Sentinel satellites, other altimetry missions, etc.) can provide additional value for water stage and flood extent (event-based) information.

Planning/Engineering applications would utilize the entire time series of SWOT observations. Over the mission life, what range of quantities or system states (e.g., minimum, maximum, and mean of given parameters) may be measured by SWOT. At a minimum, future systems should be designed to accommodate the observations. Can a model be developed or calibrated to reproduce the SWOT observations? If so, can it be used to simulate other potential scenarios or conditions? Other important considerations for this perspective include; time series data, boundary conditions for other models, and new data for many locations around the globe.

- Applied uses in this activity area include improved flood modeling, potentially providing regional and local planners, insurance organizations, and infrastructure managers to mitigate hazards to life and property. FM Global and other insurance companies develop models to predict floods (e.g., 100-year floods) and will be among the first SWOT Early Adopters.

Management applications are similar to forecasting in data latency but less focused on modeling of future conditions and more on the current status of systems that will influence decision trees. Important considerations for this perspective include data latency and observations that may be difficult to obtain from existing data.

- Applied uses in this area may include managing freshwater for urban, industrial, and agricultural consumption, reducing environmental risk, or contributing to public policy decision-making processes. SWOT will provide water supply services and distribution companies and organizations requiring information about major reservoirs, rivers, and catchment areas, enabling better planning for future stocks. Water resources managers require constant updating of available storage, particularly when multiple reservoirs are involved. SWOT will be able to facilitate the system integration of storage changes in these circumstances. Through existing relationships with the Water Management Bureaus in California, and other western U.S. states, SWOT will seek partnership from water managers to guide an early adopter effort.

The SWOT mission concept and data will provide completely new measurement, data management, and observational paradigms for the earth systems it will measure. In addition to the mission and applications objectives noted above, the SWOT data will be useful for a variety of other scientific and operational applications as the mission science and data systems mature. Although it is impossible to foresee all the applications that could be made of the SWOT data, a number of applications are natural candidates for consideration

1. SWOT data can be complementary to the operational oceanographic altimeters to improve the understanding of global and regional sea level change.
2. SWOT data can potentially be used for mapping the thickness of floating sea ice by measuring sea ice freeboard to support marine operations and shipping.
3. SWOT can collect data over the tidally affected portions of rivers, and estuaries and wetlands, to help better understand the dynamics of freshwater/marine interaction dynamics.
4. SWOT data can be used to improve Earth's mean land/ice topography, its changes and potential land-cover classifications.
5. SWOT could provide valuable information for modeling water circulation in large lakes such as the Caspian Sea, African lakes, Lake Baikal and Lake Titicaca, and the Great Lakes of North America.

These applications will bring synergistic value to the mission's science (Rodriguez 2015). A SWOT survey to determine user needs (https://www.surveymonkey.com/r/SWOTusersurvey2015) has been designed to identify ways in which the proposed SWOT data and information products may be useful to operational, private, institutional, and other individuals and organizations. The survey helps identify the capabilities and areas of interest of potential SWOT data users and guides the Early Adopter interests for the mission. The application process is described in the 'SWOT Early Adopters Guide' is accessible from swot.jpl.nasa.gov/applications. Applications will be accepted beginning in 2016.

10 Conclusion and Way Forward with Prospects of the Early Adopter Program

The NASA Early Adopter program was started so the mission could learn more about their product applications. What evolved was a mechanism that provides feedback and guidance to the utility and value of satellite data for all of NASA. NASA data have the potential to improve models, forecasts, and operational activities, and with the understanding and advocacy of Early Adopters, the science can reach this potential faster and will less cost to the taxpayer. Applied research activities and the efforts from Early Adopters support NASA science designed to inform policy, climate, education, business, and decision-making activities that impact society. The Early Adopters provide a quantifiable impact to the uses of satellite data.

SMAP will complete its mission operations for Phase E in 2018 and provide a quantitative assessment of SMAP product value for weather, agriculture, flooding, drought, health, and national security. Additionally, the feedback from Early Adopters regarding the loss of the SMAP Radar will help inform the new decadal survey and provide quantitative examples of the long-standing need for new radar mission.

ICESat-2 continues to implement its Early Adopter Program and will soon transition into operations with approximately 20 early adopters working to inform a variety of applications. Quantitative research is lined up for operational sea ice forecasting for Arctic shipping, coastal deliveries, water resources management, monitoring of high water levels, weather hazards, sea level rise monitoring, fire fuel mapping, monitoring ecosystem, and biodiversity change over large regions (Arctic Tundra, Boreal forest), and national defense environmental forecasting.

SWOT's Early Adopter program begins in 2016. The Early Adopter targeted by SWOT will champion the first set of science application questions and begin exploring how SWOT data fit into the mission science objectives. Feedback from the Early Adopters will provide detailed accomplishments and challenges to the use of the products so the science team can help improve upon the development.

The Early Adopter program has spread across NASA missions and the 2017 Decadal Survey includes a strong focus on applications. Formal and appropriately scaled EA programs will ultimately extend to future NASA EO programs impacting capacity building programs like DEVELOP, SERVIR as well as smaller satellite missions like the Earth Venture Initiatives (EVIs). EVIs, unlike decadal survey missions (SMAP, ICESAT-2, and SWOT) operate under a shorter timeline for planning the mission and launching into operation. EVI missions have two-year process effort. EVIs like CYGNSS and ECOSTRESS are actively pursuing Early Adopter programs that selects two to three specific case studies scaled to leverage the science objectives of the mission and demonstrate the impact of the science through a specific partner in a particular are of applications. Early Adopter programs across all NASA will be modified to meet needs such as this and allow opportunity for user feedback for all NASA time lines. As NASA's culture shifts to integrate more applications into their mission life cycle, EVIs will also have the benefit of integrating with stakeholders during the mission development.

Crosscutting themes between water, energy, food, and health requires there to be scientific and end user guidance. Integrating feedback on the utility of data products produces more streamlined and user-friendly products for science and stakeholders. The Early Adopter approach provides NASA insight on how science addresses the specific needs of an ever changing society. Development of strategic partnerships with industry or private sector entities should benefit both scientific and research objectives, as well as societal needs. Building upon the cooperative partnering described in the few examples here, we are in a position to leverage the combined observational power of current and future Earth observing platforms and expand the societal applications of these assets. Already funded by the tax-paying public, the innovations of new missions and maturing data and modeling systems can support a robust business model between scientific research communities and stakeholders that can support sustainable partnerships for moving into the next decade of sustained and experimental observations.

The Early Adopter Program demonstrates that there is a deep-rooted drive among several communities to ensure that science touches lives. Early Adopters present science from a perspective we can all identify with and see the impacts of soil moisture, ice, and other data and information products in a context that makes it personal. Making science applications personal helps demonstrate that satellite data provide value to society (better health, efficiency, and cost savings). The clear value of the Early Adopter program is further demonstrated by the decision to implement similar activities across other NASA missions (future missions as well as those currently in operation).

Appendix 1

See Table 12.

Table 12 SMAP early adopter table

Early Adopter PI and institution; SMAP Contact	Applied Research Topic
Weather and Climate Forecasting	
* **Stephane Bélair**, Meteorological Research Division, Environment Canada (EC); SMAP Contact: **Stephane Bélair**	Assimilation and impact evaluation of observations from the SMAP mission in Environment Canada's Environmental Prediction Systems
* **Lars Isaksen and Patricia de Rosnay**, European Centre for Medium-Range Weather Forecasts (ECMWF); SMAP Contact: **Eni Njoku**	Monitoring SMAP soil moisture and brightness temperature at ECMWF
* **Xiwu Zhan, Michael Ek, John Simko and Weizhong Zheng**, NOAA National Centers for Environmental Prediction (NCEP), NOAA National Environmental Satellite Data and Information Service (NOAA-NESDIS); SMAP Contact: **Randy Koster**	Transition of NASA SMAP research products to NOAA operational numerical weather and seasonal climate predictions and research hydrological forecasts
* **Michael Ek, Marouane Temimi, Xiwu Zhan and Weizhong Zheng**, NOAA National Centers for Environmental Prediction (NCEP), NOAA National Environmental Satellite Data and Information Service (NOAA-NESDIS), City College of New York (CUNY); SMAP Contact: **Chris Derksen**	Integration of SMAP freeze/thaw product line into the NOAA NCEP weather forecast models
* **John Galantowicz**, Atmospheric and Environmental Research, Inc. (AER); SMAP Contact: **John Kimball**	Use of SMAP-derived inundation and soil moisture estimates in the quantification of biogenic greenhouse gas emissions
◊ **Jonathan Case, Clay Blankenship and Bradley Zavodsky**, NASA Short-term Prediction Research and Transition (SPoRT) Center; SMAP Contact: **Dara Entekhabi**	Data assimilation of SMAP observations, and impact on weather forecasts in a coupled simulation environment
◊ **Steven Quiring**, Texas A&M University; SMAP Contact: **Dara Entekhabi**	Hurricane power outage prediction
Droughts and Wildfires	
* **Jim Reardon and Gary Curcio**, US Forest Service (USFS); SMAP Contact: **Dara Entekhabi**	The use of SMAP soil moisture data to assess the wildfire potential of organic soils on the North Carolina Coastal Plain
* **Chris Funk, Amy McNally and James Verdin**, USGS & UC Santa Barbara; SMAP Contact: **Susan Moran**	Incorporating soil moisture field levels into the FEWS Land Data Assimilation System (FLDAS)
* **Brian Wardlow and Mark Svoboda**, Center for Advanced Land Management Technologies (CALMIT), National Drought Mitigation Center (NDMC); SMAP Contact: **Narendra Das**	Evaluation of SMAP soil moisture products for operational drought monitoring: potential impact on the U.S. Drought Monitor (USDM)
◊ **Uma Shankar**, The University of North Carolina at Chapel Hill – Institute for the Environment; SMAP Contact: **Narendra Das**	Enhancement of a bottom-up fire emissions inventory using Earth observations to improve air quality, land management, and public health decision support
◊ **Javier Fochesatto**, University of Alaska; SMAP Contact: **John Kimball**	Soil moisture in Alaskan ecosystem soils
◊ **Amir AghaKouchak**, University of California, Irvine; SMAP Contact: **Dara Entekhabi**	Integrating SMAP into the Global Integrated Drought Monitoring and Prediction System: Toward near real-time agricultural drought monitoring
◊ **Renato D'Auria**, ALTEC S.p.A; SMAP Contact: **Randy Koster**	Satellite soil moisture accuracy evaluation for hydrological operative forecasting (SMAHF)
◊ **Rong Fu**, University of Texas; SMAP contact: **Randy Koster**	Using SMAP data to improve drought early warning over Texas and the U.S. Great Plains
Floods and Landslides	
* **Fiona Shaw**, Willis, Global Analytics; SMAP Contact: **Robert Gurney**	A risk identification and analysis system for insurance; eQUIP suite of custom catastrophe models, risk rating tools and risk indices for insurance and reinsurance purposes
* **Kashif Rashid and Emily Niebuhr**, UN World Food Programme; SMAP Contact: **Eni Njoku**	Application of a SMAP-based index for flood forecasting in data-poor regions
◊ **Konstantine Georgakakos**, Hydrologic Research Center; SMAP Contact: **Narendra Das**	Development of a strategy for the evaluation of the utility of SMAP products for the Global Flash Flood Guidance Program of the Hydrologic Research Center
◊ **Luca Brocca**, Research Institute for Geo-Hydrological Protection, Italian Dept. of Civil Protection; SMAP contact: **Dara Entekhabi**	Use of SMAP soil moisture products for operational flood forecasting: data assimilation and rainfall correction
◊ **Jennifer Jacobs**, University of New Hampshire; SMAP contact: **Narendra Das**	Satellite enhanced snowmelt flood predictions in the Red River of the North Basin

(continued)

Table 12 (continued)

◊ **Huan Wu, Xiwu Zhan, and Robert F. Adler,** University of Maryland, NASA Jet Propulsion Laboratory (JPL), and NOAA/NESDIS/STAR; SMAP contact: **Seungbum Kim**	Improving the Global Flood Monitoring System (GFMS) with GPM precipitation, SMAP soil moisture and surface water mask information
◊ **G. Robert Brakenridge,** Dartmouth Flood Observatory, University of Colorado; **SMAP contact: Seungbum Kim**	Use of SMAP data for early detection of inland flooding
Agricultural Productivity	
* **Catherine Champagne,** Agriculture and Agri-Food Canada (AAFC); SMAP Contact: **Stephane Bélair**	Soil moisture monitoring in Canada
* **Zhengwei Yang and Rick Mueller,** USDA National Agricultural Statistical Service (NASS); SMAP Contact: **Wade Crow**	US National cropland soil moisture monitoring using SMAP
* **Amor Ines and Stephen Zebiak,** International Research Institute for Climate and Society (IRI) Columbia University; SMAP Contact: **Narendra Das**	SMAP for crop forecasting and food security early warning applications
* **Jingfeng Wang, Rafael Bras, Aris Georgakakos and Husayn El Sharif,** Georgia Institute of Technology (GT); SMAP Contact: **Dara Entekhabi**	Application of SMAP observations in modeling energy/water/carbon cycles and its impact on weather and climatic predictions
* **Curt Reynolds,** USDA Foreign Agricultural Service (FAS); SMAP Contact: **Wade Crow**	Enhancing USDA's global crop production monitoring system using SMAP soil moisture products
◊ **Alejandro Flores,** Boise State University; SMAP Contact: **Dara Entekhabi**	Data fusion and assimilation to improve applications of predictive ecohydrologic models in managed rangeland and forest ecosystems
◊ **Barbara S. Minsker,** University of Illinois and sponsored by John Deere Inc.; SMAP Contact: **Wade Crow**	Comprehensive, large-scale agriculture and hydrologic data synthesis
◊ **Lynn J. Torak,** U.S. Geological Survey, Georgia Water Science Center; SMAP contact: **Dara Entekhabi and Vanessa Escobar**	Downscaling SMAP soil-moisture data to improve crop production and efficient use of energy and water resources and to assess water availability in the Apalachicola-Chattahoochee-Flint River basin
◊ **Kamal Labbassi,** Faculty of Sciences, MARSE, El Jadida, Morocco; SMAP contact: **Susan Moran**	Hydrologic models and remote sensing data to assess indicators for irrigation performance monitoring in Morocco
◊ **Shibendu Ray,** Mahalanobis National Crop Forecast Centre, New Delhi, India; SMAP contact: **Narendra Das**	Evaluation of SMAP soil moisture products for drought assessment under National Agricultural Drought Assessment and Monitoring System (NADAMS) of India
◊ **Niladri Gupta,** Tocklai Tea Research Institute; SMAP contact: **Susan Moran**	Crop management advisory service using SMAP in tea growing regions of northeast India
Human Health	
* **Hosni Ghedira,** Masdar Institute, UAE; SMAP Contact: **Dara Entekhabi**	Estimating and mapping the extent of Saharan dust emissions using SMAP-derived soil moisture data.
* **Kyle McDonald and Don Pierson,** City College of New York (CUNY) and CREST Institute, New York City Dept. of Environmental Protection; SMAP Contact: **Erika Podest**	Application of SMAP freeze/thaw and soil moisture products for supporting management of New York City's potable water supply
◊ **James Kitson, Andrew Walker and Cameron Hamilton,** Yorkshire Water, UK; SMAP Contact: **Robert Gurney**	Using SMAP L-2 soil moisture data for added value to the understanding of land management practices and its impact on water quality
◊ **Luigi Renzullo,** Commonwealth Scientific and Industrial Research Organisation (CSIRO), Australia; SMAP Contact: **Jeff Walker**	Preparing the Australian Water Resources Assessment (AWRA) system for the assimilation of SMAP data
◊ **David DuBois,** New Mexico State University; **SMAP Contact: Dara Entekhabi**	Tracking and assessment of dust storm events in Southwestern US
National Security	
* **John Eylander and Susan Frankenstein,** U.S. Army Engineer Research and Development Center (ERDC) Cold Regions Research and Engineering Laboratory (CRREL); SMAP Contact: **Steven Chan**	U. S. Army ERDC SMAP adoption for USACE civil and military tactical support
* **Gary McWilliams,** George Mason, Li Li, Andrew Jones and Maria Stevens, Army Research Laboratory (ARL); U.S. Army Engineer Research and Development Center (ERDC) Geotechnical and Structures Laboratory (GSL); Naval Research Laboratory (NRL); and Colorado State University (CSU); SMAP Contact: **Steven Chan**	Exploitation of SMAP data for Army and Marine Corps mobility assessment
◊ **Kyle McDonald,** City College of New York (CUNY); SMAP Contact: **Simon Yueh**	Integration of SMAP datasets with the NRL environmental model for operational characterization of cryosphere processes across the north polar land-ocean domain

(continued)

Table 12 (continued)

◊ Georg Heygster, Institute of Environmental Physics, University of Bremen, Germany; SMAP Contact: Simon Yueh	SMAP-Ice: Use of SMAP observations for sea ice remote sensing
◊ Lars Kaleschke, Institute of Oceanography, University of Hamburg; SMAP Contact: Simon Yueh	SMOS to SMAP migration for cryosphere and climate application
◊ Jerry Wegiel, Headquarters Air Force Weather Agency; SMAP contact: Peggy O'Neill	Optimization of NASA's Land Information System (LIS) at HQ Air Force Weather Agency (AFWA)
◊ Matthew Arkett, Canadian Ice Service; SMAP contact: Simon Yueh	Pre-launch evaluation of SMAP L-band SAR data for operational sea ice monitoring
◊ Derek Ward, Lockheed Martin Missiles and Fire Control; SMAP contact: Steven Chan	Manned and unmanned vehicle ground mobility predictions and route selection
◊ David Keith, U.S. National Ice Center, Naval Research Laboratory; SMAP contact: Simon Yueh	NIC cryospheric investigations in support of NASA ROSES arctic sea ice applications of geodetic imaging
◊ Christopher Jackson and Frank Monaldo, NOAA NESDIS Center for Satellite Applications and Research (STAR); SMAP Contact: Simon Yueh	Ocean surface wind and sea ice measurements derived from SMAP SAR data
General	
◊ Srini Sundaram, Agrisolum Limited, UK; SMAP Contact: Robert Gurney	Application of SMAP data products in Agrisolum - A bigdata social agritech platform
* Rafael Ameller, StormCenter Communications, Inc.; SMAP Contact: Randy Koster	SMAP for enhanced decision making
◊ Joey Griebel, Exelis Visual Information Solutions; SMAP Contact: Barry Weiss	Utilization of SMAP Products in ENVI, IDL and SARscape - Products L1 to L4
◊ Kimberly Peng, Africa Soil Information Service (AfSIS) and Center for International Earth Science Network (CIESIN); SMAP contact: Eric Wood	Input generator for digital soil mapping
◊ Tyler Erickson and Rebecca Moore, Google Earth Engine, Google, Inc.; SMAP contact: Narendra Das and Amanda Leon	Making SMAP data products available in the Google Earth Engine Analysis Platform
◊ Jeff Morisette, USGS and DOI North Central Climate Science Center, the National Drought Mitigation Center, SMAP Contact: John Kimball	Evaluation of SMAP data for incorporation into the USGS's Software for Assisted Habitat Modeling
◊ Benjamin White, Integra, LLC; SMAP Contact: John Kimball	Application of SMAP L4_C and related products to REDD+ monitoring reporting and validation (MRV)

NOTES:
† *Early Adopters* are defined as those groups and individuals who have a direct or clearly defined need for SMAP-like soil moisture or freeze/thaw data, and who are planning to apply their own resources (funding, personnel, facilities, etc) to demonstrate the utility of SMAP data for their particular system or model. The goal is to accelerate the use of SMAP products after launch by engaging in applied research that would enable integration of SMAP data in applications. This research would promote understanding of how SMAP data products can be scaled and integrated into policy, business and management activities to improve decision- making efforts.
* *Early Adopters selected in 2011-2012* agreed to engage in pre-launch research that will enable integration of SMAP data after launch in their application, complete the project with quantitative metrics prior to launch, and take a lead role in SMAP applications research, meetings, workshops, and related activities.
◊ *Early Adopters selected from 2013 forward* agreed to engage in pre-launch research that will enable integration of SMAP data after launch in their application, and to provide feedback to the SMAP project upon request concerning their experience in using the data.

Appendix 2

See Table 13.

Table 13 ICESat-2 early adopter table

Early adopter	Science theme	Regions	Early adopter title	End user(s)	Applications
Pamela G. Posey, Naval Research Laboratory SDT/Project Office Partner: Sinead L. Farrell	Sea Ice	Full Arctic Region	Use of ICESat-2 data as a Validation Source for the U.S. Navy's Ice Forecasting Models	U.S. Navy, POC: Bruce McKenzie, Naval Oceanographic Office U.S. National/Naval Ice Center, POC: LTJG David Keith	Navigation; Arctic shipping
Nancy Glenn, Boise Center Aerospace Laboratory (BCAL) SDT/Project Office Partner: Amy Neuenschwander	Vegetation	Semiarid regions; western U.S.	Improved Terrestrial Carbon Estimates with Semiarid Ecosystem Structure	USDA (US Forest Service and Agricultural Research Service, POC: Dr. Stuart Hardegree) DOI (BLM, POC: Anne Halford; including the Great Basin Landscape Conservation Cooperative (LCC), and USGS, POC: Dr. Matt Germino) DoD (Charles BaunIdaho Army National Guard) Regional partners (Great Basin Research and Management Partnership and Joint Fire Sciences Program)	Long-term land management. Use estimates of aboveground biomass to quantify carbon, fuel loads, and monitor change in semiarid regions

(continued)

Table 13 (continued)

Early adopter	Science theme	Regions	Early adopter title	End user(s)	Applications
Lucia Mona, National Research Council of Italy—Institute of Methodologies for Environmental Analysis (CNR-IMAA) SDT/Project Office Partner: Steve Palm	Atmospheric Sciences	Polar Region	APRIL—Aerosol optical Properties in polar Regions with ICESat-2 Lidar	NASA, policy makers at local (Polar regions) and global (climate change) scale Examples: Aerocom, WMO	Climate; Air quality (effects on health and environment); Volcanic Hazards
Greg Babonis, SUNY at Buffalo SDT/Project Office Partner: Alex Gardner	Ice Sheets; Solid Earth	Subglacial volcanic events in areas such as Antarctica, Iceland, and in southern Andean ice fields	Applications of ICESat-2 in Volcanic and Geohazards-related research	Volcano Observatories, Vhub user group, UB GMFC (Geophysical Mass Flow Group)	Volcanic hazard mitigation, monitoring, and forecasting.
Lynn Abbott, Virginia Polytechnic Institute and State University SDT/Project Office Partner: Sorin Popescu	Vegetation	Not specified (global)	Detection of ground and top of canopy using simulated ICESat-2 lidar data	American Forest Management, POC: John Welker; USDA Forest Service, POC: John Coulston	Monitor forest-related harvesting and land use

(continued)

Table 13 (continued)

Early adopter	Science theme	Regions	Early adopter title	End user(s)	Applications
Sudhagar Nagarajan, Florida Atlantic University SDT/Project Office Partner: Bea Csatho	Ice Sheets	Not specified (global)	Incorporation of simulated ICESat-2 (MABEL) data to increase the time series and accuracy of Greenland/Antarctica Ice Sheet DDEM (Dynamic DEM)	Center for Environmental Studies, POC: Dr. Leonard Berry (works with local, national, and international government organizations on sea level rise)	Sea level rise monitoring/forecasting
Andy Mahoney, Geophysical Institute, University of Alaska Fairbanks SDT/Project Office Partner: Sinead L. Farrell, Ron Kwok	Sea Ice	Alaska's Northern Coastline; Arctic Coastal System	Repeat altimetry of coastal sea ice to map landfast sea ice extent for research and operational sea ice analysts	NOAA National Weather Service Ice Desk, POC(s): James Nelson, Meteorologist in Charge, (james.a.nelson@noaa.gov) Rebecca Heim, Ice Forecaster, (Rebecca.Heim@noaa.gov)	Operational ice charts/navigation; coastal deliveries; monitoring habitat of marine wildlife; offshore oil and gas industry roads; travel/transportation;
Charon Birkett, ESSIC, University of Maryland SDT/Project Office Partner: Mike Jasinski	Hydrology	Global (requirement: observations of lakes/reservoirs at least down to 10km2 target sizes)	The Application of Altimetry Data for the Operational Water Level Monitoring of Lakes and Reservoirs	USDA/FAS (POC: Dr. Curt Reynolds, Office of Global Analysis)	Hydrological drought; agricultural drought; monitoring of high water (flood) levels; monitoring of crop condition and production

(continued)

Table 13 (continued)

Early adopter	Science theme	Regions	Early adopter title	End user(s)	Applications
Guy Schumann, Joint Institute for Regional Earth System Science & Engineering, University of California, Los Angeles (UCLA) SDT/Project Office Partner: Mike Jasinski	Hydrology	California Bay Delta; Niger Inland Delta	Assessing the value of the ATL13 inland water level product for the Global Flood Partnership	Global Flood Partnership (GFP) (POCs: Dr. Florian Pappenberger; Global Flood Service and Toolbox Pillar; Dr. Guy Schumann, member of the Global Flood Partnership)	Prediction and managing of flood disaster impacts and global flood risk
Kuo-Hsin Tseng, SUNY at Buffalo SDT/Project Office Partner: Mike Jasinski	Hydrology	Ganges-Brahmaputra-Meghna (GBM) river basin covering India, Nepal, China, Bhutan, and Bangladesh	Using ICESat-2 Ground and Water Level Elevation Data towards Establishing a Seasonal and Flash Flood Early Warning System in the lower Ganges-Brahmaputra-Meghna River Basin	Institute of Water Modelling (POC: Zahirul Haque Khan), Bangladesh Water Development Board (POC: Engr. Zahirul Islam), Bangladesh Inland Water Transport Authority (POC: Md. Mahbub Alam)	Water resource management; observation of freshwater storage change
Andrew Roberts, Naval Postgraduate School Alexandra Jahn, University of Colorado at Boulder	Sea Ice	Central Arctic analysis domain	An ICESat-2 emulator for the Los Alamos sea ice model (CICE) to evaluate DoE, NCAR and DOD sea ice predictions for the Arctic	U.S. Department of Energy (POC: Elizabeth Hunke); National Center for Atmospheric Research (POC: Marika Holland, Jennifer Kay)	Sea ice forecasting; national defense environmental forecasting; coordinated disaster response: oil spill mitigation, field

(continued)

Table 13 (continued)

Early adopter	Science theme	Regions	Early adopter title	End user(s)	Applications
Adrian Turner, Los Alamos National Laboratory SDT/Project Office Partner: Ron Kwok				U.S. Department of Defense (POC: Wieslaw Maslowski, Ruth Preller) University of Colorado Boulder (POC: John Cassano)	campaigns; improved climate projections at all latitudes
Stephen Howell, Environment Canada SDT/Project Office Partner: Ron Kwok	Sea Ice	Canadian Arctic	Use of ICESat-2 Data for Environment Canada observational applications and prediction systems	Climate Research Division (POC: Howell) Canadian Meteorological Centre (POC: Belair) Canadian Ice Service (POC: Arkett) Canadian Centre for Climate Modelling and Analysis (POC: Derksen)	Climate data records; operational sea ice forecasting for Arctic shipping; sea ice info for mariners; weather hazards; prevention/mitigation of atmospheric catastrophes
Wenge Ni-Meister, Hunter College, The City University of New York, SDT/Project Office Partner: Sorin Popescu	Vegetation	Global	Mapping Vegetation with On-Demand Fusion of Remote Sensing Data for Potential Use of U.S. Forest Service Inventories and Fire Fuel Estimates	U.S. Forest Service	Forest inventories and fire fuel mapping

(continued)

Table 13 (continued)

Early adopter	Science theme	Regions	Early adopter title	End user(s)	Applications
Birgit Peterson, USGS Earth Resources Observation and Science Center SDT/Project Office Partner: Amy Neuenschwander	Vegetation	United States	Evaluation of ICESat-2 ATLAS data for wildland fuels assessments	U.S. Forest Service's Wildland Fire Assessment System (POC W. Matt Jolly, mjolly@fs.fed.us, Project Manager)	Wildfire decisions; fire behavior modeling variables
G. Javier Fochesatto, Geophysical Institute University of Alaska Fairbanks Falk Huettmann, Institute of Arctic Biology, University of Alaska Fairbanks SDT/Project Office Partner: Lori Magruder	Vegetation	Arctic Tundra and Boreal Forest; Interior Alaska	Using ICESat-2 prelaunch data in high latitude terrestrial ecosystems to allow for continuous monitoring of boreal forests and Arctic tundra	USDA Forest Service PNW Research Station (POC: Dr. Hans-Erik Andersen)	Land Management and monitoring over large regions (Arctic Tundra, Boreal Forest)

References

Babonis, G., Ogburn, S., Calder, E., &Valentine, G. (2013). Applications of ICESat-2 in Volcanic and Geohazards-related Research. Proposal to the ICESat-2 Early Adopter program. SUNY at Buffalo, Buffalo, NY.

Bolten, J., Crow, W., Zhan, X., Reynolds, C., & Jackson, T. (2009). Assimilation of a satellite-based soil moisture product into a two-layer water balance model for a global crop production decision support system. In S. K. Park & L. Xu (Eds.), *Data Assimilation for Atmospheric, Oceanic and Hydrologic Applications* (pp. 449–464). Berlin: Springer.

Brown, M., & Escobar, V. (2014). Assessment of soil moisture data requirements by the potential SMAP data user community: Review of SMAP mission user community. *IEEE Journal Selected Topics in Applied Earth Observations and Remote Sensing, 7*, 277–283. doi:10.1109/JSTARS.2013.2261473.

Brown, M., Moran, S., Escobar, V., & Entekhabi, D. (2012). Assessment of soil moisture data requirements by the potential SMAP data user community: Review of SMAP mission user community. *IEEE Journal Selected Topics in Applied Earth Observations and Remote Sensing, 7*, 277–283. doi:10.1109/JSTARS.2013.2261473.

Escobar, V., & Arias, S. (2015). Early adopters prepare the way to use ICESat-2 data. *The Earth Observer, 27*, 31–35.

Fochesatto, G., & Huettmann, F. (2015). Using ICESat-2 prelaunch data in high latitude terrestrial ecosystems to allow for continuous monitoring of boreal forests and Arctic tundra. [Proposal to the ICESat-2 Early Adopter program]. University of Alaska Fairbanks, Fairbanks, AK.

Fu, L., Alsdorf, D., Rodriguez, E., Morrow, R., Mognard, N., Lambin, J., et al. (2009). The SWOT (Surface Water and Ocean Topography) Mission: Spaceborne Radar Interferometry for Oceanographic and Hydrological Applications, White paper submitted to OceanObs'09 Conference, March 2009.

Glenn, N., Shrestha, R., Spaete, L., Li, A., Mitchell, J. (2013). Improved Terrestrial Carbon Estimates with Semiarid Ecosystem Structure. [Proposal to the ICESat-2 Early Adopter program]. Boise State University, Boise, Idaho.

Howell, S., Derksen, C., Belair, S., Smith, G., Lemieux, J., Buehner, M., et al. (2014). Use of ICESat-2 Data for Environment Canada observational applications and prediction systems. Proposal to the ICESat-2 Early Adopter program. Environment Canada, Ontario, Canada.

Mahoney, A. (2014). Repeat Altimetry of coastal sea ice to map landfast sea ice extent for research and operational sea ice analysts. [Proposal to the ICESat-2 Early Adopter program]. Geophysical Institute, University of Alaska Fairbanks, Fairbanks, AK.

Moran, M., Doorn, B., Escobar, V., & Brown, M. (2014). Connecting NASA science and engineering with earth science applications. *Journal of Hydrometeorology, 16*, 473–483.

Moran, S., Doorn, B., Escobar, E., & Brown, M. (2015). Connecting NASA science and engineering with earth science applications. *Journal of Hydrometeorology, 16*, 473–483. doi:10.1175/JHM-D-14-0093.1.

Nagarajan, S., Csatho, B., & Schenk, T. (2013). Generation of Greenlad and Antarctic Dynamic DEMs (DDEM). Proposal to the ICESat-2 Early Adopter program. Florida Atlantic University, Boca Raton, FL.

National Research Council. (2007). *Earth Science and Applications from Space: National Imperatives for the Next Decade and Beyond* (456 pp). National Academies Press.

Ni-Meister, W. (2014). Mapping Vegetation with On-Demand Fusion of Remote Sensing Data for Potential Use of U.S. Forest Service Inventories and Fire Fuel Estimates. [Proposal to the ICESat-2 Early Adopter program]. New York, NY.

Posey, P., Li L., Gaiser P., Allard R., Hebert, D., Wallcraft, A., et al. (2013). Use of ICESat-2 data as a Validation Source for the U.S. Navy's Ice Forecasting. Proposal to the ICESat-2 Early Adopter program. Naval Research Laboratory, Stennis Space Center, MS.

Roberts, A., Jahn, A., & Turner, A. (2014). An ICESat-2 emulator for the Los Alamos sea ice model (CICE) to evaluate DOE, NCAR, and DOD sea ice predictions for the Arctic. [Proposal to the ICESat-2 Early Adopter program]. Naval Postgraduate School, Monterey, CA.

Rodriguez, E. (2015). SWOT Mission Science Requirements Document, JPL D-61923, http://swot.jpl.nasa.gov/files/swot/SRD_021215.pdf.

Srinivasan, M., Peterson, C., Andral, A., & Dejus, M. (2012). SWOT Early Adopters Guide, http://swot.jpl.nasa.gov/files/swot/SWOT%20Early%20Adopters%20Guide1.pdf.

Srinivasan, M., Peterson, C., & Callahan, P. (2013). Mission Applications Support at NASA: SWOT, *Proceedings of the Symposium on 20 Years of Progress in Radar Altimetry*, Venice, Italy, 24–29 Sep 2012. Noordwijk, The Netherlands, *ESA Publications Division*, European Space Agency, Paper-2494283.

Srinivasan, M., Peterson, C., Andral, A., Dejus, M., Hossain, F., Cretaux, J.-F., et al. (2014). SWOT Applications Plan, JPL Document ID; D-79129, Ver. 1.0, Sept. 2014, http://swot.jpl.nasa.gov/files/swot/Final_SWOT%20Applications%20Plan_D79129.pdf.

Srinivasan, M., Andral, A., Dejus, M., Hossain, F., Peterson, C., Beighley, E., et al. (2015). Engaging the Applications Community of the future Surface Water and Ocean Topography (SWOT) Mission. International Archives of the Photogrammetry, Remote Sensing and Spatial Information Sciences, XL-7/W3, 1497–1504. doi:10.5194/isprsarchives-XL-7-W3-1497-2015, http://www.int-arch-photogramm-remote-sens-spatial-inf-sci.net/XL-7-W3/1497/2015/isprsarchives-XL-7-W3-1497-2015.html.

Wynne, R., Thomas, V., Abbott, L., & Awadallah, M. (2013). Detection of ground and top of canopy using simulated ICESat-2 lidar data. [Proposal to the ICESat-2 Early Adopter program]. Virginia Polytechnic Institute and State University, Blacksburg, VA.

Application of Satellite Radar Altimeter Data in Operational Flood Forecasting of Bangladesh

Amirul Hossain and Md. Arifuzzaman Bhuiyan

Abstract This study demonstrates the applicability of satellite-based radar altimeter flood forecasting in downstream flood prone areas of Bangladesh without the use of data assimilation. The joint NASA-French satellite mission JASON-2 provides water level elevation data over the Ganges-Brahmaputra-Meghna (GBM) basin at no cost to end users. A nearly linear relationship was established between these upstream water elevation observations with the downstream water levels in Bangladesh especially in the Ganges and Brahmaputra basins. Using this remotely sensed data as boundary inflow in the current flood forecasting model developed by the Flood Forecasting and Warning Center (FFWC; see FFWC 2014) within the Bangladesh Water Development Board (BWDB), it is possible to generate a 5 day or more flood forecast in some downstream locations with considerable accuracy. This approach was validated using remotely sensed JASON-2 observations from 2014 and comparing the resulting flood forecast with recorded ground measurements of river height. This study highlights the potential of using JASON-2 data for flood forecasting on international rivers like in Bangladesh where upstream hydrologic observations are otherwise unavailable.

Keywords Altimeter · JASON-2 · Flood forecast

1 Introduction

Located on the downstream section of the GBM (Ganges-Brahmaputra-Meghna) basin, Bangladesh is vulnerable to floods due to intense rainfall during monsoon season, improper land use practices, and poor national water policy. Limited sharing of hydro-meteorological data from transboundary countries makes flood

---

A. Hossain (✉) · Md. Arifuzzaman Bhuiyan
Flood Forecasting and Warning Center, Dhaka, Bangladesh
e-mail: amirulbd63@yahoo.com

Md. Arifuzzaman Bhuiyan
e-mail: arif81_bwdb@yahoo.com

forecasting a complex process. Currently, model-based flood forecasts are generated and circulated to the people nationally by the Flood Forecasting and Warning Center (FFWC) of the Bangladesh Water Development Board (BWDB) under the Ministry of Water Resources (MoWR). The current model-based system has difficulties increasing the forecast lead time while maintaining a high level of confidence. Recently, remote sensing technology has received widespread acceptability in flood management and forecasting, especially for countries where there are transboundary issues like Bangladesh (Islam and Sado 2000; Kaku and Held 2013 and Islam et al. 2009). Some recent studies on Bangladesh monsoon flood have proved that a combination of satellite estimates of rainfall and modeling can satisfyingly forecast flooding downstream in Bangladesh (Hossain et al. 2014a; ADB 2014; Siddique-E-Akbor et al. 2014) with reasonable low cost.

Radar altimetry satellites (JASON-1, JASON-2, ALTIKA) are the latest addition of NASA which are capable of measuring river flow discharge and water level elevation of wide rivers with considerable accuracy. JASON-2 is a joint NASA-French Satellite mission that measures water levels and makes the data publicly accessible within 24 h (Hossain et al. 2014a, b and NASA website). Near real-time data on JASON-2 is made available at a publicly accessible ftp site at ftp://avisoftp.cnes.fr/. Four JASON-2 virtual stations (192, 079, 014 and 155; see Table 1 and Fig. 1) on the upstream (outside Bangladesh) region of the Ganges River and three virtual stations (166, 053 and 242; see Table 1 and Fig. 1) on the upstream region of the Brahmaputra River have the potential to provide data for downstream flow forecasting of Bangladesh. Here, 'virtual stations' are defined as the specific location on the river that are overpassed by JASON-2 (see Fig. 1 for specific location and numbering of virtual stations). Table 1 shows the locations and distance of JASON-2 pass points from the border of Bangladesh.

On the Brahmaputra River, two virtual stations (VS 166 and VS 53) are located near Tejpur and one station (VS 242) lies near Guwahati. On the Ganges River, one station (VS 155) is near the Farraka Barrage area and three stations (VS 192, 79 and 14) are found near *Varanasi*, *Patna*, and *Chappara*. Depending on the proximity to

Table 1 JASON-2 pass locations and distance from Bangladesh border

	JASON-2 pass name	JASON-2 pass location	Approximate distance from border of Bangladesh (KM)
Brahmaputra	VS 166	Lat = 26.215; Lon = 91.026	130
	VS 53	Lat = 26.764; Lon = 93.48	385
	VS 242	Lat = 26.725; Lon = 93.618	395
Ganges	VS 155	Lat = 25.289; Lon = 87.112	100
	VS 14	Lat = 25.51; Lon = 85.689	250

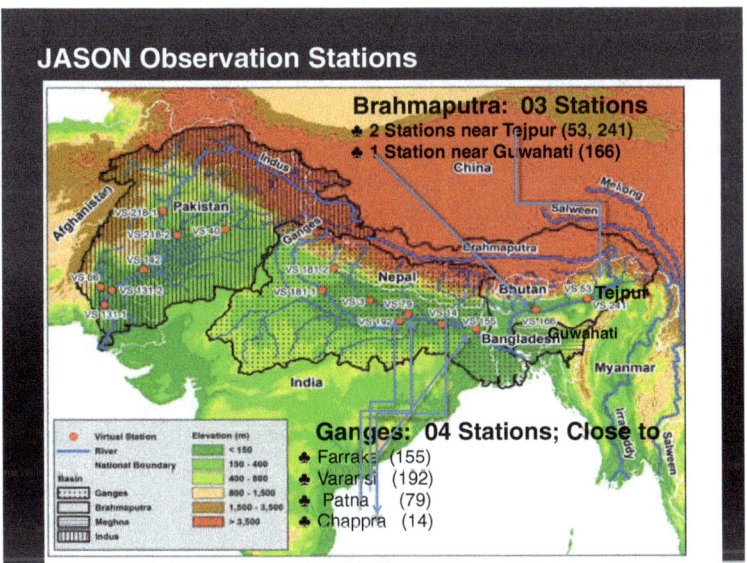

Fig. 1 Figure JASON-2 pass stations over Ganges and Brahmaputra river

Bangladesh, these seven stations on the Ganges and Brahmaputra Rivers are close to Bangladesh and hence data of these stations was considered to be used for downstream flood forecasting in Bangladesh. The flood risk information can be provided more efficiently to the affected people as well as policy makers, researchers, mass media, etc.

2 Objective

The main objective of this study is to make real-time operational flood forecast based on data found from radar altimeter satellite JASON-2 (Source: NASA) by using GIS-based customized MIKE 11 super model (FFWC 2014). Forecast performance was compared with observed river levels and performance was evaluated with some well-known statistical parameters. A secondary objective was to check the applicability of altimeter satellite forecast in the downstream of Bangladesh without the use of upstream flow data.

3 Study Area

Bangladesh is located at the downstream end of floodplain delta of three major river basins: the Ganges, the Brahmaputra, and the Meghna (combinedly referred to as the GBM Basin). The major cause of the monsoon flood relies on high magnitude of the inflow from upper GBM catchment associated with the intense rainfall inside the country during monsoon period, i.e., from June to October.

Based on the availability of JASON-2 data, a major part of monsoon flood prone area of Bangladesh situated on the lower end of the Ganges and Brahmaputra basin was chosen as a study area in this research. Simulation was done over the selected area and forecasted water levels at nine locations along the Brahmaputra–Jamuna, Ganges and Padma Rivers were found. Figure 2 shows the entire study area and

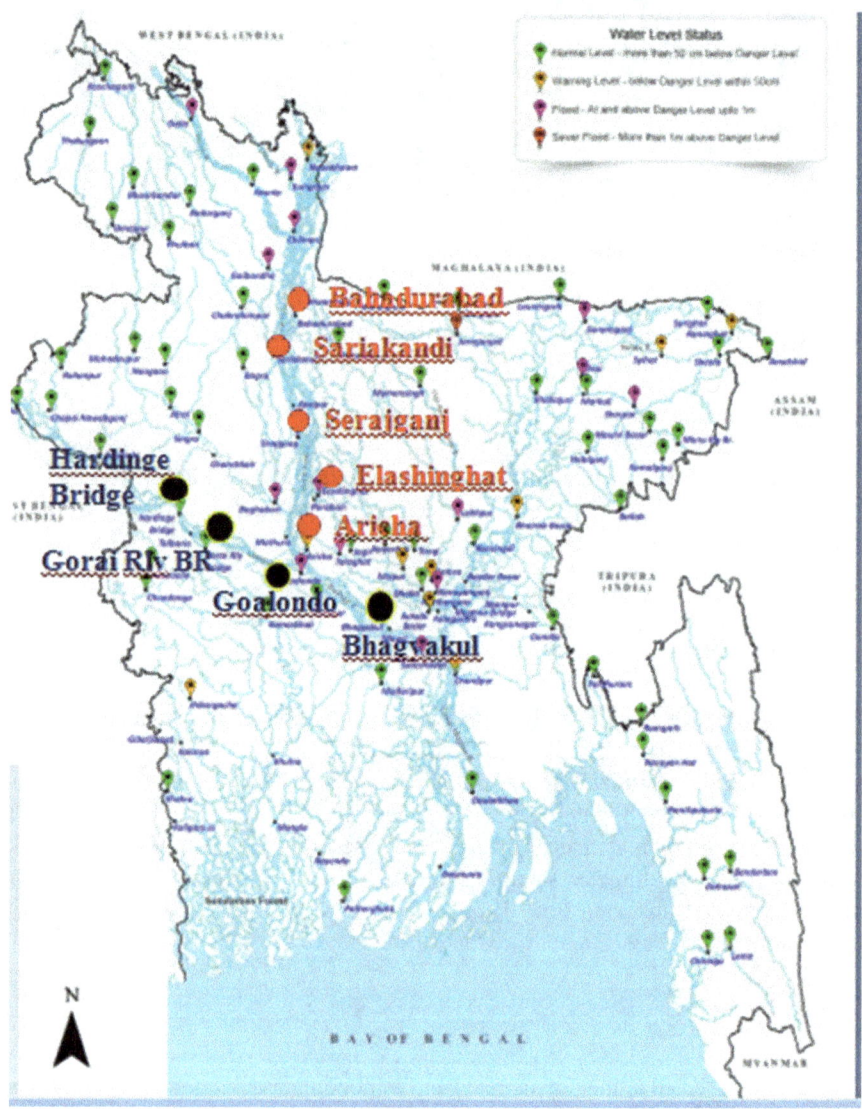

Fig. 2 Study area and locations of 9 flood forecast stations

locations of the nine forecasting stations. Among the nine stations, five (Bahadurabad, Sariakandi, Serajgonj, Aricha, and Elasin Ghat) stations are on the Brahmaputra; two stations (Goalondo and Bhagyakul) are on the Padma, and two stations (Hardinge Bridge and Gorai railway Bridge) are on the Ganges. The north eastern portion, the south eastern hilly areas, and the coastal areas of Bangladesh are beyond the scope of this study. This research work only focuses on monsoon flooding in Bangladesh lasting from mid-June to mid-October.

4 Data and Methodology

4.1 JASON-2 Data Download and Forecast Rating Curve (FRC)

JASON-2 data was downloaded from an ftp site and water level elevations at the respective seven sites (three in Brahmaputra and four in Ganges river) shown in Table 1 and Fig. 1 were extracted using the mathematical software platform, MATLAB. Based on JASON-2 water elevation data, a relationship between upstream (outside Bangladesh) water elevation data and downstream water level on Brahmaputra and Ganges river were established to find downstream water level with a lead time of 1–8 days during monsoon season. These relationships were called Forecast Rating Curves (FRC) and they were established at Bahadurabad station on the Brahmaputra River and Hardinge Bridge station on the Ganges River (Hossain et al. 2013). The 5- and 8-day lead time FRCs for Bahadurabad and Hardinge bridge are shown in Fig. 3. These FRCs are based on trend forecast derived initially for a period spanning from October, 2008 to November, 2009 and later upgraded for a period spanning 2008–2013 (6 years) during which JASON-2 flew over the Ganges-Brahmaputra basin. Here, only 4 out of 56 FRC curves are shown. Interested readers are requested to see Hossain et al. (2013, 2014a, b) for details of preparation of FRC curves. While producing forecasts using this data, some human judgment was also applied. The water level data found in this way were transferred to the boundary Noonkhawa at Brahmaputra and Pankha at Ganges considering lagtime due to travel of flow.

4.2 Setup of MIKE 11 Flood Simulation Model

MIKE11-GIS coupled flood model in FFWC was used to conduct simulation over study area. This model has 13 water level boundary inflow stations (shown in Fig. 4). The boundary estimate form provides a list of all rainfall and water level boundaries during the period. As guidance for the forecaster, it also includes water level and rainfall records at each boundary point during the past 7 days. Now, using available information, the forecaster has to fill out this boundary estimation form,

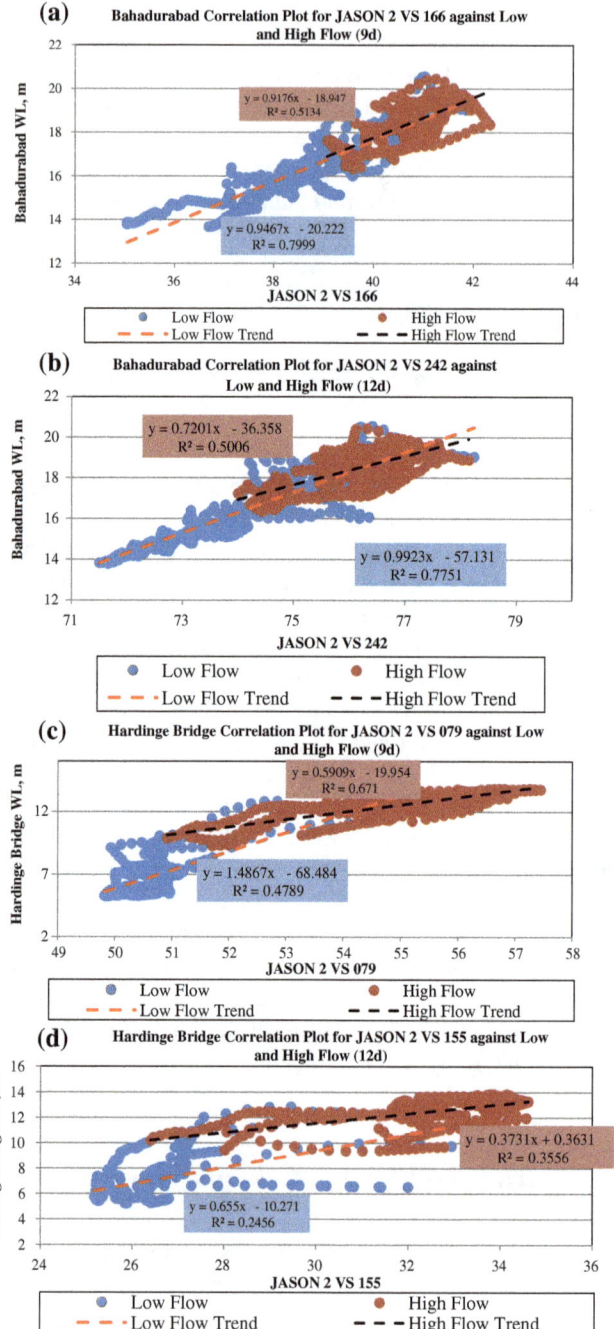

Fig. 3 Forecast Rating Curve (FRC) of Bahadurabad (**a** and **b**) and Hardinge Bridge Stations (**c** and **d**) for 2 particular days. **a** 5 day lead. **b** 8 day lead. **c** 5 day lead. **d** 8 day lead

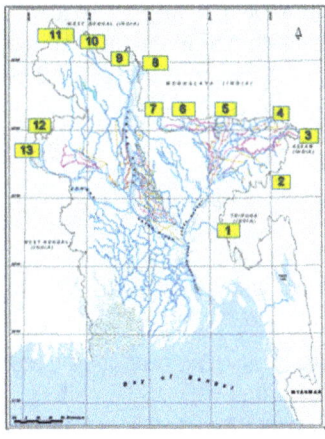

Significant Cross border Inflow points

1. Comilla, Gumti
2. Manu RB, Manu
3. Amalshid, Barak
4. Sarighat, Sariaowain
5. Lourergoar, Jadukata
6. Durgapur, Shibaanidhal
7. Nakuagoan, Bhuaai
8. Noonkhawa, Brahmaputra
9. Kurigram, Dharala
10. Dalia, Teesta
11. Panchagarh, U-Karatova
12. Rohanpur, Mohananda
13. Pankha, Ganges

Fig. 4 Boundary locations of MIKE11 super model

i.e., he has to make rainfall and water level estimates during the forecast period (next desired hours). In this research, boundary estimates were provided using JASON-2 data. About 90 % of flood flow into Bangladesh during monsoon comes through two mighty rivers—Brahmaputra and Ganges (Immerzeel et al. 2010). Hence, estimation of boundary inflow/discharge through these two rivers are of great importance in simulation of flood flow and inundation for Bangladesh. Noonkhawa (8) and Pankha (13) are two most important boundary stations in MIKE11-GIS super model.

Estimation of boundary inflow of Noonkhawa and Pankha are already described above. Other 11 out of 13 boundary inflow were estimated based on the water level trend of Brahmaputra and Ganges found in Bahadurabad and Hardinge bridge from FRC curve. The 13 boundary locations and station names are shown in Fig. 4. After setting all boundary inflow, MIKE11 model simulation was performed. Daily simulation was conducted from mid-July to mid-October and forecasted water level of 8-day lead time at nine locations were found. In this research, only monsoon flood was focused.

4.3 Evaluation of Forecast

Forecasted water levels with 8-day lead time at nine locations were compared with observed water level during the period of mid-July to mid-October. Root Mean Square Error (RMSE) and Mean Absolute Error (MAE) were calculated for the nine specific stations for the months of August and September only. RMSE and MAE were also calculated for peak flood condition. It is to be mentioned here that in 2014 monsoon, there were two major floods in Bangladesh; one in August and the other in September. The performance of simulation results were checked for these two months as well as two peak flood conditions.

5 Results and Discussions On

5.1 Simplification in Forecasting Techniques

The present flood forecasting system is based on real time and forecasted hydrometeorlogical data (water level, rainfall, etc.) that need to be collected from different sources like BWDB hydrological network, Bangladesh Meteorological Department, Indian Meteorological Department, Central Water Commission India, etc., and feed to model after proper check. Here, boundary estimation depends on some assumptions which are not always reliable. On the other hand, new method, i.e., JASON-2 data-based forecasting system simplifies the entire procedure by reducing data gathering from different sources. The only usable data is JASON-2 water height data that can be freely downloaded from internet and after some simple processing, it can be directly fed to the model. Also, this eliminates the uncertainty in boundary estimation by using forecast rating curves (FRC) that were developed with observed time series water level data collected from BWDB.

5.2 Comparison of Flood Hydrograph for 8-Day Lead Time

JASON-2 data-based monsoon flood forecasts using customized MIKE11-GIS model results were compared with the observed value as a first basis of evaluation. The simulated water level results of 8-day lead time for nine stations were compared with the observed water levels and are shown in Fig. 5.

2014 experienced two major periods of flooding, one in August and one in September. At the Bahadurabad, Aricha, Serajgonj and Sariakandi stations, during August, forecasted water level trends that show good agreement with the observed except during the initial rising period. In September, during the second spell of flood, there was also a good match initially for those stations but during the rising phase, the forecasts overestimated the 1, 2, and 3 days lead time. For the Hardinge bridge, Goalondo, and Bhagyakul stations, simulated hydrographs are similar to observed hydrographs during August, but overestimation was observed for lead times of 1, 2, and 3 days during September.

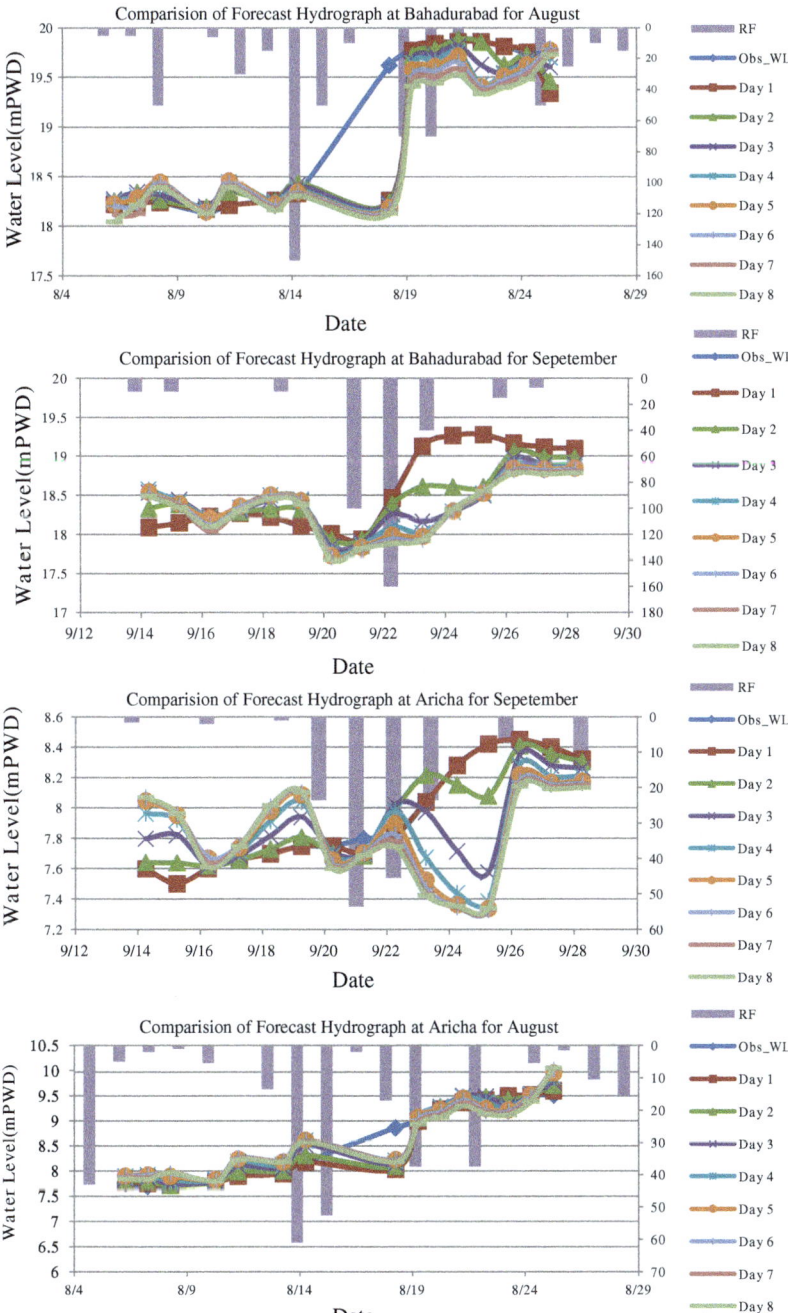

Fig. 5 Observed and simulated hydrographs for 1–8 day lead time

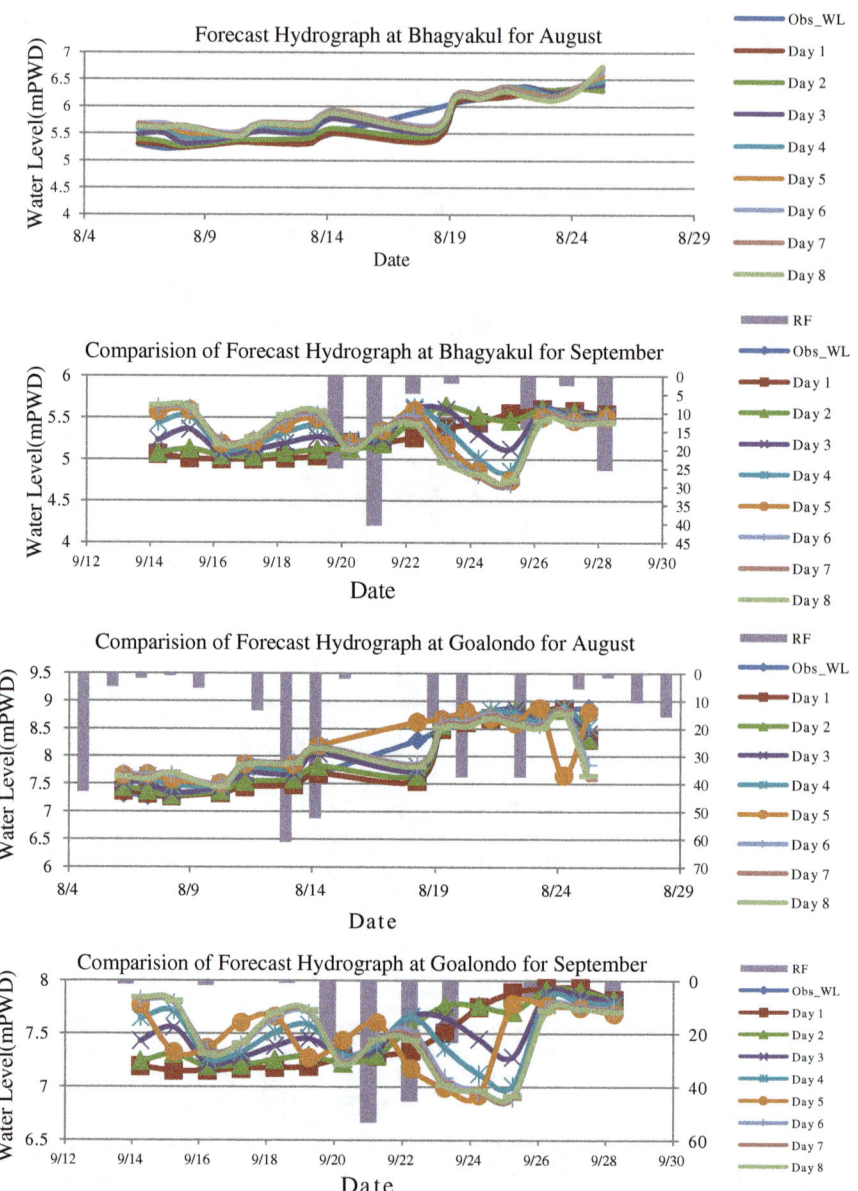

Fig. 5 (continued)

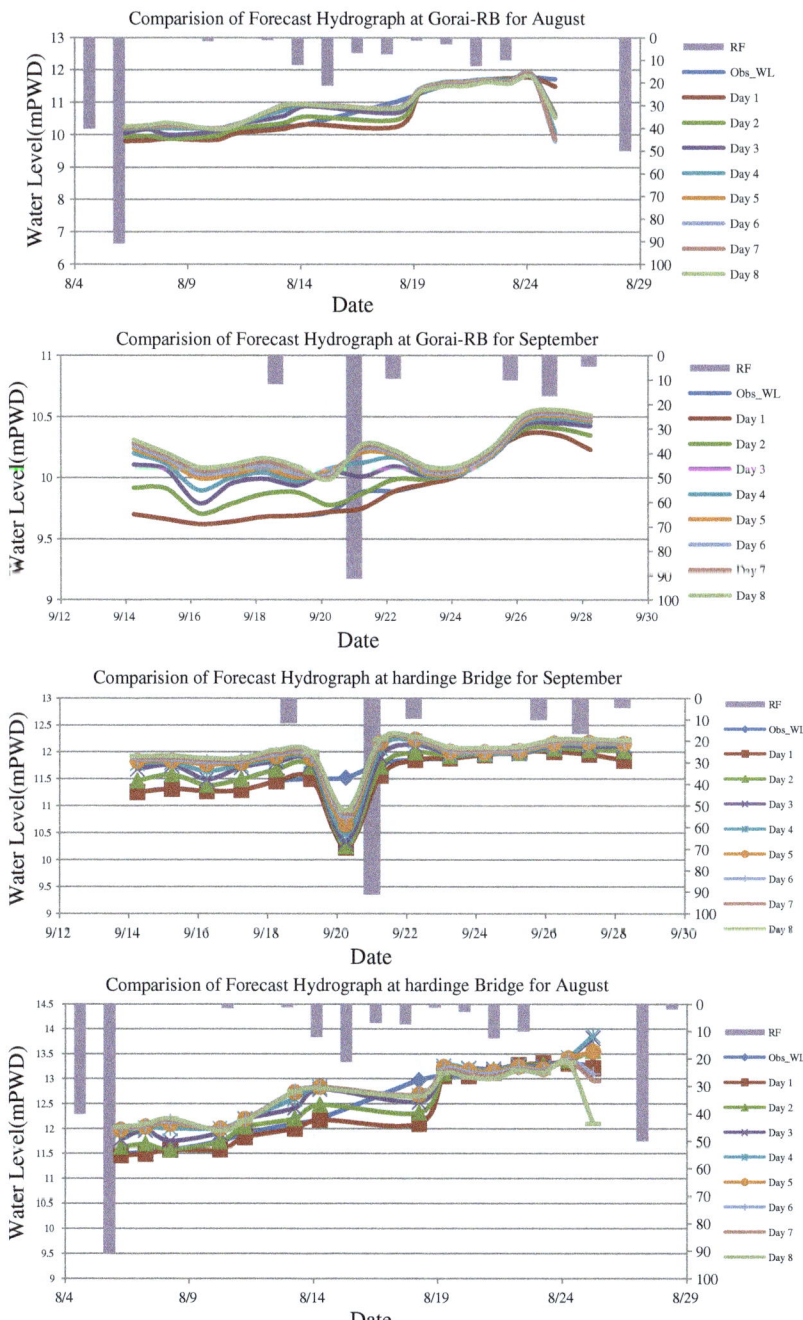

Fig. 5 (continued)

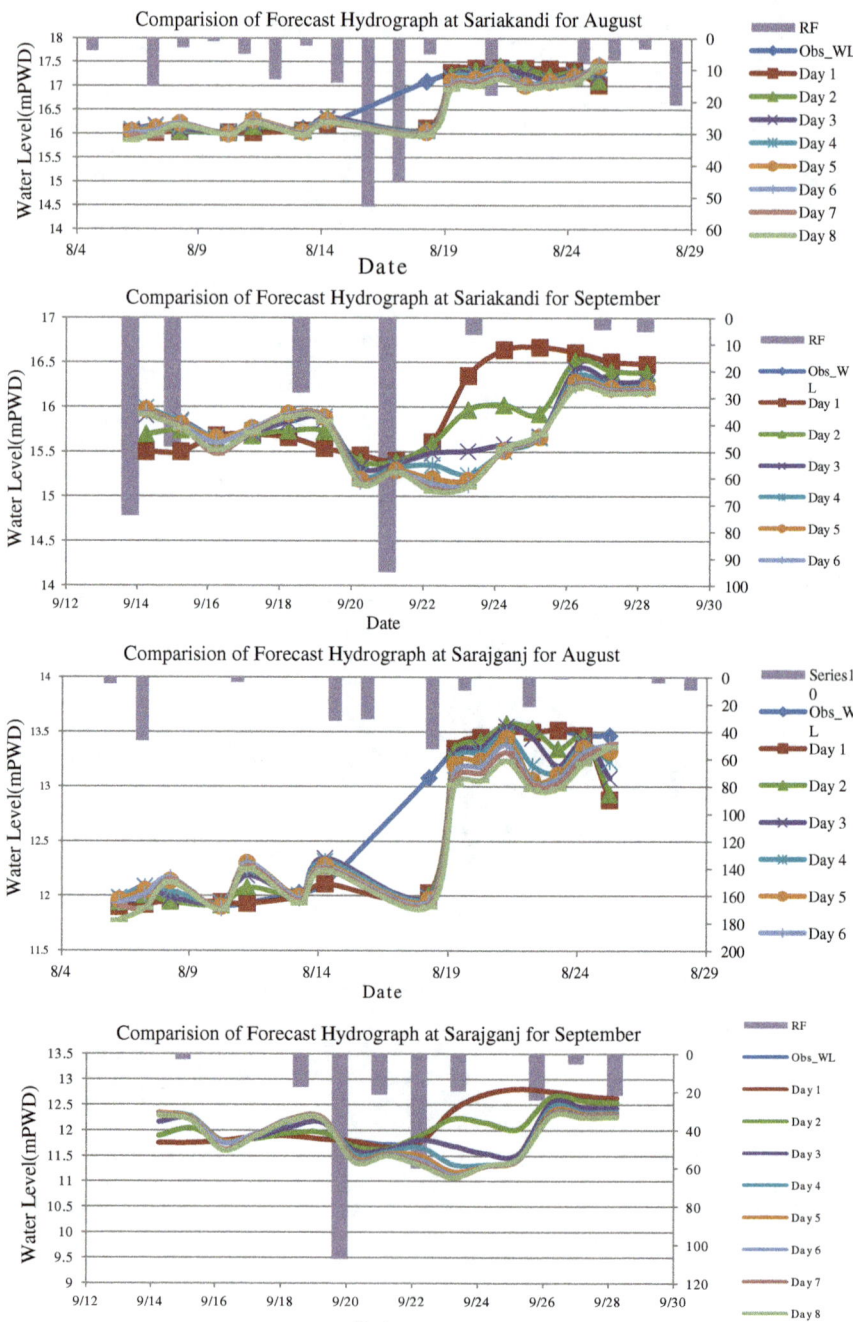

Fig. 5 (continued)

5.3 Comparison of RMSE, Mean Value Error

The Root Mean Square Error (RMSE) and Mean Absolute Error (MAE) values for lead time 1–8 days were calculated for nine stations. This is shown in Fig. 6.

RMSE values for nine stations are shown in Table 2a. The values range from 0.32–0.96 m. RMSE values are found higher for the stations along the Brahmaputra-Jamuna River (Bahadurabad, Sariakandi, Serajgonj, and Aricha stations) and the Hardinge Bridgestation on the Ganges River. The lower RMSE values are found in Goalondo, Bhagyakul, and Elasinghat. Similarly, Mean

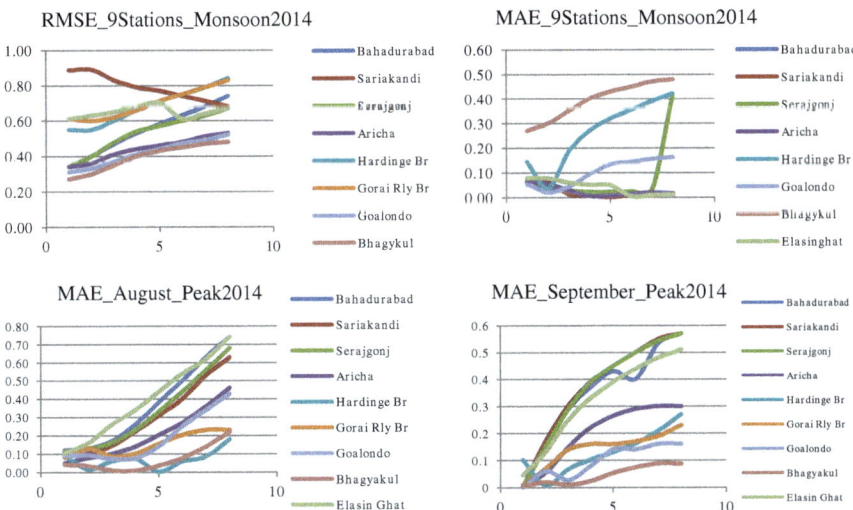

Fig. 6 RMSE and MAE values for nine stations (*Top*) for whole period (June–October/2014) and (*Bottom*) for Peak flood condition (August–September/2014) only

Table 2a RMSE of 9 forecast stations

Station	Day 1	Day 2	Day 3	Day 4	Day 5	Day 6	Day 7	Day 8
Bahadurabad	0.50	0.52	0.54	0.63	0.73	0.81	0.88	0.96
Sariakandi	0.42	0.44	0.46	0.52	0.60	0.65	0.73	0.80
Serajganj	0.49	0.53	0.53	0.57	0.66	0.73	0.8	0.88
Aricha	0.52	0.52	0.53	0.52	0.57	0.6	0.64	0.72
Hardinge Br	0.77	0.69	0.61	0.55	0.51	0.51	0.53	0.57
Gorai Rly Bridge	0.86	0.81	0.76	0.71	0.66	0.63	0.65	0.67
Goalondo	0.45	0.46	0.44	0.42	0.38	0.45	0.47	0.53
Bhagyakul	0.32	0.33	0.33	0.32	0.34	0.35	0.36	0.4
Elashinghat	0.49	0.54	0.54	0.64	0.66	0.73	0.79	0.85

Table 2b MAE of 9 forecast stations

Station	Day 1	Day 2	Day 3	Day 4	Day 5	Day 6	Day 7	Day 8
Bahadurabad	0.18	0.20	0.24	0.33	0.45	0.57	0.69	0.81
Sariakandi	0.15	0.17	0.21	0.29	0.37	0.46	0.59	0.69
Serajganj	0.18	0.201	0.23	0.292	0.4	0.51	0.624	0.75
Aricha	0.182	0.211	0.234	0.268	0.34	0.41	0.49	0.6
Hardinge Br	0.28	0.21	0.13	0.12	0.164	0.191	0.21	0.24
Gorai Rly Bridge	0.281	0.261	0.2	0.174	0.21	0.25	0.28	0.32
Goalondo	0.16	0.16	0.14	0.143	0.141	0.253	0.33	0.414
Bhagyakul	0.11	0.1	0.079	0.075	0.102	0.143	0.21	0.29
Elashinghat	0.17	0.22	0.26	0.34	0.42	0.51	0.6	0.71

Absolute Error (MAE) values range from 0.11 to 0.81 m and are shown in Table 2b. The RMSE and MAE values for peak flood condition during August and September 2014 were also calculated and shown in Fig. 5.

The RMSE values for peak conditions in August range from 0.16 to 0.91 m and in September from 0 to 0.88 m. The stations on the Brahmaputra-Jamuna River and Hardinge bridge in Ganges river show acceptable results whereas more downstream stations, i.e., stations on the Padma (Goalondo, Bhagykul), Elasin Ghat, and Gorai Rivers show higher values of RMSE and MAE.

6 Conclusion and Recommendations

Flood forecasting in Bangladesh using JASON-2 altimeter data has the advantage over conventional flood forecasting techniques that upstream water level data can be easily obtained free of cost at any time 24 h after observation. This method eliminates the necessity of using river information from upstream countries during forecast calculation which was a big challenge. In 2014, FFWC used JASON-2 data in a customized MIKE 11 GIS coupled modeling environment for forecasting floods with 8-days lead time during monsoon. Two major boundary inflows in two major rivers (at Noonkhawa on the Brahmaputra and at Pankha on the Ganges) were directly estimated using JASON-2 data and others were calculated based on the trend water level of those two boundaries. The simulation was done from June to October. The overall model performance was found satisfactory during monsoon. The simulated water levels accurately followed the observed values except for some minor discrepancies. The stations on the Brahmaputra River (Bahadurabad, Serajgonj, Sariakandi, and Aricha) and on the Ganges River (Hardinge Bridge) 8-day lead forecasted hydrographs have similar trend which is observed from 4th to 8th day; a dissimilarity was found for lead time 1st–3rd day. The first major period of flooding in 2014 was observed during the 3rd week of August followed by another major period of flooding during the 3rd week of September. The simulated hydrographs were quite capable of capturing both of the peaks.

Among the RMSE and MAE values for nine forecasting stations for the whole simulation period, upstream stations (Bahadurabad, Serajgonj, Sariakandi, Hardinge Bridge) on the Ganges and Brahmaputra Rivers showed acceptable values whereas more downstream stations showed higher errors due to tidal effects and influence of more river streams. Those parameters were also checked for the month of August and September separately which provided reasonable values. This methodology of forecasting relieves the FFWC forecaster from accumulating upstream flow measurements, which was a big challenge until now. Although 2 out of 13 boundary stations are directly found from JASON-2 data, with the improvement of quality of satellite data, other boundaries can also be calculated. In addition, surrounding weather condition during the time of forecast need to be analyzed to improve the quality of forecast. The flood forecast information using JASON-2 data during 2014 monsoon was satisfactory. As experimental scheme, the results were disseminated nationally to the government, private bodies, mass media, and other related stakeholders actively involved in flood disaster management through FFWC website (www.ffwc.gov.bd/ForecastandWarning) and received enormous praise from both home and abroad. This is a unique example in flood management of Bangladesh where satellite information has been made available for societal needs.

Given the first year of assessment, we observe some interesting possibilities moving forward with a satellite-assisted flood forecasting system. On the one hand, if qualitative trends are the major concern (i.e., "will the river rise or fall?"), then the JASON-2 forecast (based on the two boundary points that explain 80 % of Monsoon flow) can be relied on from days 4 to 8. However, if quantitative decision-making is required (i.e., "How much will the river rise?") the RMSE is quite high for the higher lead time days for such an exercise. On the other hand, the trend of forecast (rising or receding) is not as consistent during the lead times of day 1–4 even though the RMSE appears reasonably low. It may be possible to harness JASON-2-based system in such a way that the use of the forecast is targeted for the nature of the forecast quality.

It should also be mentioned that the forecasting scheme could not cover the north eastern part of Bangladesh—a flash flood prone area because of narrow width of the rivers like Meghna and Surma (compared to Ganges and Brahmaputra).Although this JASON-2-based forecasting scheme was experimental and covered few flood vulnerable areas, it (along with other satellite altimeter data) has wide scope to generate flood forecast information during both monsoon and flash flood with more area coverage of places where application or introduction of flood forecasting system is still challenging like coastal and hilly areas. The frequency of data collection by satellite altimeters will need to be higher to prepare good forecasts with increased lead time and more area coverage. JASON-2 is a sensor with a 10 day repeat which means that the virtual stations are not observed daily. Even when considering as a collection of stations on a river, the most recent JASON-2 overpass can be as old as 3 days. Next generation satellite altimeter like (JASON-3, Altika—already flying, Sentinels 3A/3B) may be able to capture river height data at higher frequency and even for smaller width rivers. We hope to investigate these opportunities in our effort to continually improve the flood forecasting system for downstream countries like Bangladesh.

References

ADB. (2014). A report on "Space Technology and Geographic Information Systems Application in ADB Projects". Published by Asian Development Bank (ADB) in 2014.
FFWC. (2014). Annual Flood Report Bangladesh Water Development Board. http://www.ffwc.gov.bd/. Annual Flood Report.
Hossain, F., Siddique-E-Akbor, A. H., Mazumder, L. C., ShahNewaz, S. M., Biancamaria, S., Lee, H., et al. (2013). Proof of concept of an altimeter-based river forecasting system for transboundary flow inside Bangladesh. *IEEE Journal of Selected Topics in Applied Earth Observations and Remote Sensing, 7*(2), 587–601 doi:10.1109/JSTARS.2013.2283402.
Hossain, F., Maswood, M., Siddique-E-Akbor, A. H. M., Yigzaw, W., Mazumder, L. C., Ahmed, T., Hossain. M., et al. (2014a). A promising radar altimetry satellite system for operational flood forecasting in flood-prone Bangladesh. *IEEE Magazine on Geosciences and Remote Sensing.* doi:10.1109/MGRS.2014.2345414.
Hossain, F., Yigzaw, W., Siddique-E-Akbor, A. H. M., Shum, C. K, Turk, F. J., Biancamaria, S., et al. (2014b). A guide for crossing the valley of death: Lessons 1 learned from making a satellite based flood forecasting system operational and independently owned by a stakeholder agency. *Bulletin of American Meteorological Society (BAMS)*. doi:10.1175/BAMS-D-13-00176.1.
Immerzeel W. W., van Beek, L. P. H, & Bierkens, M. F. P. (2010). Climate change will affect Asian Water Towers. *Science, 328.*
Islam, M. M., & Sado, K. (2000). Development of flood hazard maps of Bangladesh using NOAA-AVHRR images with GIS. *Hydrological Sciences-Journal-des Sciences Hydrologiques, 45*(3).
Islam, A. S., Bala, S. K., & Haque, A. (2009) Flood Inundation Map of Bangladesh using surface reflectance data. *2nd International Conference on Water and Flood Management (ICWFM-2009).*
Kaku, K., & Held, A. (2013). Sentinel Asia: A space-based disaster Management support system in the Asia-Pacific Region. *International Journal of Disaster Risk Reduction, 6*(2013), 1–17.
NASA Website. www.nasa.gov/missionpages/ostm/overview, ftp://avisoftp.cnes.fr//AVISO/pub/jason-2.
Siddique-E-Akbor, A. H. M., Hossain, F., Shum, C. K., Sikdar, S., Turk, F. J., Lee, H., et al. (2014). Satellite precipitation data driven hydrological modeling for water resources management in the Ganges, Brahmaputra and Meghna Basins. *Earth Interactions.* doi:10.1175/EI-D-14-0017.1.

CPSIA information can be obtained
at www.ICGtesting.com
Printed in the USA
LVHW050053280119
605463LV00004B/119/P